# EUTROPHIC SHALLOW ESTUARIES AND LAGOONS

*Edited by*

## Arthur J. McComb

*Institute for Environmental Science*
*Murdoch University*
*Murdoch, Australia*

**CRC Press**
**Boca Raton   London   Tokyo**

**Library of Congress Cataloging-in-Publication Data**

Eutrophic shallow estuaries and lagoons/edited by Arthur J. McComb.
     p.    cm.
    Includes bibliographical references and index.
    ISBN 0-8493-6839-1
    1. Eutrophication.   2. Eutrophication—Control.   3. Estuarine
ecology.   4. Lagoon ecology.     I. McComb, A. J.
    QH96.8.E9E93   1995
    594.5′26365—dc20
    for Library of Congress                                94-37997
                                                                 CIP

No claim to original U.S. Government works
International Standard Book Number 0-8493-6839-1
Library of Congress Card Number 94-37997
Printed in the United States of America    2  3  4  5  6  7  8  9  0
Printed on acid-free paper

# ACKNOWLEDGMENTS

I am indebted to a number of people who have been involved in various ways in preparing this volume for publication. In particular, of course, I am indebted to the authors of the chapters for accepting the invitation to prepare contributions for the volume, and for carrying the work to completion. I am grateful for their promptness in answering queries and for their valuable contributions to the field of knowledge about eutrophication in these shallow systems. I am especially indebted to a number of colleagues who have helped with the editing, proofreading and refereeing of the different chapters, and here I thank especially Dr. Karen Hillman and Mr. Qiu Song. Particular thanks go to Professor Ian Potter for picking up the progress of the volume when I became seriously ill for some weeks. Without him, this work would not have been brought to completion. Special thanks to Mrs. Susan Flay for her expert help in standardizing the format of the chapters and dealing with correspondence.

**Arthur J. McComb**
Perth, Western Australia

# THE EDITOR

**Arthur J. McComb, Ph.D.,** is Professor of Environmental Science at Murdoch University, Perth, Western Australia. He earned his B.Sc. and M.S. degrees from the University of Melbourne, and his Ph.D. from the University of Cambridge. He is a Fellow of the Institute of Biology. In 1963, he joined the staff of the Botany Department at the University of Western Australia and later became Head of Department, and co-director with Professor Jorg Imberger of the University's Centre for Water Research. In 1989, he moved to Murdoch University and his section of the Centre for Water Research moved to Murdoch with him.

His research interests include the ecology of estuarine, marine and freshwater systems, and especially the consequences of nutrient enrichment for plant production. He has worked on wetlands throughout Western Australia but has had particular interest in the ecology of Shark Bay, Cockburn Sound and the Peel-Harvey Estuary.

Dr. McComb has been a member of the Scientific Advisory Committee of the World Wide Fund for Nature Australia, the Australian Society for Limnology and has been president of the Royal Society of Western Australia and the Ecological Society of Australia. He was chairman of the State's National Parks and Nature Conservation Authority from 1991–1994.

He has published some 150 papers and has edited or authored books on biology of Australian native plants, seagrasses, and Australian wetlands.

# CONTRIBUTORS

**H. Beekman**
City of Cape Town
City Engineer's Department
Sewerage Branch
Cape Town, South Africa

**G. Boddington**
City of Cape Town
City Planner's Department
Town Planning Branch
Cape Town, South Africa

**R. Dick**
City of Cape Town
City Engineer's Department
Scientific Services Branch
Cape Town, South Africa

**W. D. Harding**
City of Cape Town
City Engineer's Department
Scientific Services Branch
Cape Town, South Africa

**Marilyn M. Harlin**
Department of Botany
The University of Rhode Island
Kingston, Rhode Island

**Clifford J. Hearn**
Department of Geography and Oceanography
University College
University of New South Wales
Australian Defence Force Academy
Canberra, Australia

**I. J. Hodgkiss**
Department of Ecology and Biodiversity
The University of Hong Kong
Hong Kong

**Bruce R. Hodgson**
Pacific Power Services
Sydney, Australia

**Zhang Jia-ping**
Graduate School of Oceanography
The University of Rhode Island
Narragansett, Rhode Island

**Victor N. de Jonge**
National Institute for Coastal and
  Marine Management
Rijkswaterstaat
Haren, The Netherlands

**Robert J. King**
School of Biological Science
University of New South Wales
Sydney, Australia

**R. C. J. Lenanton**
Western Australian Department
  of Fisheries
Western Australian Marine Research
  Laboratories
North Beach, Australia

**M. Lief**
City of Cape Town
City Engineer's Department
Sewerage Branch
Cape Town, South Africa

**Rod J. Lukatelich**
Department of Environmental Sciences
University of Venice
Venice, Italy

**A. Marcomini**
Department of Environmental Sciences
University of Venice
Venice, Italy

**I. R. Morrison**
City of Cape Town
City Engineer's Department
Scientific Services Branch
Cape Town, South Africa

**Arthur J. McComb**
Institute for Environmental Science
Murdoch University
Murdoch, Australia

**A. A. Orio***
Department of Environmental Sciences
University of Venice
Venice, Italy

* Deceased

**B. Pavoni**
Department of Environmental Sciences
University of Venice
Venice, Italy

**I. Potter**
School of Biological and Environmental
  Science
Murdoch University
Perth, Australia

**A. J. R. Quick**
City of Cape Town
City Planner's Department
Town Planning Branch
Cape Town, South Africa

**Wim van Raaphorst**
Netherlands Institute for Sea Research
Dem Burg
Texel, The Netherlands

**Timothy O'Riordan**
School of Environmental Science
University of East Anglia
Norwich, England

**S.-O. Ryding**
Federation of Swedish Industries
Stockholm, Sweden

**A. Sfriso**
Department of Environmental Sciences
University of Venice
Venice, Italy

**R. A. Steckis**
Western Australian Department of
  Fisheries
Western Australian Marine Research
  Laboratories
North Beach, Australia

**J. A. Thornton**
Environmental Planning Division
Southeastern Wisconsin Regional
  Planning Commission
Waukesha, Wisconsin

**W. W.-S. Yim**
Earth Science Unit
The University of Hong Kong
Hong Kong

**Qi Yu-Zao**
Chairman of Department of Biology
Director of Institute of Hydrobiology
Guangzhou, China

# TABLE OF CONTENTS

Chapter 1
Introduction ....................................................................................................... 1
**Arthur J. McComb**

Chapter 2
The Peel-Harvey Estuarine System, Western Australia .......................................... 5
**Arthur J. McComb and Rod J. Lukatelich**

Chapter 3
Tuggerah Lakes System, New South Wales, Australia ........................................ 19
**Robert J. King and Bruce R. Hodgson**

Chapter 4
Shenzhen Bay, South China Sea ..................................................................... 31
**Qi Yu-Zao and Zhang Jia-ping**

Chapter 5
A Case Study of Tolo Harbour, Hong Kong ..................................................... 41
**I. J. Hodgkiss and W. W.-S Yim**

Chapter 6
Eutrophication of the Lagoon of Venice: Nutrient Loads and Exchanges ................ 59
**A. Marcomini, A. Sfriso, B. Pavoni, and A. A. Orio**

Chapter 7
The Ems Estuary, The Netherlands ................................................................. 81
**Victor N. de Jonge**

Chapter 8
The Ecology and Management of Zandvlei (Cape Province, South Africa), an Enriched
Shallow African Estuary ................................................................................ 109
**J. A. Thornton, H. Beekman, G. Boddington, R. Dick, W. R. Harding, M. Lief,**
**I. R. Morrison, and A. J. R. Quick**

Chapter 9
Eutrophication of the Dutch Wadden Sea (Western Europe), an Estuarine Area Controlled
by the River Rhine ...................................................................................... 129
**Victor N. de Jonge and Wim van Raaphorst**

Chapter 10
Water Exchange between Shallow Estuaries and the Ocean ................................ 151
**Clifford J. Hearn**

Chapter 11
Changes in Major Plant Groups Following Nutrient Enrichment .......................... 173
**Marilyn M. Harlin**

Chapter 12
The Commercial Fisheries in Three Southwestern Australian Estuaries Exposed to Different
Degrees of Eutrophication ............................................................................. 189
**R. A. Steckis, I. C. Potter, and R. C. J. Lenanton**

Chapter 13
The Role of the Sediments ............................................................................................................ 205
**J. A. Thornton, A. J. McComb, and S.-O Ryding**

Chapter 14
The Sustainable Economics of Eutrophication Control ...................................................... 225
**Timothy O'Riordan**


Index....................................................................................................................................................233

# EUTROPHIC SHALLOW ESTUARIES AND LAGOONS

# Chapter 1

# Introduction

## *Arthur J. McComb*

## CONTENTS

I. Overview ............................................................................................................................ 1
II. Nutrient Enrichment ........................................................................................................ 1
III. Shallow Estuaries and Lagoons ..................................................................................... 2
IV. Management Implications ............................................................................................... 3
V. The Structure of the Volume .......................................................................................... 3
References ................................................................................................................................ 4

## I. OVERVIEW

This book is about the consequences and management of nutrient enrichment in shallow estuaries. The importance of estuaries in general has been reviewed on many occasions, and scarcely needs addressing here — suffice it to say that their values include the provision of sheltered waters for port facilities and recreation; the support of commercial and amateur fishing for fish, prawns, crabs and molluscs; they offer routes for transport involving boats which pass through them or vehicles which pass around them, often on reclaimed marsh; they provide for the discharge of effluents and stormwater; and they have less tangible assets including the values of landscape, conservation and cultural attributes. For these various reasons, estuaries have acted as foci for human settlement, and much of the world's population lies in cities centered on them.

Attempts to take advantage of these diverse values can lead to conflicts in planning and management, because the various purposes for which estuaries can be exploited are not all necessarily compatible. The need for resolution of these conflicts becomes more urgent as the pressure of human numbers increases, techniques for fishing improve, the numbers of boats and vehicles rises, and as public appreciation of natural values becomes more sharply focused and more demanding of managers. In the context of this book, which addresses eutrophication, conflicts may arise through attempts to achieve a balance between the cost of reducing nutrient loading and the maintenance of those estuarine values which depend upon water quality.

It is useful to remind ourselves at the outset that water quality problems and their resolution involve not just the waters of an estuary, but the use of land both in the immediate surrounds and, more broadly, on catchments which may be quite remote from the estuary itself. The streams which drain those catchments bring to the estuary fresh water, suspended sediments and dissolved and suspended organic matter and nutrients. A marked change in any of these components of stream transport may have profound effects on an estuary downstream. Recognition of the fundamental relationship between estuaries and their catchments leads to an appreciation of the complexities not only of devising management solutions for estuarine problems, but in allocating institutional responsibility for those problems and the implementation of management solutions.

## II. NUTRIENT ENRICHMENT

Estuaries are generally regarded as highly productive systems even under pristine conditions, as they typically contain a large biomass of algae, seagrasses and phytoplankton, and may support relatively large fish and bird populations. This high biomass of plants and animals suggests that the biota and sediments represent very significant nutrient pools and that there occurs significant nutrient cycling between them; direct evidence for these suggestions comes from a large literature, some of it reviewed later in the volume. It might be argued that estuaries are likely to be so nutrient rich that they would not react in a sensitive way to increased nutrient loading resulting from human activities; that is, that their biomass may not be nutrient limited. In contrast, examples given in the subsequent chapters of this volume, and elsewhere, clearly illustrate marked consequences for estuarine organisms of an increase in nutrient loading.

The nutrient enrichment of estuaries may lead to the generation of large masses of macroalgae which impede boat progress and decompose in offensive masses on the shoreline. They can include the loss of seagrasses from estuaries, and the occurrence of unacceptable phytoplankton blooms, especially of blue-green algae which affect food webs, reduce opportunities for contact recreation and may lead to nauseous odors and the production of toxins. These and other consequent changes seriously impair the ability of people to use and enjoy estuaries.

Although plant growth in estuaries might in principle be affected by a range of factors including light, temperature and salinity, an increase in nutrient load has in many instances been the major reason for an increase in plant biomass. It must be concluded that at least in a gross sense, biomass has been nutrient limited. On reflection, it should be recalled that under pristine conditions the vegetation of a catchment is very effective at trapping and recycling nutrients, so that the streams which leave them have waters with low nutrient concentrations; it is these generally nutrient-poor waters which reached pristine estuaries. So too, the oceanic waters which exchanged with the waters of estuaries were generally of low nutrient concentration. The estuarine biota were able to trap these scarce nutrient resources and play a role in their recycling, and it is easy to speculate that biomass may have increased to a level controlled by nutrient availability; from this perspective it is not so surprising to find that increased nutrient loading has such profound consequences.

We can conclude that although estuaries may contain a relatively large biomass of plants and animals they nevertheless respond to nutrient enrichment. One consequence of an already high biomass under natural conditions is that it can be difficult to define estuarine cultural eutrophication and detect its onset. What exactly is a eutrophic estuary? These are matters to which we return later in the volume.

The significance of eutrophication for estuaries is already clearly recognized and widely appreciated, and a great deal has been written about the topic. By 1979, a bibliography of 543 references could be assembled on the response of estuaries to nutrient enrichment in a survey which did not claim to be exhaustive,[1] by 1981 a symposium volume treated the subject,[2] and a great deal of further information has accumulated in journals, reports, symposium volumes and books about estuaries. It was not our intention to review this material here, but rather to deal with a particular type of estuary.

## III. SHALLOW ESTUARIES AND LAGOONS

This volume is particularly concerned with eutrophication of shallow, broad estuarine basins such as occur on coastal plains eroded to low relief, and where rivers may flood behind coastal dune systems. Such estuaries are often of the 'bar-built' type,[3] in which water exchange between the basin and the ocean is impeded by a sand bar across a narrow inlet channel, which may close the estuary off either seasonally or for years at a time. Because the seasonal broaching of a sand bar is a consequence of seasonal river discharge, one might expect such systems to be especially characteristic of areas with highly seasonal rainfalls and low tidal amplitudes.

The term 'lagoon' is used in the title to the book because broad estuarine basins behind sand dunes are sometimes referred to by this term, especially if they flood laterally along an interdunal depression. Day[4] has argued that a lagoon should continue to be thought of as any 'expanse of sheltered, tranquil water' near to or communicating with another, larger water body; that it should not be used to describe a particular kind of estuary; and that it should not be restricted to estuaries. The water bodies dealt with in this volume are best referred to technically as estuaries.

Why devote a book to these systems? As we show below, they are scattered in different parts of the world, but their eutrophication has not been generally well studied or integrated, perhaps because of their geographical locations and the emphasis which has been given to generally larger, deeper systems. Because of their distinctive features, and because eutrophication has been studied in reasonable depth in some of them, it seemed opportune to draw the threads together and seek for common features which relate to estuarine processes and management.

While these systems share with other estuaries the values we have enunciated above, they have a number of common features which are different from other estuaries, and which are likely to exacerbate the effects of nutrient loading and to have management significance. One of the most common of these features is poor exchange with the open ocean because of a narrow inlet channel which may be seasonally closed. For this reason, the salinity of the estuarine basin may at times differ very significantly from that of the ocean; it may be relatively fresh during river flow, but become marine when flow ebbs, and even hypersaline if evaporation rates are high. The change in salinity emphasizes and is a measure of poor water exchange, and importantly for our purposes, this poor exchange means that river water may be impounded to some extent in the system, enhancing opportunities for nutrient entrapment; and once river flow has ebbed, a low rate of water exchange with the ocean provides little opportunity for the loss of nutrient-rich water through the mouth of the estuary.

Another important feature of these systems is that they are relatively shallow, and this has two important consequences. The first is that a relatively large proportion of the basin floor may be in the photic zone, which predisposes the system to the growth and accumulation of seagrasses and benthic macroalgae, which may in consequence be more significant to the ecology of the system than are phytoplankton. The second consequence is that, because there is a low ratio between basin volume and sediment surface area, there is the potential for an especially close interaction between sediment and water; this interaction is further enhanced in systems which are so shallow that the surface sediments can be readily mixed into the water column during high winds.

## IV. MANAGEMENT IMPLICATIONS

Much of this volume is concerned with the causes, consequences and management of eutrophication problems in shallow systems, and by way of introduction it is useful to reflect here on some of the management implications of the points which have been raised so far. They will form a basis for later discussion.

1. It would be useful to better define the symptoms of eutrophication in these systems.
2. An estuary is likely to contain a large pool of nutrients in plant biomass and surface sediments. Management options to reduce eutrophication could therefore involve removal of this nutrient pool, or methods for reducing nutrient loss from sediments to water column.
3. Nutrient loading to an estuary may be derived from catchments via streams, drains and groundwater. Modification of land use and effluent disposal might significantly reduce estuarine eutrophication.
4. Nutrient loss to the ocean may be physically impeded by a channel of low cross-sectional area. Reduction of this barrier to water exchange, for example through dredging, may enhance nutrient loss from the system.
5. It is important to recognize the administrative difficulties of supporting responsibility for nutrient enrichment and for the management of its consequences.

## V. THE STRUCTURE OF THE VOLUME

The 'case study' approach has been adopted for the book, which begins by examining eutrophication in a number of estuaries, dealt with in alphabetic order by country. The case study approach was emphasized so that it would be possible to understand what problems, investigations and management options have been adopted in different countries. In each case authors were invited to use the same headings in their presentations, so that the information could be more readily compared between chapters. Of course the coverage is far from comprehensive; we have selected a range of systems which have been at least reasonably well documented. The examples include for comparative purposes some which lie outside the type of systems emphasized above; some for example are deeper, or are more open and have better exchange with oceanic waters.

After the case studies chapters follow which address specific aspects of eutrophication in these shallow systems, though most of them also adopt a 'case study' approach, rather than present general reviews. More specifically, one chapter addresses the changes which take place in the primary producers of these systems, in response to nutrient enrichment. Another examines the consequences of eutrophication for fisheries. There is a chapter which illustrates the state of our knowledge concerning the measurement of exchange with the ocean, and there is one which is concerned with the eutrophication of a marine area outside an estuary. The role of sediments in eutrophication is explored, and a final chapter in the series is concerned with an economic basis for management.

The final overview chapter attempts to draw together the threads of the volume by integrating information from the other chapters, using the matters raised in this introduction as a framework, and concentrating especially on implications for management.

# REFERENCES

1. **Neilson, B. J. and Cronin, L. E., Eds.,** *Estuaries and Nutrients,* Humana Press, Clifton, NJ, 1981.
2. **Webb, K. L., Hayward, D. M., Baker, J. M., and Murray, B.,** *Estuarine Response to Nutrient Enrichment, a Bibliography,* Chesapeake Research Consortium Publication 68, Special Scientific Report Number 95, Virginia Institute of Marine Science, 1979.
3. **Pritchard, D. W.,** What is an estuary: physical viewpoint, in *Estuaries,* Lauff, C. H., Ed., American Association for the Advancement of Science Publications, Washington, DC, 1967, 3–5.
4. **Day, J. H., Ed.,** *Estuarine Ecology with Particular Reference to Southern Africa,* Balkema, Rotterdam, 1981.

# Chapter 2

# The Peel-Harvey Estuarine System, Western Australia

*Arthur J. McComb and Rod J. Lukatelich*

## CONTENTS

I. Introduction ....................................................................................................................5
II. Physical Properties .........................................................................................................5
III. Catchments ....................................................................................................................7
IV. Water Movement and Exchange ....................................................................................9
V. Nutrient Concentrations ...............................................................................................11
VI. Sediment Properties ......................................................................................................11
VII. Symptoms of Nutrient Enrichment ..............................................................................13
VIII. Consequences of Eutrophication ..................................................................................14
IX. Discussion ....................................................................................................................15
References .................................................................................................................................16

## I. INTRODUCTION

Peel Inlet and Harvey Estuary are two adjoining estuarine basins in southwestern Australia, which in recent years have become highly eutrophic. One of the basins, Peel Inlet, supports a large biomass of green macroalgae, while the other, Harvey Estuary, has dense summer blooms of the blue-green alga (Cyanobacterium) *Nodularia spumigena* Mert.

This estuarine system is fed by three rivers, and communicates with the Indian Ocean through a common inlet channel (Figure 1). Beside the inlet channel lies the city of Mandurah (population 30,000), which is within 70 km of the capital city of the state, Perth, and its port of Fremantle. Once a quiet weekend retreat and a place for retirement, the Mandurah area has become rapidly urbanized because of the increasing population of the state and the construction of good access roads to Perth. Canal estates have been established at several points along the shores, especially near the inlet channel and along the Murray River. Depending on the seasons the shallow basins are used for sailing, water-skiing, windsurfing, fishing, crabbing and prawning; for example during a 5-day survey in January 1978, 1314 boats were found to be engaged in crabbing, 427 in fishing, 344 in sightseeing, 76 in water skiing, and 71 in 'miscellaneous activities'.[1] There is a commercial fishery (usually some $A2M per annum) which, although small in relation to marine fisheries of the western coast, is nevertheless one of the largest estuarine-based fisheries in Australia.

The estuaries lie on a sandy coastal plain of low relief, drained for agriculture since the 1920s. The coastal plain is separated from uplands in the catchment by a major fault line, the Darling Escarpment, which runs north–south for approximately 300 km, and is some 300 m tall (Figure 2). The escarpment marks an abrupt change in land use from agriculture on the plain to forested catchment on the immediate uplands; much of the water supply for urban areas is from reservoirs located along the escarpment. Further inland rainfall is reduced, and forested country gives way to land which has been cleared for grazing and wheat farming.

## II. PHYSICAL PROPERTIES

The region has a Mediterranean climate, with hot dry summers and cool wet winters (mean temperature for January 30.1°C, for August 18.3°C). The resulting highly-seasonal flow, together with the relief of a sandy, coastal plain, results in the structure of the estuary being of the 'bar-built' type,[2] in common with some 80 other estuaries along 1200 km of coast in southwestern Australia.[3] Of these, 64 are open to the sea for a period every year or less frequently, and 7 are permanently barred coastal lagoons; Peel Inlet is one of only 9 which are open continuously.

0-8493-6839-1/95/$0.00+$.50

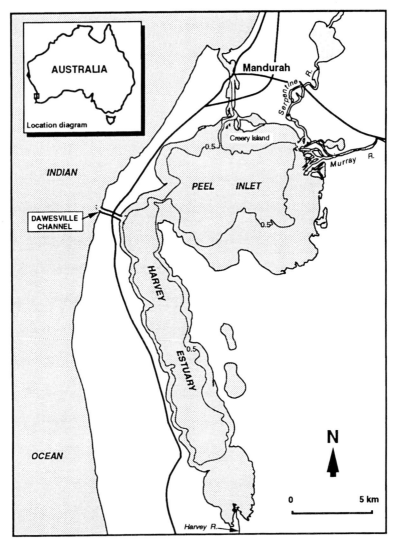

**Figure 1** The Peel-Harvey estuarine system. Note the two estuarine basins, the three rivers, the narrow inlet channel at Mandurah, between the Indian Ocean and Peel Inlet, and the site of the newly constructed channel at Dawesville, between the Indian Ocean and the northern end of Harvey Estuary. Prominent lines represent main roads.

The inlet channel is long (5 km) and narrow (200 m), and although open all year, sand has to be dredged from it to allow the passage of small boats during late summer. Harvey Estuary and Peel Inlet communicate across a shallow sill through which a small channel has been dredged to allow the passage of boats.

The basins have approximately the same area (Peel Inlet, 72 km$^2$; Harvey Estuary, 56 km$^2$) and water depth (mean in Peel Inlet, 0.8 m; in Harvey Estuary, 1.0 m); maximum depths reach 2.5 m in each basin. Peel Inlet in particular is characterized by large, shallow marginal platforms which are less than 0.5 m deep and occupy more than half the total area. The mean volumes of the two basins are $61 \times 10^6 \, \text{m}^3$ for Peel and $56 \times 10^6 \, \text{m}^3$ for Harvey.

The tides of southwestern Australia have very small amplitudes.[4] The maximum daily range due to astronomic effects is about 1.0 m and is mainly diurnal, with a small, semi-diurnal component. Superimposed on this pattern are water level changes of a similar magnitude but longer time scale (circa 7–10 days), attributed in the main to atmospheric effects and referred to for convenience as a 'barometric' tidal component. The total oceanic water level change in the ocean near Mandurah is therefore some 1.5 m. The effect of the long inlet channel is to filter out the astronomic signal but to

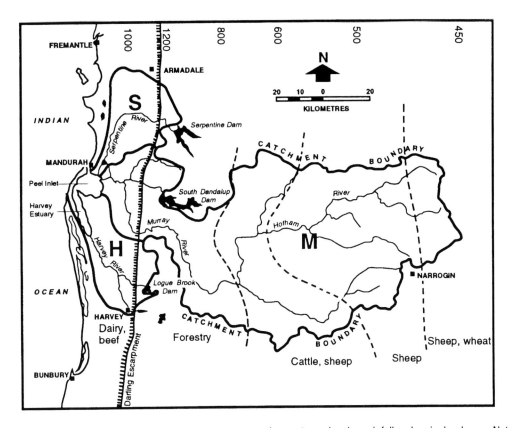

**Figure 2** The three catchments of the Peel-Harvey estuarine system, showing rainfall and major land uses. Note the north-south oriented Darling Escarpment (shaded line), which separates the sandy, coastal-plain soils from the largely lateritic uplands. The figures above, and the broken lines, indicate rainfall (mm year[-1]). (Redrawn from McComb and Humphries.[9])

leave the barometric component essentially unchanged (Figure 3). There is, therefore, little diurnal change in water level in the system, and even less as one moves far from the inlet channel into Harvey Estuary or along the tidal reaches of the Murray River. Such estuaries have been referred to as 'microtidal'.

## III. CATCHMENTS

The total catchment area is 11,378 km², and can be divided into three basins (Figure 2): that of the Harvey River and associated drains which enter the southern end of Harvey Estuary; and those of the Serpentine and Murray Rivers which enter the eastern side of Peel Inlet.

The Harvey and Serpentine catchments are on sandy coastal plain soils cleared for grazing by cattle, sheep and horses. There are also potentially significant point sources of nutrient in these two basins, consisting of stock holding yards used in connection with live sheep export to the Middle East, and piggeries of which the largest, in the Serpentine Catchment, has some 20,000 animals. Both the Harvey and Serpentine rivers have been dammed, and their truncated catchments consist only of the coastal plain sections of what were larger catchments with forested uplands. Water from the Harvey Reservoir is used in part for channel irrigation on the coastal plain, but excess water from this irrigated land passes directly to the ocean through a drainage canal and does not enter Harvey Estuary. Despite the construction of the dams, it is estimated that the total volume of water entering the estuary from the Harvey River has not been greatly affected by dam construction; what was lost from the upper catchment has been approximately replaced by flow from agricultural drains on the coastal plain.[1]

The large catchment of the Murray River (the main channel of which is undammed) ranges from high to low rainfall zones. The higher rainfall area is covered by *Eucalyptus* forest, mainly *Eucalyptus marginata* (jarrah) and *Eucalyptus calophylla* (marri). This forested catchment has a low water yield by world standards (less than 10% of precipitation may be accounted for by streamflow), but the water is

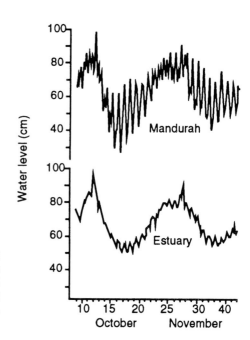

**Figure 3** A portion of tide trace from (above) the ocean beyond the inlet channel at Mandurah, and (below) inside Peel Inlet, at the northwest (see Figure 1). (Redrawn from McComb, A.J. and Lukatelich, R.J., *Limnology in Australia*, Williams, W.D. and deDeckker. P., Eds., Dr. W. Junk, Dordrecht, 1986, 433–455. With permission.)

of high quality and rainfall is relatively reliable. Further inland the catchments have been cleared, and this leads to increased salinity (up to 3 ppt is recorded during peak flow of the Murray River where it crosses the escarpment)[5] as well as increased nutrient loads. The clearing has taken place for orchards, wheat and sheep farming, depending approximately on rainfall. The lower rainfall sections of the catchment have less reliable rainfall than do the catchments further west.

River flow is highly seasonal, more than two thirds in the 5 months of June to October, often with half of the flow in July and August; on the other hand there is little or no flow from December to April inclusive. There is also large year-to-year variation; over the period 1977–1986 mean streamflow from the Murray River was 595 $m^3 \times 10^6$ with a standard deviation of 185 (71% of the mean). Over the same period, flow from the Harvey catchment, which is nearer to the coast and with more reliable rainfall, had a mean flow of 221 $m^3 \times 10^6$ with a standard deviation 35% of the mean.[1] The truncated catchments of the Serpentine and Harvey rivers show rapid changes in hydrographs in response to individual winter storm events.

Forested subcatchments in the jarrah forest are very effective at trapping nutrients, and the loads of nitrogen and phosphorus leaving microcatchments can be less than those which arrive in rainfall. The clearing of forested subcatchments in the Murray River basin has been accompanied by increasing concentrations of nitrogen in streamflow; there is a linear relationship between flow-weighted mean nitrogen concentration and percentage of catchment cleared, (for a range of land uses, including orchards, sheep farms and even a piggery).[6] Most of the increased nitrogen in streamflow can be attributed to increased fertilizer usage and the fixation of atmospheric nitrogen by pasture legumes. Much of the nitrogen in the Murray River is in the form of nitrate, which can reach a concentration of 6 mg $L^{-1}$. The situation is very different for phosphorus, which is retained in the gravelly and loamy soils of the upper Murray catchments.

In contrast, soils of the coastal plain show little ability to retain phosphorus. The natural soils are highly leached and phosphorus deficient, and the native vegetation is adapted to retain and recycle what little is available of this scarce resource. The establishment of agriculture has only been possible through the use of phosphatic and other fertilizers and the development of strains of legumes efficient in nitrogen fixation under the particular climatic conditions of the region. The use of phosphatic fertilizers on soils with a low phosphorus-retaining capacity has led to increases in phosphorus in streamflow. The increase has been exacerbated because it is the practice to spread superphosphate onto paddocks before the winter rains, when the pastures are dry and plants dormant; it is not convenient to spread the fertilizer later, because machinery is readily bogged down in the wet soil. Thus the rains occur when there is little capacity for plants to retain phosphate, and it is estimated from experimental catchments on pasture that

**Table 1** Annual River Discharge and Phosphorus and Nitrogen Loads to the Peel-Harvey Estuarine System 1977–1988[9]

| River | Discharge load (billion liters/year) | | | Total phosphorus load (tonnes/year) | | | Total nitrogen load (tonnes/year) | | |
|---|---|---|---|---|---|---|---|---|---|
| | Min. | Max. | Mean | Min. | Max. | Mean | Min. | Max. | Mean |
| Harvey River and drains | 86 | 370 | 225 (36%) | 25 | 133 | 82 (57%) | 138 | 1,115 | 430 (35%) |
| Serpentine River | 50 | 190 | 129 (21%) | 14 | 70 | 43 (30%) | 110 | 629 | 250 (20%) |
| Murray River | 62 | 756 | 264 (43%) | 3 | 69 | 18 (13%) | 46 | 2,012 | 553 (45%) |
| Total | | | 618 (100%) | | | 143 (100%) | | | 1,233 (100%) |

*Note:* The rivers and their catchments are shown in Figure 2.

some 30% of phosphorus lost from catchments on deep gray sands and 15% of the loss from sands over clays derive directly from fertilizer application, because of poor retention of the high content of water-soluble phosphorus in superphosphate.[1,7]

Mean flows and nutrient loads from the three catchments are presented in Table 1. While there are large differences between years, because of variability in rainfall (especially on the inland section of the Murray Catchment), in an 'average' year 43% of the flow is contributed by the Murray, but only 13% of the phosphorus; in contrast the Harvey and its associated drains contribute 36% of the flow but 57% of the phosphorus.[8,9] In years when there is little flow from the Murray River, the Harvey may contribute up to 75% of the phosphorus. The Serpentine Catchment, which has a much lower drainage density than the Harvey, is somewhat intermediate in behavior. For these reasons attempts to control phosphorus concentrations in drainage have concentrated on the sandy, coastal-plain catchments.

## IV. WATER MOVEMENT AND EXCHANGE

The highly seasonal river flow, the narrow channel to the ocean, and the large surface area of the water bodies combine to bring about marked seasonal changes in salinity, and the documentation of salinity behavior has been critical to understanding water exchange. In Peel Inlet the salinity is at a minimum (typically 5 ppt) in late winter and then rises because of the combined effects of ebbing river flow, exchange with the ocean and surface evaporation; eventually the salinity increases beyond that of the ocean (35 ppt), rising to 45 ppt in some years.[10] The salinity reached in late summer is affected by the occurrence of unseasonal summer storms — thunderstorms or tropical cyclones which rarely move as far south as this system. In late summer the rate of increase in salinity is arrested; insolation falls and water exchange begins to dominate the equation. Salinity is depressed when rain falls directly on the estuary, but especially when rivers respond to rainfall as their catchments become saturated. The minimum salinity reached in the basin in winter is related to the volume of river flow.

Harvey Estuary shows similar seasonal changes (Figure 4), but they are more extreme because of the greater distance from the ocean; salinities in summer may reach 52 ppt.[10] Estuary flushing by river flow is clearly more rapid in winter, when the volume of water entering by flow is estimated at about 5.8 volumes of the estuary.[1] It is estimated that, including tidal and wind forcing, the gross annual flushing rate for the system is about 10.7 volumes, equivalent to a residence time of about 34 days.[1]

The factors which affect water exchange through the inlet channel to this system have been detailed elsewhere[11] and have been the subject of modeling exercises.[12] On one occasion direct measurements were attempted for exchange during a period of 5 days in winter,[13] and the results illustrate the complexity of the interactive processes involved. Over the whole exercise there was a net water inflow because of a barometric increase in water level (see above), even though this was a period of river discharge. The salinity at a sampling station in the middle of the channel changed dramatically in response to diurnal tidal excursions and illustrated that water leaving the estuary was not the same as that which had entered the estuary during the previous flood tide; and that water entering the estuary was 'new' marine water, not estuary water previously lost from the system. Nutrient concentrations in surface water, especially nitrate,

10

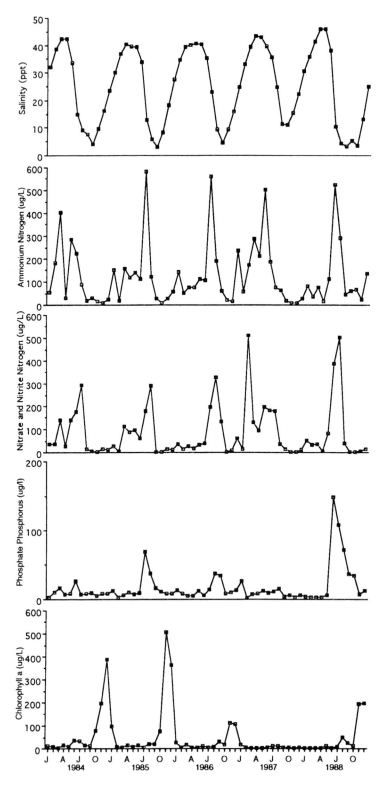

**Figure 4**   The salinity and inorganic nutrient concentrations of Harvey Estuary. The points are mean surface data for three representative sites along the length of the estuary.[10,14]

showed dramatic changes with the ebb and flow of the tides. A budget for the 5 days based on these figures showed massive loss of nitrogen in surface water (23,500 kg), despite a net inflow of water ($3.2 \times 10^7$ m$^3$) entering from the ocean.

Water lost from the system is swept northwards by prevailing wind-induced ocean-water movement.[13] Marine water entering the system behaves as a front which can be detected in Peel bottom water. Marine water from Peel Inlet penetrates Harvey Estuary and can be detected as density-driven intrusions during periods of low wind speed; the system is mixed vertically during strong wind. Thus, in Harvey Estuary there are two states; wind-forced with the water column vertically mixed, and calm with the water column stratified. The constant oscillation between the two states ensures that there is always a horizontal density differential along the length of Harvey Estuary to induce density-driven salt wedge intrusions.[11]

## V. NUTRIENT CONCENTRATIONS

Because of eutrophication problems in the system, there has been an ongoing study of nutrient concentrations and their implications for management. These investigations began in 1977, but were intensified in 1978 in a collaborative, 2-year study designed primarily to discover interrelationships between concentrations at different sites in the inlet, and this study, carried out at seven sites at fortnightly intervals, established that river flow was the only significant exogenous source of nutrients for the inlet.[8] Monitoring of water quality then continued at weekly, fortnightly and (more recently) at somewhat longer intervals in summer. It was found that mean nutrient concentrations for three sites in each inlet did not differ significantly from means calculated using data from eight and seven sites, for Peel Inlet and Harvey Estuary, respectively,[14] and it is the mean for the three sites which is described below.

Concentrations of nutrients differ greatly at different times of year, and between years (Figure 4). Nitrate rises sharply at the time of river flow, and especially during heavy flows of the Murray River, when nitrate concentrations ($NO_3$-N) may reach 33 mg L$^{-1}$ in surface water in Peel Inlet,[10] when it is stratified. At other times of year, nitrate may rise sharply because of nitrification following the decomposition of organic matter in the estuary. Ammonia concentrations are not high in river water, but are mainly generated in the system through decomposition, for example in the estuary in winter or following the collapse of blue-green algal blooms in summer.

Phosphate concentrations ($PO_4$-P) are generally low (<10 µg L$^{-1}$) in the inlet, until the time of river flow; concentrations then increase, especially in Harvey Estuary. However, concentrations then fall rapidly to very low figures and generally remain low throughout the summer period.

Total phosphorus and total nitrogen concentrations are generally high and erratic, but are especially high during the time of *Nodularia* blooms because of the large amounts of nutrients associated with the high biomass. Much of the variability at other times is due to wind stirring, which, as explained further below, readily brings the surface sediments into suspension in this shallow system.

One of the striking features of the system is that nutrient concentrations are generally low until times of river flow, yet the macroalgae and blue-green algae grow most rapidly in the warmer months. This provides circumstantial evidence that nutrients must be trapped in winter and recycled via the sediments to support the development of a high biomass in summer. This recycling is reviewed in greater detail below.

## VI. SEDIMENT PROPERTIES

The sediments range from organic silts to sands.[15] The surface few centimeters have relatively high nutrient concentrations, and are bioturbated through the activities of amphipods and annelids. They are also readily resuspended through wind-induced wave action, and the effect is well seen for shallow sites with a long fetch (Figure 5). Rates of deposition into sediment traps reach 234 g m$^2$ day$^{-1}$ in Harvey Estuary, and this is very largely because of the deposition of resuspended sediment.[16] Such resuspension occurs frequently, and is relatively more significant in Harvey Estuary than in Peel Inlet because of the orientation of the long axis of Harvey Estuary in relation to the prevailing wind direction, and also because of the less dense, organic-rich sediments there. The turbidity brought about by resuspension in Harvey Estuary, together with dense algal blooms for part of the summer, combine to reduce light at the sediment surface to the point that growth of benthic macroalgae are essentially precluded except at the far north of the estuary.[17] In contrast, sufficient light penetrates to the sediment surface in Peel Inlet to allow the growth of macroalgae, especially on the broad, shallow marginal platforms.[18]

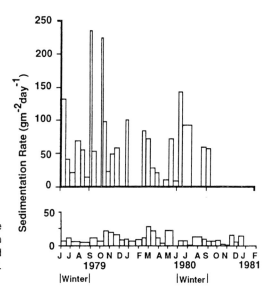

**Figure 5** Sediment resuspension at two sites in the Peel-Harvey estuarine system, one (above) in the north of Harvey Estuary, the other (below) at a more sheltered site in Peel Inlet. Data are in grams per day dry weight. (Data from Gabrielson and Lukatelich.[16])

Another feature of the sediment surface is the presence of relatively high populations of microphytobenthos, as measured by chlorophyll *a* concentrations.[19] These algae offer the potential for high rates of benthic photosynthesis and could affect nutrient exchange between the sediment surface and overlying water. Resuspension of benthic algae, especially pennate diatoms, accounts for much of the chlorophyll recorded in the water column in late summer in both basins.

The sediments are both a source and sink for nutrients. Wind-induced resuspension of sediments provides an opportunity for sediment phosphate adsorption to occur, but in our experience a subsequent period of stratification and consequent oxygen reduction is sufficient to allow phosphate release from the sediments once again. Diatom blooms in winter are responsible for the trapping and sedimentation of nutrients; it was calculated for a single winter river inflow event that of the phosphorus arriving over a period of 18 days (some 80% of which is in the form of filterable reactive phosphate), 71% was trapped in a rapidly developing diatom bloom.[20]

The winter diatom bloom has two effects — one is to trap phosphorus, and the other is to load the sediments with a source of respirable carbon which drives the sediment surface anaerobic and allows subsequent phosphate release. If sediment cores collected from the estuary are incubated in darkness in the laboratory, and phosphate concentrations are measured in the water above, then large seasonal and between-year differences are observed (Figure 6). These differences are attributed to differences in the nutrient loading between years and the consequent occurrence of winter diatom blooms. In years with

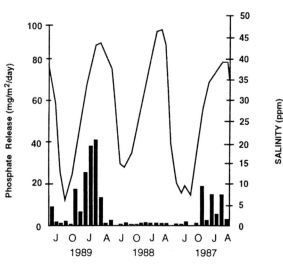

**Figure 6** Release of phosphorus from the sediments of Harvey Estuary. The data are release rates, recorded under standard conditions in the laboratory, for cores removed from the same site in the estuary at different times of year. The background line shows estuarine salinity at the time. (Data of Hill and McComb, not previously published.)

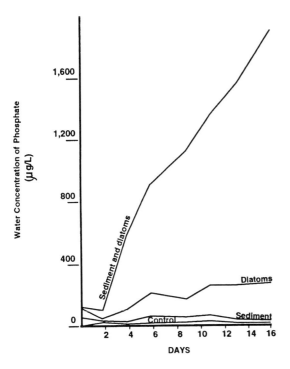

**Figure 7** The effect of diatom detritus on the release of phosphate from samples incubated in the dark under laboratory conditions. Control = estuary water with no sediment; + diatoms = estuary water plus diatom detritus; Sediment = estuary water above a sediment core; Sediment + diatom = estuary water, sediment core, and added diatom detritus. (Data of Lukatelich.[1])

little nutrient loading and no diatom bloom, there was no subsequent nutrient release. In other years there was considerable nutrient release (up to 190 mg P m$^{-2}$ day$^{-1}$), and release continued during and after the occurrence of blue-green blooms in summer.[10,21] The effect of the diatoms on nutrient release has been demonstrated experimentally in the laboratory (Figure 7).

There are large year-to-year differences in the magnitude of the summer *Nodularia* blooms, and these differences are related closely to the volume of river flow in each previous winter, and so to the nutrient loading in that winter.[10,14] This is attributed to the magnitude of diatom blooms and nutrient cycling from the sediments. Thus the size of a *Nodularia* bloom in summer depends not so much on the total nutrient pool accumulated over the long term in the sediments, but whether or not conditions are favorable for nutrient release in a particular year. This has important management implications.

Like the blue-green algae, the macroalgae in Peel Inlet (see below) grow mainly in the summer months, when ambient nutrient concentrations are low. They too depend on the recycling of nutrients from sediments and decaying masses of algae. The anaerobic conditions beneath banks of algae favor nutrient release, and the nutrients promote macroalgal growth where light is not limiting.[22]

## VII. SYMPTOMS OF NUTRIENT ENRICHMENT

Since the mid 1960s massive populations of green macroalgae have been reported from Peel Inlet and from the northern end of Harvey Estuary. The algae foul what were once clean, sandy beaches, onto which they are driven by wind-induced water movement, and may smother marsh plants fringing the estuary. They also accumulate through water movement as dense, offshore banks, in which all but the surface layer becomes a decomposing mass.[24] The first dominant species was *Cladophora montagneanna*, which had not been recorded previously from Australia.[25] This occurred as beds of unattached spheres of radiating filaments (diameter 2–3 cm) in the deeper waters of Peel Inlet, beds from which spheres buoyed up by bubbles of oxygen floated to the water surface and were driven to the shore in long windrows; alternatively, spheres brought into resuspension by vertical mixing would be transported in surface currents. Arrival on the beaches was not seasonal, but laboratory and field studies confirmed that growth occurs mainly in the summer months, and that in winter light and temperature become limiting.[18,25] Salinity has little effect over the range encountered in the estuary. Of the nutrients, experiments using laboratory culture, together with a knowledge of field-measured concentrations, showed that phosphorus was often at growth-limiting concentrations when light was available.[25,27,28] This was confirmed by analysis of tissue concentrations in the field, in relation to critical tissue nutrient concentrations measured in the labora-tory.[25]

Over the years *Cladophora* has been replaced as a dominant alga by other green algae in the genera *Chaetomorpha, Enteromorpha* and *Ulva*.[17] This has occurred as light conditions have generally deterio-rated in Peel Inlet; species of these genera have a different growth form than *Cladophora*, and are higher in the water column. The final demise of *Cladophora* dominance corresponded with a massive storm in 1979, which disrupted the deep algal beds, and which have never reestablished.[17] *Chaetomorpha* and *Enteromorpha* have lower requirements for phosphorus than does *Cladophora*, a conclusion based on laboratory uptake determinations and the concentrations of nutrients in the tissues of plants collected from the field.[17,22,23] Nevertheless, light is of critical importance, and there is a close relationship between the amount of algal biomass found at the end of the growth period in Peel Inlet and the light attenuation coefficient during the growth period.[14,17]

In Harvey Estuary there is a different symptom of eutrophication: the development of massive blooms of the blue-green cyanobacterium *Nodularia spumigena*. These first became obvious in 1978, though earlier reports suggest that large blooms had occurred in some previous years. Since 1978, the blooms have occurred in many but not all summers and, as explained above, blooms only occur after years of substantial river flow. The *Nodularia* blooms are initiated when temperatures rise sufficiently to allow the 'germination' of akinetes in the sediments, in late spring or early summer; the bloom then increases to a level dictated by the amount of sediment phosphorus available, the organisms fixing massive amounts of dinitrogen dissolved in the water.[29,30] The blooms finally collapse as salinity rises to 30 ppt in midsummer.[10]

There is considerable variation in the peak chlorophyll *a* concentrations reached by *Nodularia* blooms in different years, and the size of the bloom is related to the minimum salinity reached in the estuary in the previous winter, which is a reflection of river volume and hence, nutrient loading.[9,10]

A deterioration in the condition of the estuarine system has therefore occurred since the mid-1960s in Peel Inlet, and since at least the late 1970s in Harvey Estuary. The total biomass of macroalgae has differed between years but has in general drifted downwards in recent years, with a deterioration in light climate.[17] In Harvey Estuary there have been few years without blooms recently, and the presence of post-*Nodularia* diatom blooms has become more common in summer, suggesting an increase in nutrient cycling.[21] In some years there have been benthic accumulations of *Oscillatoria* and relatively high populations of the coccoid blue-green *Synechococcus*. Nevertheless, the available evidence suggests that a reduction of nutrient loading will be met with a corresponding reduction in *Nodularia*, rather than a continuation of large blooms sustained by long-term accumulations in the sediments.[1,10,14]

## VIII. CONSEQUENCES OF EUTROPHICATION

The macroalgal populations impact on amateur and commercial fisheries in different ways, which are detailed elsewhere in this volume. The catches of fish 'per unit effort' have increased somewhat with increased macroalgal biomass, and because the amount of primary production is so high in the system, the increase in fish catch has been attributed to other factors, such as additional shelter for juveniles.[31] Prawns and crab populations have been affected in ways which depend on the species of macroalgae, and this is explained in detail in Chapter 13. However, despite this general increase in catch, fish kills have occasionally been reported where large amounts of organic material have decomposed rapidly in a local area and led to oxygen depletion. In addition the presence of large amounts of macroalgae leads to problems in the use of boat motors and to net fouling.[32]

Winter diatom blooms are grazed by large populations of the calanoid copepods *Sulcanus* and *Gladioferans*. While these might in turn be grazed and support food webs, they are rapidly eliminated when *Nodularia* blooms become established; this may be because they are unable to cope with the filaments of the blue-green.[20] Fish may leave the estuary during severe blooms.[33] While severe oxygen depletion is not a general feature of eutrophication in Harvey Estuary, there are occasional local events of severe oxygen depletion at depth during *Nodularia* blooms; high mortality of invertebrates have been reported in the Estuary and are attributed to deoxygenation. There is also the possibility of toxicity. Early classic work linked stock deaths with drinking water containing *Nodularia* in Lake Alexandrina in South Australia,[34] and the strain of *Nodularia*, which occurs in Harvey Estuary, has been shown to produce the toxin under laboratory conditions.[35] While government concern has led to warnings against swimming in the Estuary during the blooms, there have been no reported human responses to the blooms, nor of death of pets and other animals; the occasional deaths of fish and invertebrates referred to above can be attributed to oxygen depletion.

The occurrence of large populations of macroalgae and *Nodularia* have led to the production of nauseous odors which offend local residents and visitors. While there are no quantitative data, a reasonable case has been made that these odors, the appearance of the estuaries and the need to walk through rotting algae on what were once sandy beaches, must have reduced potential visitor numbers and affected real estate values near to the shore.

## IX. DISCUSSION

The nutrients responsible for eutrophication in the Peel-Harvey system clearly derive from catchments cleared for agriculture, especially on the sandy coastal plain. One possible management option is, therefore, to control fertilizer usage on the catchments.

A survey of some 120 possible management options[36,37] confirmed that fertilizer control should be a key component of a general management strategy. Attention is being directed particularly at control in the Harvey Catchment, where several approaches have been adopted. There was an immediate ban on the further clearing of perennial native bushland in the catchment. Improved predictions were made of the levels of fertilizer needed to obtain high plant yield, based on experimentation in the catchment.[1,38] Detailed, paddock-by-paddock testing of soil phosphorus levels were undertaken, and it was found that many soils were already phosphate saturated. Recommendations were made about the timing of fertilizer application, traditionally at the end of summer before the break of season. And a new formulation of phosphatic fertilizer was marketed, with a smaller proportion of readily water-soluble phosphate.[38,39] With the strong support of farm advisers, these modifications were widely accepted by many farmers, and it is estimated that a 30% reduction in phosphorus application had taken place by 1991.[9] Longer-term options are also being promoted, such as altered land use to the growing of tree crops, which require less fertilizer and, probably more important, increase water use and so reduce the volume of water entering drainage.

Already there is some evidence that phosphorus loading from Harvey River is decreasing.[9] Nevertheless, while we can expect a fall in the amount of phosphorus reaching drainage, even if all phosphorus application ceased there remains a phosphorus store in the catchment soils which would continue to be lost to drainage over a long period, and for this reason *Nodularia* might not be satisfactorily controlled in the short term by the reduction in phosphorus loading in the catchment.[1] In addition there are point sources of nutrient addition, including some large piggeries, sheep-holding yards, and increasing urbanization; these are also potential sources of nutrients to the system.

For these reasons, in addition the suggestion was made that a new channel be cut to the ocean at Dawesville (Figure 1), in order to increase water exchange. It was calculated that this increased loss, coupled with a reduction in fertilizer usage, has the potential to control *Nodularia* blooms. After a considerable period of modeling the effects of such a channel on water exchange, tidal amplitude and salinity,[1] the decision has been made to proceed with construction; this has begun, and should be complete by mid-1994. The channel would have two major effects. One is to increase the loss of nutrient-enriched water from the system (the volume exchange through the channel will, according to modeling studies, be three times the present exchange through the Mandurah channel);[1] the other is to increase the salinity in early spring to levels which are not favorable to *Nodularia* growth. There will be a number of other effects, including a reduction of salinity in late summer, an increase in the magnitude of the daily tidal signal, and changes in invertebrate and fish populations in the direction of more marine populations. Some of the secondary effects are difficult to predict with accuracy (for example, effects of tidal regimes on fringing marshes), but the possibility of relatively minor changes in these components of the ecosystem does not outweigh the benefits of reducing eutrophication.

The control measures suggested should lead to a reduction in the populations of *Nodularia* in Harvey Estuary, but the effects on macroalgae are more difficult to predict. Because of the relationship between macroalgal biomass and light attenuation, which is attributed to material from *Nodularia* blooms entering Peel Inlet, it is reasonable to predict a rise in macroalgal populations as water clarity increases. However, providing nutrient loading can be reduced, algal biomass should only remain high until surface sediment stores have been depleted, after which there should be a fall in biomass. These changes are more difficult to model in quantitative terms and remain somewhat more speculative than the conclusions for *Nodularia*. It will clearly be a valuable exercise to document the changes which take place in the system after construction of the channel, as they will be important not only to the management of this estuary but to similar systems in other regions.

The Peel-Harvey estuarine system remains a key area for recreation in the expanding urban area near to the capital. It is also an important area for conservation and has been listed under the Ramsar Convention as an important habitat for waterbirds, including migratory waders. It supports a key recreational and commercial fishery. Its informed management will remain a major challenge for research workers and for those authorities responsible for management implementation.

# REFERENCES

1. **Anon.,** *Peel Inlet and Harvey Estuary Management Strategy,* Environmental Review and Management Programme, Stage 2, prepared for the Department of Agriculture and the Department of Marine and Harbours by Kinhill Engineers Limited, Perth, Western Australia, 1988, 192.
2. **Pritchard, E. J.,** What is an estuary: physical viewpoint, in *Estuaries,* 83, Lauff, G. H., Ed., American Association for the Advancement of Science, Washington, D.C., 1967, 3–5.
3. **Hodgkin, E. P. and Lenanton, R. C. J.,** Estuaries and coastal lagoons of southwestern Australia, in *Estuaries and Nutrients,* Nielsen B. J. and Cronin, L. E., Eds., Humana Press, Clifton, NJ, 1981, 307–321.
4. **Hodgkin, E. P. and Di Lollo, V.,** The tides of southwestern Australia, *J. R. Soc.,* 41, 42–54, 1958.
5. **Collins, P. D. K.,** Murray River basin surface water survey, Technical Note 45, Water Resources Section, Public Works Department, Perth, Western Australia, 1974, 52.
6. **Loh, I. C., Gilbert, C. J., and Browne, K. P.,** Nutrient concentration of streamflow in the Murray River Basin, Western Australia. Report No. WRB 17, Water Resources Branch, Public Works Department, Perth, Western Australia, 1981.
7. **Birch, P. B.,** Phosphorus export from coastal plain drainage into the Peel-Harvey estuarine system of Western Australia, *Aust. J. Mar. Freshwater Res.,* 33, 23–32, 1982.
8. **Hodgkin, E. P., Birch, P. B., Black, R. E., and Humphries, R. B.,** The Peel-Harvey estuarine system study (1976–1980), Report 9, Department of Conservation and Environment, Perth, Western Australia, 1980, 82.
9. **McComb, A. J. and Humphries, R.,** Loss of nutrients from catchments and their ecological impacts in the Peel-Harvey estuarine system, Western Australia, *Estuaries,* 15, 529–537, 1992.
10. **Lukatelich, R. J. and McComb, A. J.,** Nutrient levels and the development of diatom and blue-green blooms in a shallow Australian estuary, *J. Plankton Res.,* 8, 597–618, 1986.
11. **Hearn, C. J. and Lukatelich, R. J.,** Dynamics of Peel-Harvey Estuary, Southwest Australia, in *Residual Currents and Long-Term Transport,* Vol. 1, Cheng, R. T., Ed., Springer-Verlag, New York, 1990, 431–450.
12. **Hearn, C. J., Lukatelich, R. J., and McComb, A. J.,** Coastal lagoon ecosystem modelling, in *Coastal Lagoon Processes,* Kjerfve, B., Ed., Elsevier, Amsterdam, 1994, 471–506.
13. **Black, R. E., Lukatelich, R. J., McComb, A. J., and Rosher, J. E.,** The exchange of water, salt, nutrients and phytoplankton between Peel Inlet, Western Australia and the ocean, *Aust. J. Mar. Freshwater Res.,* 11, 27–38, 1981.
14. **McComb, A. J. and Lukatelich, R. J.,** Interrelations between biological and physiochemical factors in a database for a shallow estuarine system, *Environ. Monitoring Assessment,* 14, 223–238, 1990.
15. **Hill, N. A., Lukatelich, R. J., and McComb, A. J.,** A comparative study of some of the physical and chemical characteristics of the sediments from three estuarine systems in southwestern Australia, Report 23, Waterways Commission, Perth, Western Australia, 1991, 45.
16. **Gabrielson, J. O. and Lukatelich, R. J.,** Wind related resuspension of sediments in the Peel-Harvey estuarine system, *Estuarine Coastal Shelf Sci.,* 20, 135–145, 1985.
17. **Lavery, P. S., Lukatelich, R. J., and McComb, A. J.,** Changes in the biomass and species composition of macroalgae in a eutrophic estuary, *Estuarine Coastal Shelf Sci.,* 33, 1–22, 1991.
18. **Gordon, D. M., Birch, P. B., and McComb, A. J.,** The effect of light, temperature and salinity on photosynthetic rates of an estuarine *Cladophora, Bot. Mar.,* 23, 749–755, 1980.
19. **Lukatelich, R. J. and McComb, A. J.,** Distribution of abundance of benthic microalgae in a shallow southwestern Australian estuarine system, *Mar. Ecol. Prog. Ser.,* 27, 287–297, 1986.
20. **Lukatelich, R. J.,** Nutrients and Phytoplankton in the Peel-Harvey Estuarine System, Western Australia, Ph.D. thesis, University of Western Australia, 1987.
21. **Lukatelich, R. J. and McComb, A. J.,** Nutrient cycling and the growth of phytoplankton and macroalgae in the Peel-Harvey estuarine system, report to Waterways Commission, Environmental Dynamics Report ED-85-118, University of Western Australia, Perth, 1985, 75.
22. **Lavery, P. S. and McComb, A. J.,** Macroalgal-sediment nutrient interactions and their importance to the nutrition of macroalgae in eutrophic estuaries, *Estuarine Coastal Shelf Sci.,* 32, 281–295, 1991.
23. **Lavery, P. S. and McComb, A. J.,** The nutritional ecophysiology of *Chaetomorpha linum* and *Ulva rigida* in Peel Inlet, Western Australia, *Bot. Mar.,* 34, 251–260, 1991.

24. **McComb, A. J., Atkins, R. P., Birch, P. B., Gordon, D. M., and Lukatelich, R. J.,** Eutrophication in the Peel-Harvey estuarine system, Western Australia, in *Estuaries and Nutrients,* Nielson, B. and Cronin, E., Eds., Humana Press, Clifton, NJ, 1981, 323–342.

25. **Gordon, D. M., van den Hoek, C., and McComb, A. J.,** An aegagropeloid form of the green alga *Cladophora montagneana* Kutz, (Chlorophyta, Cladophorales) from southwestern Australia, *Bot. Mar.,* 27, 37–65, 1985.

26. **Gordon, D. M. and McComb, A. J.,** Growth and production of the green alga *Cladophora montagneana* in a eutrophic Australian estuary and its interpretation using a computer program, *Water Res.,* 23, 633–645, 1989.

27. **Birch, P. B., Gordon, D. M., and McComb, A. J.,** Nitrogen and phosphorus nutrition of *Cladophora* in the Peel-Harvey estuarine system, Western Australia, *Bot. Mar.,* 24, 381–387, 1981.

28. **Gordon, D. M., Birch, P. B., and McComb, A. J.,** Effects of inorganic phosphorus and nitrogen on the growth of an estuarine *Cladophora* in culture, *Bot. Mar.,* 24, 93–106, 1981.

29. **Huber, A. L.,** Factors affecting the germination of akinetes of *Nodularia spumigena, Appl. Environ. Microbiol.,* 49, 73–78, 1985.

30. **Huber, A. L.,** Nitrogen fixation by *Nodularia spumigena* Mertens (Cyanobacteriaceae), 1. Field studies and the contribution of blooms to the nitrogen budget of the Peel-Harvey estuary, Western Australia, *Hydrobiologia,* 131, 193–203, 1986.

31. **Potter, I. C., Loneragan, N. R., Lenanton, R. C. J., Chrystal, P. J., and Grants, C. G.,** Abundance, distribution and age structure of fish populations in a Western Australian estuary, *J. Zool. (London),* 200, 21–50, 1983a.

32. **Hodgkin, E. P., Black, R. E., Birch, P. B., and Hillman, K.,** The Peel-Harvey estuarine system. Proposals for management, Report 14, Department of Conservation and Environment, Perth, Western Australia, 1985, 54.

33. **Potter, I. C., Loneragan, N. R., Lenanton, R. C. J., and Chrystal, P. J.,** Blue-green algae and fish population changes in a eutrophic estuary, *Mar. Pollut. Bull.,* 14, 228–233, 1983b.

34. **Francis, G.,** Poisonous Australian Lake, *Nature,* 2, 11–12, 1878.

35. **Runnegar, M. T. C., Jackson, A. R. D., and Falconer, I. R.,** Toxicity of the cyanobacterium *Nodularia spumigena* Mertens, *Toxicon,* 143–151, 1988, 26.

36. **Humphries, R. B. and Croft, C. M.,** Management of the eutrophication of the Peel-Harvey estuarine system, final report, Bulletin 165, Department of Conservation and Environment, Perth, Western Australia, 1984, 55.

37. **Hodgkin, E. P.,** Potential for management of the Peel-Harvey estuary, Proceedings of a Symposium at the University of Western Australia, Bulletin 160, Department of Conservation and Environment, Perth, Western Australia, 1984, 261.

38. **Anon.,** Land use and the environment (six papers on catchment management in the Peel-Harvey system), *J. Agric. West. Aust.,* 30, 83–107, 1989.

39. **Anon.,** Peel Harvey estuarine study, (eleven papers concerned with catchment management.) *J. Agric. West. Aust.,* 25, 83–107, 1984.

40. **McComb, A. J. and Lukatelich, R. J.,** Nutrients and plant biomass in Australian estuaries with particular reference to southwestern Australia, in *Limnology in Australia,* Williams, W. D. and deDeckker, P., Eds., Dr. W. Junk, Dordrecht, 1986, 433–455.

# Chapter 3

# Tuggerah Lakes System, New South Wales, Australia

*Robert J. King and Bruce R. Hodgson*

## CONTENTS

I. Introduction ........................................................................................................... 19
   A. General ........................................................................................................... 19
   B. Lake Usage .................................................................................................... 20
      1. History ....................................................................................................... 20
      2. Fisheries .................................................................................................... 20
      3. Power Generation ................................................................................... 20
II. Physical Properties ............................................................................................... 20
III. Catchment Description .......................................................................................... 22
IV. Water Movement and Exchange .......................................................................... 22
   A. General ........................................................................................................... 22
   B. Water Quality ............................................................................................... 23
      1. Temperature ............................................................................................. 23
      2. Salinity ...................................................................................................... 23
      3. Turbidity ................................................................................................... 23
      4. Dissolved Oxygen and pH ...................................................................... 23
V. Nutrients ................................................................................................................ 23
   A. Input from the Catchment ........................................................................... 24
   B. Urban Inflow ................................................................................................. 24
   C. Sediments ...................................................................................................... 25
   D. Ocean Exchange ........................................................................................... 26
VI. Symptoms of Nutrient Enrichment ..................................................................... 26
   A. Phytoplankton .............................................................................................. 26
   B. Macroalgae .................................................................................................... 26
   C. Seagrasses ..................................................................................................... 27
VII. Discussion .............................................................................................................. 27
Acknowledgments .......................................................................................................... 28
References ....................................................................................................................... 28

## I. INTRODUCTION

### A. GENERAL

Barrier estuaries or estuarine lagoons are characteristic of much of the temperate southeast coastline of Australia.[1] They range in size from large estuaries such as Lake Macquarie and the Wallis Lakes which are over 100 km[2] to small lakes such as Lakes Wallaga, Pambula and Merimbula on the southern New South Wales coast. All are characterized by narrow, elongated entrance channels sometimes only intermittently open to the ocean, and broad tidal and backbarrier sand flats.[2] Away from the active channels the lakes are shallow, low energy environments with the margins often densely covered with seagrass communities.

Three of these barrier estuary systems (Lake Illawarra near the industrial city of Wollongong, south of Sydney, and Lake Macquarie and the Tuggerah Lake system on the New South Wales Central Coast) have been studied extensively. These lakes all support some commercial fishing and prawning, but are in regions of rapid urbanization and are subject to nutrient input from sewage and catchment runoff. They are also subject to high and increasing levels of recreational use. The proximity to major population centers and the coal deposits of the Sydney basin has made these lakes suitable sites for coal-fired power stations. Condenser cooling water is drawn from the lakes and the thermally enhanced water discharged back into the lakes, which then act as a cooling field. This review considers the Tuggerah Lakes only (Figure 1), although similar observations have been made in all three systems.

## B. LAKE USAGE

### 1. History

The first evidence of human habitation in the coastal environment of the Tuggerah Lakes is provided by aboriginal midden sites, shell and other food remains and artifacts. In the coastal escarpment, rock engravings and painting sites provide further clues to the hunter and gathering activities of the pre-European population.[3]

European settlers moved into the area in the 1820s, initially for cattle grazing and dairying, but by the late 1820s logging commenced, with consequent large scale land clearance. Subsequently dairying and citrus growing became important, as they are to some extent today. Commercial fishing on the lakes was well established before the Sydney & Newcastle Railway was officially opened in 1889. Prior to the railway, timber and produce were transported to Sydney from ports outside the Tuggerah estuary, but early in the 20th century there was a thriving ferry service on Lake Tuggerah. As roads were built this was replaced by land transport.[3]

The population of the Wyong Shire has increased markedly in recent years: in 1954 it was 13,097 but had increased to 32,967 in 1971, 47,362 in 1976, and 102,000 in 1990. Employment is now largely in manufacturing and wholesale industries, with power generation, coal mining and primary industry next most important.[4] Recreation and tourism are also of economic significance.

### 2. Fisheries

The Tuggerah Lakes today support a large commercial fishery. In addition, recreational fishing is important with an estimate of 505,778 fishing hours per annum (approximately 1400 hours per day) in 1978/80. The annual commercial fish catch was around 250 tonnes from the late 1950s to the mid-1970s, but increased to approximately 350 tonnes in the early 1980s, due largely to an increase in sea-mullet catches. The average commercial fish catch composition over the period 1955/56 to 1983/84 was sea-mullet 47.6% (by 1980 approximately 60%); luderick, 19.8%; bream, 10.3%; dusky flathead, 6.4%, with all others 15.9% (by 1980 approximately 10% only). The average prawn catch over the same period was 69,700 kg but with a large interannual variation, characteristic of highly fecund species and single-year class fisheries. There appears to be little conflict between commercial fishing and amateur angling since the two groups in large part fish different areas, using different methods, at different times, and the bulk of commercial catch is mullet, a non-angling species. The Munmorah Power Station has influenced the distribution of fish in Tuggerah Lakes; in particular an increase in fish near the heated water discharge has benefitted recreational fishing.

### 3. Power Generation

Munmorah Power Station commenced operation in 1967 and was fully commissioned in 1969 with a maximum generating capacity of 1400 MW. The flow rate at full capacity is 54 $m^3$ $s^{-1}$, but the average capacity factor over the period 1985–1989 was approximately 55%. It operates using coal-fired steam-driven generators, with lake water continuously drawn into an intake canal in Lake Munmorah and discharged into Lake Budgewoi (Figure 1). The flow rate is sufficient to reverse the normal hydraulic flow between the two lakes.

## II. PHYSICAL PROPERTIES

The Tuggerah Lakes system (Figure 1), approximately 100 km north of Sydney on the east Australian coast (33°17'S, 151°30'E), consists of three interconnected coastal lagoons: Lake Munmorah, Lake Budgewoi and Lake Tuggerah.[3] The physical characteristics of the lake system are given in Table 1. The lakes are a barrier estuarine system at a relatively youthful stage[2] and are thought to have formed by longshore currents building a series of sand bars across an indentation in the coastline. The eastern foreshores are consequently of a sand dune nature.[5] The lakes have a narrow opening to the ocean at The Entrance, and this is blocked by sand bars at infrequent intervals. Tidal exchange with Tuggerah Lake has been estimated at less than 1%.[3]

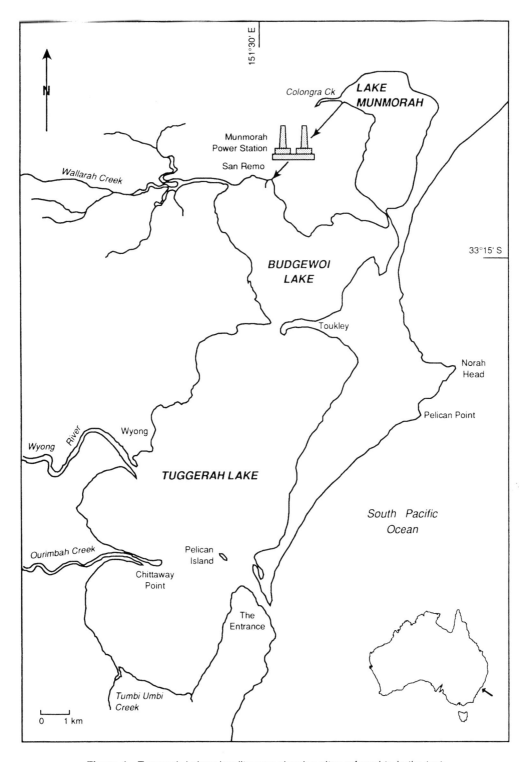

**Figure 1** Tuggerah Lakes: locality map showing sites referred to in the text.

**Table 1** Physical Characteristics of the Tuggerah Lakes and the Catchment[3]

| Lake | Surface (km²) | Shoreline (km) | Volume (10⁶ m³) | Mean depth (m) | Catchment area (km²) |
|---|---|---|---|---|---|
| Tuggerah | 58 | 60 | 91 | 1.6 | 599.5 |
| Budgewoi | 11 | 30 | 18 | 1.6 | 42 |
| Munmorah | 8 | 15 | 14 | 1.8 | 28.5 |
| Total | 77[a] | 105 | 123 | — | 670 |

[a] West et al.[15] gave a value of only 70.29 km², based on aerial photographs. This reflects the shallow topography such that there can be a large change in surface area with a small fall in lake level.

From Inter-Departmental Committee, *Tuggerah Lakes Study Report,* Ministry for Public Works, Sydney, 1979.

The climate is mild with daily mean minimum and maximum temperatures in the range 9–20°C. Mean annual rainfall is approximately 1160 mm, and this is distributed throughout the year with rainfall generally in 130–140 days.[3]

## III. CATCHMENT DESCRIPTION

The catchment of the lakes is only 670 km² (Table 1) and practically coincides with the Wyong Shire. The major streams flowing into the lakes are the Wyong River and Ourimbah Creek into Lake Tuggerah, and Wallarah Creek into Lake Budgewoi. Colongra Creek, the only major stream flowing into Lake Munmorah, has been dammed to form the Munmorah ash dam. Water from the catchment runoff is diverted from the dam into the inlet canal of the power station and then pumped into Lake Budgewoi. Catchment runoff is determined largely by rainfall, with the main runoff events occurring in summer and winter. The larger part of the catchment is either agricultural land or natural vegetation, but as discussed in the introduction, agriculture is no longer an important activity. The considerable increase in the population that has occurred in the last 40 years is located in medium density housing surrounding the lake shores.

## IV. WATER MOVEMENT AND EXCHANGE

### A. GENERAL

The hydrodynamics of the system are controlled largely by catchment runoff. For example salinity from 1973–1988 can be correlated with catchment runoff. Nutrient levels are however lower when catchment flow increases, but this is presumably due to flushing of the system.

Within the lakes, water circulation is affected by the limited tidal exchange, irregular and variable river discharge, and density gradients. In these shallow lakes most of the water circulation results from wind-induced currents and the water column is well mixed. The exception to this is that some areas behind fringing seagrass beds are partially isolated from the main water body, and this is reflected in high nutrients from local runoff, more extreme temperature regimes, and increased macroalgal growth.

The effect of Munmorah Power Station operation is that Lake Budgewoi and Lake Munmorah can in some ways be regarded as a single hydrological system, but differences in nutrients, turbidity and chlorophyll *a* between these lakes and Lake Tuggerah are not due to the increase in water circulation caused by power station operation. Nutrient levels in these lakes are almost always higher than in Lake Tuggerah, which can be attributed to their small catchment and resultant low rates of flushing. During runoff periods there is a net loss of nutrients from these lakes to Lake Tuggerah and thence to the ocean. There is no evidence from seagrass and algal growth, nor from turbidity measures, that the reversal of flow between Lake Budgewoi and Lake Munmorah has had a detrimental effect on the latter. Phytoplankton chlorophyll *a* in these lakes is, however, higher reflecting the higher nutrient levels.

## B. WATER QUALITY

Water quality characteristics of the lakes have been assessed largely on the basis of data routinely collected by the Electricity Commission of New South Wales (ECNSW, now trading as Pacific Power).

### 1. Temperature

The average daily natural surface temperature is about 12°C in winter and 28°C in summer; 13–27°C for bottom waters. Considerable attention has been paid to the temperature in the Tuggerah Lakes, especially in relation to the discharge of cooling water from Munmorah Power Station. Data for the period 1973–1979 show that within 500 m of the outfall the surface temperature is elevated about 7°C on average. Mean surface temperature in Lake Budgewoi is raised 2 to 5°C and the extent of the temperature plume depends on power station output and wind direction.[4] Warmer temperatures are confined to the surface layers. Temperatures in Lake Munmorah are uniform and about 0.8°C higher than in Lake Tuggerah, indicating that the cooling field is mostly confined to Lake Budgewoi.[4]

Temperature measurements in inshore macroalgal beds, effectively isolated from the main body of the lakes by seagrass beds, show significant diurnal and day-to-day variation. These variations are a natural phenomenon. Records from Chittaway Point at the mouth of Ourimbah Creek from 1986/87 show summer temperatures up to 37°C and diurnal variation as high as 15°C in summer and 5°C in winter.[6] A maximum temperature of 40°C was recorded inshore at San Remo.

### 2. Salinity

Salinity in the lakes is affected by freshwater input from catchment runoff, groundwater flow and precipitation, ocean exchange and evaporation, with catchment runoff being the main factor. Minimum salinities occur in late summer and early winter. Salinity variation is normally in the range 17–28 ppt. There are local variations due to proximity to freshwater inflow or ocean influence. These variations are exacerbated with changes in catchment inflow so that after heavy rainfall, salinities as low as 5 ppt are experienced and values up to 49 ppt have been recorded during drought.

### 3. Turbidity

Analysis of Secchi depth data over the period 1973–1988 show that turbidity increases after catchment inflow but also increases with wind mixing and sediment suspension, especially when lake levels are low. The northern lakes are more turbid than Lake Tuggerah. Over the period since 1963 there has been a general increase in turbidity (Table 2) and this, along with nutrient changes, has been invoked as a possible reason for the loss of seagrasses, especially *Ruppia,* from the deeper parts of Lake Budgewoi and southern Lake Tuggerah.[7]

### 4. Dissolved Oxygen and pH

All three lakes were well oxygenated apart from low values in 1974 associated with stratification following flooding. Slightly higher values in spring/early summer and to a lesser extent in autumn are consistent with increased phytoplankton activity, but this is not confirmed. Similarly seasonal decreases in pH were observed. Average values for pH were between 7.8 and 8.0.

## V. NUTRIENTS

Analysis of the long-term Electricity Commission data (Table 2) shows that on the basis of water column nutrients the lakes can be considered to be oligomesotrophic in the classification scheme of Vollenweider.[8] A broad overview of sediment nutrients is given in Table 3. Nutrient concentrations appear to be strongly influenced by catchment runoff events. Seasonal trends in orthophosphate concentration indicate that *in situ* nutrient uptake and release by the sediment and biota may be more important for phosphorus than for nitrogen and silicate. There has been no systematic collection of nutrient data to address the total nutrient budget of the Lakes system and even the data for the water column nutrients are limited. For example, ammonium ion concentrations were not monitored so that attempts at correlating phytoplankton levels with nitrate plus ammonium are not possible.

**Table 2** Water Quality Data for Tuggerah Lakes, 1963–1991 (Surface/Bottom Water)

| | Lake Munmorah | | Lake Budgewoi | | Lake Tuggerah | |
|---|---|---|---|---|---|---|
| **1963–1966[a]** | | | | | | |
| Salinity (ppt) | 29.3 | 29.8 | 32.7 | 33.0 | 32.7 | 33.2 |
| Temperature (°C) | 18.1 | — | 17.9 | — | 18.4 | — |
| Dissolved $O_2$ (mg $L^{-1}$) | — | — | 8.9 | — | — | — |
| pH | — | — | 7.8 | — | — | — |
| Phosphate ($\mu$g $L^{-1}$) | — | — | 6 | 11 | — | — |
| Nitrate ($\mu$g $L^{-1}$) | — | — | 15 | 20 | — | — |
| Secchi depth (m) | | 1.3 | | 1.4 | | 1.6 |
| **1973–1979** | | | | | | |
| Salinity (ppt) | 18.9 | 19.1 | 19.0 | 20.0 | 20.0 | 21.8 |
| Temperature (°C) | 20.0 | 19.7 | 21.2 | 20.1 | 18.8 | 18.7 |
| Dissolved $O_2$ (mg $L^{-1}$) | 7.8 | 7.7 | 7.9 | 7.8 | 8.2 | 7.6 |
| pH | 7.9 | 7.9 | 7.9 | 7.9 | 7.9 | 7.9 |
| Phosphate ($\mu$g $L^{-1}$) | 2 | 2 | 2 | 2 | 2 | 3 |
| Nitrate ($\mu$g $L^{-1}$) | 25 | 38 | 23 | 22 | 19 | 18 |
| Secchi depth (m) | | 1.3 | | 1.2 | | 1.4 |
| **1980–1984** | | | | | | |
| Salinity (ppt) | 26.8 | 27.1 | 26.9 | 27.4 | 26.1 | 27.1 |
| Temperature (°C) | 21.0 | 20.4 | 21.4 | 20.6 | 19.7 | 19.5 |
| Dissolved $O_2$ (mg $L^{-1}$) | 7.4 | 7.3 | 7.7 | 7.6 | 8.0 | 7.8 |
| pH | 7.8 | 7.8 | 7.9 | 7.9 | 7.9 | 7.9 |
| Phosphate ($\mu$g $L^{-1}$) | 3 | 3 | 4 | 3 | 3 | 3 |
| Nitrate ($\mu$g $L^{-1}$) | 13 | 15 | 12 | 12 | 9 | 11 |
| Secchi depth (m) | | 1.0 | | 1.0 | | 1.3 |
| **1985–1991[b]** | | | | | | |
| Salinity (ppt) | 18.6 | 19.7 | 18.6 | 19.1 | 20.1 | 22.6 |
| Temperature (°C) | 20.1 | 19.0 | 24.1 | 21.9 | 19.3 | 19.2 |
| Dissolved $O_2$ (mg $L^{-1}$) | 8.1 | 8.0 | 7.3 | 7.3 | 8.1 | 7.7 |
| pH | 7.9 | 8.0 | 8.0 | 8.0 | 8.0 | 8.0 |
| Phosphate ($\mu$g $L^{-1}$) | 3.7 | 2.8 | 2.6 | 3.3 | 4.7 | 4.2 |
| Nitrate ($\mu$g $L^{-1}$) | 7.9 | 7.0 | 9.1 | 8.1 | 6.7 | 7.7 |
| Secchi depth (m) | | 0.8 | | 0.6 | | 0.8 |

[a] 1963–1966 data, except temperature, salinity and secchi depth from Higginson.[7]

[b] Water quality data for Lake Budgewoi (1985–1991) were taken in the outfall region where dissolved oxygen and turbidity are higher than elsewhere in the lake. The Lake Munmorah data for this period were collected near the inlet to the power station. All other data are from the center of each lake.

## A. INPUT FROM THE CATCHMENT

There are inadequate data to qualify the catchment input for the system but CSIRO estimated an annual input of 4.3–20 tonnes of orthophosphate and 110 tonnes for ammonia plus nitrate for Lake Tuggerah: comparable values for Lake Budgewoi and Lake Munmorah are not available. Particulate nitrogen and phosphorus inputs were estimated to be an order of magnitude higher.

## B. URBAN INFLOW

For urban runoff there are no direct data, and estimates vary widely. There is an urgent need to quantify this factor since estimates based on other studies of urban runoff suggest that values for urban input of dissolved inorganic N and P may be as great as 10–100% of the values for catchment runoff. Estimates of urban runoff have been made on the basis of the areas of urban foreshore, rainfall, an assumed runoff coefficient of 0.4, and the measured concentrations of total nitrogen and phosphorus in stormwater drains.[3] These estimates are given in Table 4. While the proportion of urban runoff to total input is small, it may impact in local areas and also accumulate in the inshore areas of the lakes, resulting in growth of nuisance macroalgae and seagrass. The values for input of nutrients from sewage given in the CSIRO report[4] are extremely low, 145 kg N y$^{-1}$ and 14 kg P y$^{-1}$ from the Toukley treatment

**Table 3** Sediment Characteristics in Tuggerah Lakes

| Characteristic | Lake Munmorah | Lake Budgewoi | Lake Tuggerah |
|---|---|---|---|
| **Center of Lakes (1988)** | | | |
| Total N (mg kg$^{-1}$) | 3400 | 2500 | 2400 |
| Total P (mg kg$^{-1}$) | 220 | 250 | 240 |
| TOC (%) | 5.0 | 2.7 | 3.7 |
| Silt:sand ratio | 98.6:1.4 | 92.3:7.7 | 92.4:7.6 |
| **Inshore Muddy Sand[a] in the Seagrass/Macroalgal Zone (1982–1988)** | | | |
| Total N (mg kg$^{-1}$) | 2111 | 1800–2100 | 382–1555 |
| Total P (mg kg$^{-1}$) | 112 | 125–170 | 96–262 |
| TOC (%) | 0.2–3.8 | 1.6–3.2 | 1.0–5.0 |
| Silt:sand ratio | | 60.5:39.5 | 42.4:57.6 |

[a] Values for inshore sediments given by Higginson[7] for the period 1963–1966 indicate much lower nutrient status (total N 350 mg kg$^{-1}$, total P 50 mg kg$^{-1}$), and a ratio of silt to sand of 8.9:88.1. This may be due to the unrepresentative nature of the data as well as changing conditions.

Data from Sweaney,[9] Cheng,[21] Wilson and King,[6] and ECNSW.

works, and are probably not representative of the situation that prevailed prior to the diversion of sewage to the ocean outfall at Norah Head. There are no estimates for groundwater, though the contribution may be considerable as a result of the legacy of septic tanks.

## C. SEDIMENTS

A summary of sediment data from various sources is provided in Table 3 but it must be borne in mind that these data do not result from a detailed or systematic survey. Estimates of sediment particulate matter inputs are given in Table 4. Sources of sediments include soil erosion from the immediate lake surrounds and the catchment, ash in the power station outfall, and deposition of atmospheric particulate matter. Although the exact contribution of power station operation to the greater sedimentation observed in the northern lakes cannot be fully assessed, the amounts estimated from these two sources are low in relation to total sedimentation.[4]

The lakes are particularly susceptible to siltation and appear to retain most of the silt washed in after heavy rainfall in the catchment.[7] Reports of shoaling of areas at the mouths of Wyong River and Ourimbah Creek occur in NSW Fisheries annual reports from the late 1800s, so that the creation of such

**Table 4** Estimated Nutrient Sources for Tuggerah Lakes (tonnes year$^{-1}$)

| Sources | Lake Munmorah N | Lake Munmorah P | Lake Budgewoi N | Lake Budgewoi P | Lake Tuggerah N | Lake Tuggerah P | Total N | Total P |
|---|---|---|---|---|---|---|---|---|
| Catchment[a] | | | | | | | | |
| Particulate | 436 | 112 | 626 | 102 | 3,240 | 162 | 4,302 | 376 |
| Dissolved | 0.6 | 0.3 | 0.8 | 0.5 | 11.7 | 7.0 | 13.1 | 7.8 |
| Urban runoff | 8 | 0.3 | 33 | 1.4 | 57 | 2.4 | 98 | 4.1 |
| Power Station[4] | | | | | | | | |
| Sawdust | | | 0.25 | 0.008 | | | 0.25 | 0.008 |
| Ash[b] | | | – | 2.0 | | | – | 2.0 |
| Sewage | | | 0.2 | 0.12 | | | 0.2 | 0.12 |
| Sewage (Toukley Sewage Works, 1985) | | | | | | | | |
| Dissolved | | | | | 0.145 | 0.014 | 0.145 | 0.014 |

[a] Data from ECNSW.

[b] The values available for the phosphorus content of the ash are very variable and this upper estimate may be in error by at least an order of magnitude. Less than 1% of the total P in ash is soluble.

silt fans is not a unique feature of urbanization.[3] Long-term estimates of the siltation rates, by the Geological Survey of NSW, based on borehole interpretation, gave an estimated sediment rate of 1.4 mm per year[3] but the rate has increased considerably in the last 20–30 years. Recent estimates based on trace elements in the sediments give average annual sedimentation rates of 11, 10, and 4 mm in the center of Lake Munmorah, Lake Budgewoi, and Lake Tuggerah respectively.[4] Silt and clay introduced to the lakes from the catchment areas by the main creeks appears to be transported to the center of the lakes and not be transported to the inshore areas.[9] This may account for the relatively high sedimentation rates in central lake areas. Silt derived from urban foreshore areas remains in the inshore region because of the lack of inshore water currents, caused by the presence of seagrass beds.

Sawdust, added to the condenser cooling water as a temporary measure to control condenser tube leaks, has been identified as another source of nutrients, but the amounts of nitrogen and phosphorus added to the lakes in this way are minor in relation to natural fluxes. Over the period 1971–1988 the use of sawdust averaged 250 tonnes per annum, but alterations in operations since 1982 have substantially decreased this, and by 1991 this figure was less than 12 tonnes per annum. Batley et al.[4] suggested that the organic carbon sources may stimulate nitrogen fixation, or if used as an energy source for sulfate-reducing bacteria, stimulate anaerobic processes and thereby increase mobilization of phosphate in surficial sediments. However, in the quantities involved, sawdust would be an insignificant carbon source when compared with phytoplankton sedimentation, seagrass and macroalgal production and catchment input of organic matter. Also the quantities used would have little effect on biological oxygen demand as the sawdust is widely dispersed.

## D. OCEAN EXCHANGE
On the basis of limited data the estimated annual import of phosphorus from oceanic water is 3 tonnes, but this would be exceeded by the loss due to oceanic exchange.

## VI. SYMPTOMS OF NUTRIENT ENRICHMENT

Changes over time in the populations of aquatic angiosperms, macroalgae and phytoplankton are often indicators of nutrient enrichment. The available information for Tuggerah Lakes can be interpreted as showing that the system is not markedly eutrophic. The relatively few data on phytoplankton would place the lakes in the oligo-mesotrophic category. The dominance of seagrass in the biomass compared with that of the algae also points to the health of the system. There has been an increase in macroalgae in some inshore areas since the early 1980s.

## A. PHYTOPLANKTON
Phytoplankton biomass and growth have not been extensively studied in the Tuggerah Lakes. Watson's[10] preliminary investigation of phytoplankton showed that diatoms were an important component in all three lakes. The nanoplankton (<15 $\mu$m) fraction was often significant and on one occasion represented 94% of the chlorophyll $a$ measured. Hodgson[11] measured chlorophyll $a$ values in the lakes over the period 1973/74 and noted a coincidence between high chlorophyll values and the influx of nutrients from the catchment with rain. While there is some spatial variation in the mean chlorophyll $a$ levels recorded, in Watson's survey the range was 2 $\mu$g L$^{-1}$ in spring to 4.4 $\mu$g L$^{-1}$ in autumn, with individual values from 1.6–9.1 $\mu$g L$^{-1}$. These values are low for estuarine waters and indicate oligo-mesotrophic conditions and are at levels generally too low to affect lake water turbidity.[12]

## B. MACROALGAE
There is a wide variety of macroalgae represented, but the most abundant are filamentous green algae such as *Chaetomorpha, Rhizoclonium,* and *Cladophora,* along with *Enteromorpha.*[6] These sometimes grow unattached as dense mats inshore of the seagrasses and can accumulate due to the effects of wind. Such algae have been reported as troublesome, causing offensive smells and degrading the recreational amenity at particular sites. In a few inshore localities, values of 200 gdw m$^{-2}$ have been recorded, but if the measured values are considered in relation to the total area of the lakes, then the value would be less than 10 gdw m$^{-2}$. Values up to 400 gdw m$^{-2}$ have been reported in eutrophic estuaries in Western Australia.[13] In the Tuggerah Lakes the offshore seagrass beds not only buffer the main water body of the lake from nutrient-rich urban runoff, but also appear to physically retain the algae in that region.

## C. SEAGRASSES

Three aquatic angiosperms occur in quantity in Tuggerah Lakes: *Halophila ovalis* (R. Brown) Hooker f., *Ruppia megacarpa* Mason and *Zostera capricorni* Ascherson. Their broad distribution has been mapped by Higginson[14] and West et al.[15] King and Holland[16] provided a more detailed survey, plotting seagrasses in terms of subjective estimates of abundance and sociability. In that survey, undertaken in summer 1985, the total area occupied by seagrass was 19.11 km$^2$ or about 25% of the lake area. *Zostera* was most widely distributed, occurring over 12.26 km$^2$, with *Halophila* over 10.40 km$^2$ and *Ruppia* over 8.24 km$^2$. King and Hodgson[17] documented changes in seagrass distribution in the lakes from 1980–1985, and compared these with values from earlier surveys of Higginson in the 1960s.[14,18] In addition, estimates of *Zostera* biomass, based on the methods outlined in King and Barclay,[19] were made. In the three surveys from May 1963 to August 1966 the total area of seagrasses was in the range 28.2–41.9% of the total lake area, whereas by 1980–1985 it was only 17.1–25.0%.[17] The major decrease between the two periods of survey appears to be due to the progressive disappearance of *Ruppia* from the central parts of Budgewoi and Tuggerah Lakes. At the same time, seagrasses have increased in the inshore region. The results over 1980–1985 showed wide variation in both the area occupied by seagrass (13.13–19.11 km$^2$) and total biomass of *Zostera* (840–1888 tonnes), but there were no time-related trends or predictable changes. More recent data are given in Table 5. These again emphasize the wide variation between years, especially for *Ruppia*, which fluctuates greatly and even disappears and reappears, perhaps in response to salinity changes.[4]

## VII. DISCUSSION

The existing information on nutrients and sedimentation, supported by data on seagrasses and macroalgae, all indicate a degradation in the ecosystem of the Tuggerah Lakes. The system is not showing symptoms of serious eutrophication such as massive macroalgal, cyanobacterial and phytoplankton blooms over the whole lake: nutrient loading has not yet reached the 'threshold' limit. Excessive growths of macroalgae in some inshore areas are a response to nutrients in urban runoff though this is poorly documented. Lake management is therefore vitally important at this stage.

Most of the environmental and biological data available on the Tuggerah Lakes have resulted from studies undertaken or initiated by the Electricity Commission of New South Wales (now trading as Pacific Power), and therefore relate directly or indirectly to power station operation. There is now a specific need for a nutrient budget for the Tuggerah Lakes in which the quantitative contribution from all sources, catchment and urban runoff, groundwater input, point sources, and oceanic exchange as well as the dynamics of the system are measured and interpreted.

In 1970, an Inter-Departmental Committee on Tuggerah Lakes was established by the State Government[3] to investigate matters affecting the lakes, and as a response to concerns about siltation, pollution, weed control and the perceived environmental deterioration of the lakes. The composition of that committee emphasized the wide range of authorities with some responsibility for the Tuggerah Lakes: NSW Department of Public Works, the Electricity Commission of NSW, the Health Commission of NSW, NSW State Fisheries, the Maritime Services Board of NSW, the Department of Lands, the NSW Planning and Environment Commission, and the Wyong Shire Council, with the later addition of the State Pollution Control Commission as an observer. The terms of reference were broad, but the notion of a comprehensive development plan for the lakes and adjacent areas was paramount.[3] Specific terms of reference required that a program of works be drawn up and sources of funding be considered.

The present Tuggerah Lakes Restoration Project resulted from a State Government ministerial task force established in 1988. It is directed by a project manager from the Wyong Shire Council, who is the

**Table 5** Areas (km$^2$) of Seagrasses in Tuggerah Lakes

|  | 1980–1985[17] | 1986[a] | 1988[a] | 1991[a] |
|---|---|---|---|---|
| *Zostera* | 8.66–16.69 | 9.69 | 8.97 | 12.31 |
| *Halophila* | 4.10–13.36 | 6.24 | 8.98 | 5.38 |
| *Ruppia* | 1.76–8.24 | 10.82 | 6.81 | 9.10 |

[a] Unpublished data: R. J. King.

executive member of a management advisory committee consisting of one representative from each of the Public Works Department, Department of Lands, State Pollution Control Commission, Electricity Commission and the Soil Conservation Service. In addition, there are two Shire Councillors and a further two members of the local community. The major works being undertaken in this project include:

1. Reduction of sediment and nutrient inputs to the lakes through: the protection of existing wetlands and the creation of sediment traps off urban drains; the installation of gross pollutant traps on major drains, with sediment pits, energy dissipation, trash racks, and wetlands as appropriate at each site; dredging of Ourimbah Creek mouth to redirect the channel, and hence sediment-laden flood flows, towards the Entrance; and aspects of catchment management. The fact that the lakes catchment is all within a single local government area makes the latter easier than might be anticipated. The entire catchment is now sewered, all housing approvals must include proposals for sediment control, and there is a broadly based program for community education.
2. Reduction of sediment and nutrients in the lakes by: the removal of algae and seagrass, and inshore black organic ooze from 56 km, or more than 50%, of the lakes' shoreline (the spoil from this operation is being used to reclaim or create up to 30 m of foreshore reserve); beach cleaning for the removal of stranded weed and litter.
3. Dredging of the Entrance channel to improve tidal exchange, and possibly control flooding in low-lying areas.

CSIRO in conjunction with Wyong Shire Council have installed a satellite monitoring station in the restored area at Long Jetty, which measures automatically at 1–4 hour intervals water depth, surface and bottom water temperatures along with salinity, pH and Eh of bottom, mid and surface waters. The aims of this are to obtain early indication in restored areas of progression towards a well-oxygenated water column, low algal growth and near zero sulfide concentrations in the upper 20 cm of sediment.[20]

Despite the recommendations in the Inter-Departmental Committee reports[3] and the CSIRO review of the Lakes[4] there is as yet no broadly funded study of the ecology of the Lakes such that the long-term effects of the present modification of the lakes environment can be determined. While the present 'restoration' activities may improve the amenity in the short term, the biological consequences have not been assessed, particularly the potential effects of the removal of inshore seagrass habitat in reclaimed areas on fish and prawns. Major areas of seagrass habitat near The Entrance and between Chittaway Point and the mouth of Wyong River are to be left undisturbed.

## ACKNOWLEDGMENTS

The research that forms the background to much of what is written here has been supported by the Electricity Commission of New South Wales (Pacific Power) as part of its environmental monitoring program (Grant No. B200 444 to R. J. King, University of NSW). We are grateful to Pacific Power staff, especially Dr. N. Marshman, for access to data and comments on the manuscript.

## REFERENCES

1. **Barnes, R. S. K.,** *Coastal Lagoons*, Cambridge University Press, Cambridge, U.K., 1980.
2. **Roy, P. S.,** New South Wales estuaries: their origin and evolution, in *Coastal Geomorphology in Australia*, Thom, B. G., Ed., Academic Press, Sydney, 1984, 91.
3. **Inter-Departmental Committee,** Tuggerah Lakes Study Report, Ministry for Public Works, Sydney, 1979.
4. **Batley, G., Body, N., Cook, B., Dibb, L., Fleming, P. M., Skyring, G., Boon, P., Mitchell, D., and Sinclair, R.,** The Ecology of the Tuggerah Lakes System. A review: with special reference to the impact of the Munmorah Power Station, Report prepared for the Electricity Commission of New South Wales, Wyong Shire Council, and the State Pollution Control Commission by CSIRO, Sydney, 1990.
5. **Bird, E. C. F.,** *Coasts — An Introduction to Coastal Geomorphology*, 3rd ed., ANU Press, Canberra, 1984.
6. **Wilson, N. C. and King, R. J.,** Studies on Macroalgae in the Tuggerah Lakes with particular reference to the effects of heated water discharged from Munmorah Power Station, Report to the Electricity Commission of New South Wales, Sydney, 1988.
7. **Higginson, F. R.,** Ecological effects of pollution in Tuggerah Lakes, *Proc. Ecol. Soc. Aust.*, 5, 143, 1971.

8. **Vollenweider, R. A.,** Scientific Fundamentals of the Eutrophication of Lakes and Flowing Waters, with Particular Reference to Nitrogen and Phosphorus as Factors in Eutrophication, Re. OAS/CSI/68-97, OECD, Paris, 1968.

9. **Sweaney, K. A.,** Studies of the Seagrass-Sediment Interrelationship of Tuggerah Lakes, New South Wales, B.Sc. (Honors) thesis, University of New South Wales, Kensington, 1982.

10. **Watson, L. G.,** Phytoplankton Studies of the Tuggerah Lakes, B.Sc. (Honors) thesis, University of New South Wales, Kensington, 1983.

11. **Hodgson, B. R.,** The Hydrology and Zooplankton Ecology of Lake Macquarie and the Tuggerah Lakes, New South Wales, Ph.D. thesis, University of New South Wales, Kensington, 1979.

12. **Kirk, J. T. O.,** *Light and Photosynthesis in Aquatic Ecosystems*, Cambridge University Press, Cambridge, U.K., 1983.

13. **Hillman, K., Lukatelich, R. A., Bastyan, G., and McComb, A. J.,** Distribution and biomass of seagrass and algae, and nutrient pools in water, sediments and plants in the Princess Royal Harbour and Oyster Harbour, Environment Protection Authority, Western Australia, Technical Series No. 33, 1990.

14. **Higginson, F. R.,** The distribution of submerged aquatic angiosperms within the Tuggerah Lakes System, *Proc. Linn. Soc. N. S. W.*, 90, 328, 1965.

15. **West, R. J., Thorogood, C. A., Walford, T. R., and Williams, R. J.,** An estuarine inventory for New South Wales, *N. S. W. Dept. Agric., Fish. Bull.*, 2, 1985.

16. **King, R. J. and Holland, V. M.,** Aquatic angiosperms in coastal saline lagoons of New South Wales. II. The vegetation of Tuggerah Lakes, with specific comments on *Zostera capricorni* Ascherson, *Proc. Linn. Soc. N. S. W.*, 109, 25, 1986.

17. **King, R. J. and Hodgson, B. R.,** Aquatic angiosperms in coastal saline lagoons of New South Wales. IV. Long term changes, *Proc. Linn. Soc. N. S. W.*, 109, 51, 1986.

18. **Higginson, F. R.,** The ecology of Submerged Aquatic Angiosperms in the Tuggerah Lakes System, Ph.D. Thesis, University of Sydney, Sydney, 1968.

19. **King, R. J. and Barclay, J. B.,** Aquatic angiosperms in coastal saline lagoons of New South Wales. III. Quantitative Assessment of *Zostera capricorni, Proc. Linn. Soc. N. S. W.*, 109, 41, 1986.

20. **Skyring, G. W. and Johns, I. A.,** An automatic monitoring and communications system for unattended survey of water quality at Long Jetty, Tuggerah Lake: NSW, a progress report (July 1990 to March 1991), Report to Wyong Shire Council, Wyong, 1991.

21. **Cheng, D.,** A Study on sediment-bound plant nutrients, Long Jetty, Tuggerah Lakes, Report to Wyong Shire Council, Wyong, 1987.

# Chapter 4

# Shenzhen Bay, South China Sea

## Qi Yu-Zao and Zhang Jia-ping

## CONTENTS

I. Introduction .................................................................................................................31
II. Physical Properties ......................................................................................................31
    A. Salinity ................................................................................................................32
    B. Temperature ........................................................................................................32
    C. pH ........................................................................................................................32
    D. Dissolved Oxygen (DO) .....................................................................................33
III. Catchment Description .................................................................................................33
IV. Water Movement and Exchange ...................................................................................34
V. Nutrient Concentrations ...............................................................................................34
    A. Dissolved Inorganic Nitrogen (DIN) .................................................................34
    B. Dissolved Inorganic Phosphorus (DIP).............................................................35
    C. DIN:DIP Atomic Ratios .....................................................................................35
    D. Reactive Si(OH)$_4$ ...............................................................................................36
    E. Chemical Oxygen Demand (COD) .....................................................................36
VI. Sediment Properties .....................................................................................................37
VII. Symptoms of Nutrient Enrichment ..............................................................................37
    A. Species Component .............................................................................................37
    B. Phytoplankton Biomass ......................................................................................38
VIII. Conclusion ...................................................................................................................39
Acknowledgments ..................................................................................................................39
References................................................................................................................................39

## I. INTRODUCTION

Shenzhen Bay is located in the eastern part of the Pearl River Estuary, between 113°53′–114 E and 22°23′–22°30′ N, surrounded by the Shenzhen Special Economic Zone of China and Hong Kong (Figure 1). Shenzhen Bay has a well-developed aquaculture industry, producing fish and the well-known Shajing Oyster. It is also an excellent recreation and sightseeing area. Shenzhen Bay receives water from the Pearl River, Shenzhen River and Yuanland River (Hong Kong). Large amounts of nutrients, organic matter and trace metals enter the Bay with the river water, and exchange with oceanic water is poor. As a result, Shenzhen Bay has become eutrophic, and has had several red tides caused by *Noctiluca scintillans, Gymnodinium* spp. and *Skeletonema costatum* in recent years.

## II. PHYSICAL PROPERTIES

Shenzhen Bay has an area of 126 km$^2$ and a water capacity of $3.78 \times 10^8$ m$^3$, with an average depth of about 3 m. It is located in the subtropic zone and has an oceanic climate, high temperature and heavy rainfall. The annual average air temperature is 22.4°C. The water temperature is affected by air temperature, the yearly average being 21.1°C. The annual average recipitation is 1948 mm, with a maximum of 2662 mm. The annual average salinity of Shenzhen Bay is 22.2‰, with a gradient high in the outer part of the Bay. Influenced by monsoons, the main wind direction over the Bay is southeastern.

    Physicochemical and biological data presented in this chapter were obtained from samples taken in cruises during May 1986, February, March, April, July, August and October of 1987, April 1988 and March, May and June 1989. Each cruise covered 10 sampling sites (Figure 1), with sites 1–3 in shallow nearshore areas and sites 4–10 in offshore areas.

0-8493-6839-1/95/$0.00+$.50

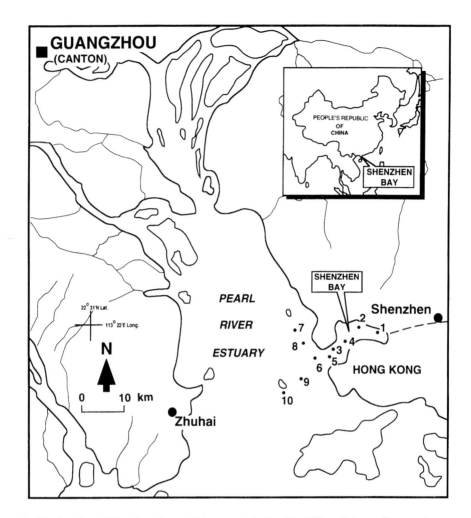

**Figure 1** The locality of Shenzhen Bay, which opens into the Pearl River Estuary. The numbers represent sampling sites referred to in the text.

## A. SALINITY

Salinity in the Bay changes seasonally according to river inflow and other catchment runoff. During the rainy season (April to August), the amount of water from the Pearl River increases greatly and the water salinity is about 20‰. The light freshwater floats on the surface over the denser salty water, forming a salinity gradient of about 10‰ (Figure 2). During dry seasons (September to March), on the other hand, the salinity rises to 30‰ due to the decreased input from the Pearl River and other catchment runoff, and the surface and bottom water is well mixed, with no obvious salinity gradient. The salinity gradient in rainy seasons is important to the vertical distributions of nutrients and occurrence of red tides.

## B. TEMPERATURE

The seasonal range of temperature in Shenzhen Bay is 16.9–30.9°C. The temperature is lowest in February and March and highest in July and August. The vertical temperature difference is slight and within 1°C in most sites, emphasizing that waters are usually well mixed vertically.

## C. pH

The water pH in Shenzhen Bay is slightly alkaline. As recorded, the highest pH is 8.16, and the lowest 7.57. The water environment with such a pH is suitable for the growth of diatoms and beneficial to the uptake of nitrogen and phosphorus.

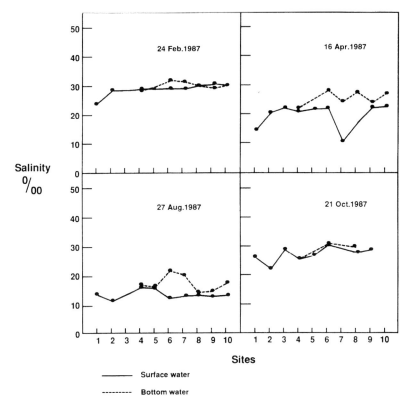

**Figure 2** The salinity of Shenzhen Bay, recorded at the sites shown in Figure 1. The broken line represents bottom water, the solid line surface water. [From Qi, Y. Z., Zhang, J. P., Wu, K. D., Li, J. R., and Qian, H. L., *J. Jinan Univ. (Red Tide Res. Vol. 10)*, 1989 (in Chinese). With permission.]

## D. DISSOLVED OXYGEN (DO)

DO ranges from 4.86 to 10.4 mg $L^{-1}$. Oxygen supersaturation was recorded in cruises during October 1987 and April 1988. The reason may be that the large amount of phytoplankton, *Skeletonema costatum*, present in the Bay during the research period, produced high concentrations of oxygen through photosynthesis.

## III. CATCHMENT DESCRIPTION

In 1990, Shenzhen City had a population of over 1 million. Industries included electronics, food manufacture, textile and machinery. Light industry and especially electronics made up 80% and 45.3% of the total industry. Assessed using the total output value of 1983, about 230 million tons of industrial waste water was discharged into Shenzhen Bay. Shenzhen City occupies 28,600 ha of land, among which agricultural use makes up 37.7%, animal husbandry, 23.5%, urban development, 30.9% and aquaculture, 6.7%. The main agricultural products are rice, sugar cane, peanuts, fruits, pigs, fowls and aquatic products. Agricultural wastes discharged into the Bay are mainly organic manure and pesticides. The amount of waste discharged every year from Shenzhen City was 2840 tons in biological oxygen demand (BOD5), raw sewage comprising 61.2%.

New Territory, Hong Kong, located on the south side of Shenzhen Bay and Shenzhen River, has well-developed livestock farms. The excrement of pigs and fowl are directly discharged into Shenzhen River and Shenzhen Bay. In addition, the sewage of the residents in New Territory and partial waste water of light industry are also directly discharged into the Bay. The pollutants from Hong Kong are mainly wastes from livestock farms, about 22,115 tons of BOD5 a year, of which organic matter comprised a major portion. Overall, Shenzhen City contributes 11% and New Territory, Hong Kong 89% of the total waste input.

## IV. WATER MOVEMENT AND EXCHANGE

Shenzhen Bay is a semi-closed trumpet-like bay, located on the eastern side of the Pearl River Estuary and connected with Hong Kong basin. There are islands on the north, south and west sides, and Shenzhen River is on the east. The Shenzhen River is 31 km long and its water runs into the Shenzhen Bay.

Shenzhen Bay has an irregular half-day tide pattern. The main tide direction is northwest when tides are rising and southeast when tides are falling. The current speed is higher when tides are falling than rising. The average tide range is 1.5 m. Water movement is mainly affected by the runoff of the Pearl River, Shenzhen River and tides.

## V. NUTRIENT CONCENTRATIONS

### A. DISSOLVED INORGANIC NITROGEN (DIN)

The average DIN concentration of the surface waters of the ten sites was 404.3 μg L$^{-1}$, but great seasonal changes were recorded. DIN in rainy seasons was 2–3 times higher than that in dry seasons. The average DIN of April and August 1987 was 676.7 μg L$^{-1}$, in May 1989 it was 697.8 μg L$^{-1}$, while the average for February and October was 95.0 μg L$^{-1}$ (Figure 3).

From the DIN horizontal distribution, there was no evident difference between the inner and outer parts of the Bay in all cruises. The vertical gradient in rainy seasons, however, reflected salinity, with higher concentrations in the surface, with a maximum difference of about 220 μg L$^{-1}$ and an average of 91.6 μg L$^{-1}$. This DIN gradient may be the result of the influence of the existing salinity gradient. Large volumes of freshwater, rich in DIN, formed a surface layer which did not mix well with denser seawater poor in DIN. There was no DIN gradient in dry seasons, which may indicate that the large quantity of DIN brought in with freshwater is the reason for high DIN concentrations in rainy seasons.

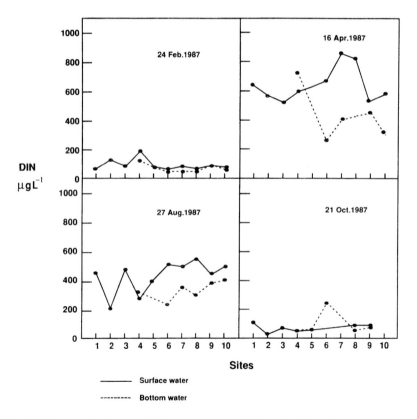

**Figure 3** Dissolved inorganic nitrogen (DIN) concentrations in Shenzhen Bay. Sampling sites are shown in Figure 1. The broken line represents bottom water, the solid line surface water. [From Qi, Y. Z., Zhang, J. P., Wu, K. D., Li, J. R., and Qian, H. L., *J. Jinan Univ. (Red Tide Res. Vol. 10)*, 1989 (in Chinese). With permission.]

## B. DISSOLVED INORGANIC PHOSPHORUS (DIP)

DIP concentrations were high at the nearshore sites (1–3) and low at offshore sites (4–10), with average concentrations of 602 μg L$^{-1}$ and 47.4 μg L$^{-1}$, respectively (Figure 4). DIP was quite constant throughout the year. Obvious DIP differences between the nearshore and offshore sites suggest that the main source of DIP was runoff from Shenzhen and Hong Kong. Unlike DIN, DIP had no vertical gradient. One possible explanation for this is that in the shallow waters, phosphorus from the sediment is released to the lower part of the water column through remixing or mineralization, thus obscuring any original vertical difference.

Statistical analyses suggested that the community distribution of phytoplankton in the cruise of May 1986 was related to DIP and zooplankton biomass, but after an examination of all cruises, DIN was generally considered more important, as is discussed below.

## C. DIN:DIP ATOMIC RATIOS

N/P atomic ratios are useful indicators for determining the relative importance of nitrogen and phosphorus to eutrophication and phytoplankton blooms. The normal N/P ratio in coastal water is about 15. The N/P ratios for the 10 cruises of this study are presented in Figure 5. All N/P ratios in nearshore sites (1–3) in all cruises were low, most lower than 10, almost all lower than 16. Although DIN concentrations were quite high in some cruises (e.g., April and August 1987, Figure 3), very high DIP concentrations in nearshore sites kept N/P ratios low (Figures 4 and 5). This result indicated that DIN was the more limiting factor for the phytoplankton growth. DIP concentrations at other sites (in the middle and mouth of the Bay) were quite stable year-round and far lower than that of nearshore sites, while the DIN concentrations were similar in all sites within the same cruise and sharply different in different cruises (e.g., much higher in rainy seasons than in dry seasons). As a result, N/P ratios in the middle and mouth of the Bay in rainy

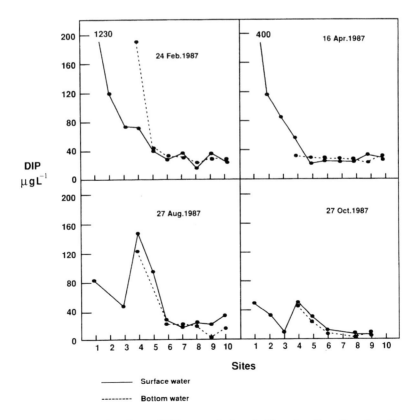

**Figure 4**  Dissolved inorganic phosphorus (DIP) concentrations in Shenzhen Bay. Sampling sites are shown in Figure 1. The broken line represents bottom water, the solid line surface water. [From Qi, Y. Z., Zhang, J. P., Wu, K. D., Li, J. R., and Qian, H. L., *J. Jinan Univ. (Red Tide Res. Vol. 10)*, 1989 (in Chinese). With permission.]

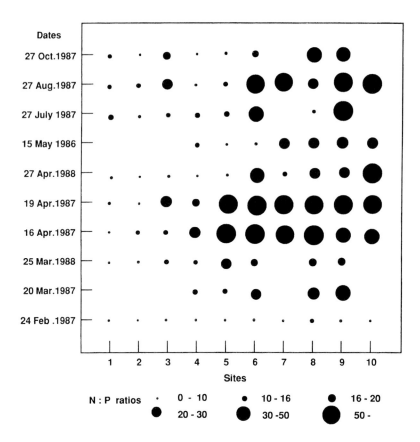

**Figure 5** Nitrogen-to-phosphorus ratios in Shenzhen Bay, at sites shown in Figure 1. Data are for dissolved inorganic nitrogen and phosphorus, and are expressed as atomic ratios. [From Qi, Y. Z., Zhang, J. P., Wu, K. D., Li, J. R., and Qian, H. L., *J. Jinan Univ. (Red Tide Res. Vol. 10)*, 1989 (in Chinese). With permission.]

seasons were over 50, with the maximum of 90.6, suggesting that in this area in rainy periods DIP was more limiting than DIN. N/P ratios in dry seasons were below 16 and DIN became relatively more limiting again.

Generally, DIP was two times higher than the 'eutrophication criteria' and relatively constant throughout the year. On the other hand, DIN underwent seasonal changes, and was just over the eutrophication standard only in rainy seasons. These may indicate that in Shenzhen Bay DIN is more limiting than DIP to the growth of phytoplankton.

## D. REACTIVE Si(OH)$_4$

An essential element for the growth of diatoms is silicon. Soluble silicon [reactive Si(OH)$_4$] concentrations can reach 1250 μg L$^{-1}$ in this Bay. The distribution of Si(OH)$_4$ was like that of DIN, in that a stratification of Si(OH)$_4$ existed in the Bay in rainy seasons, high at the surface and low at the bottom, and disappeared in dry seasons (Figure 6). It was noticeable that very low or negligible Si(OH)$_4$ concentrations were found in the cruise of October 1987, during which time a massive bloom of red tide diatom *Skeletonema costatum* occurred in the Bay. The maximum biomass was 173 gww m$^{-2}$ (Figure 7). Silicon is one of the major components of diatoms, consisting of 6.8–30% of their biomass. The massive growth of *Skeletonema costatum* obviously resulted in short term silicon depletion of the water column.

## E. CHEMICAL OXYGEN DEMAND (COD)

COD in the Bay is between 0.23 and 3.19 mg L$^{-1}$ and generally above 1 mg L$^{-1}$. In 1988, the average COD of all samples was 0.6 mg L$^{-1}$ in March, 1.37 mg L$^{-1}$ in July, 1.30 mg L$^{-1}$ in August and down to 0.44 mg L$^{-1}$ in October, showing a distinct seasonal fluctuation. COD concentrations were generally lower in dry months.

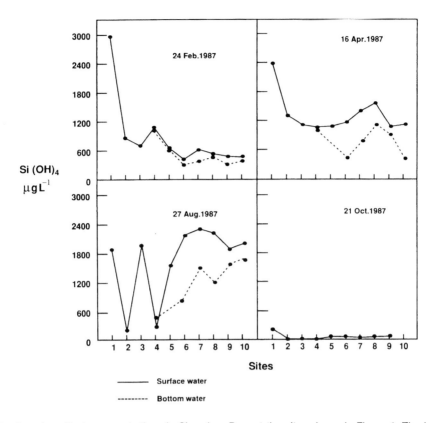

$Si(OH)_4$
$\mu g L^{-1}$

Sites

——— Surface water

--------- Bottom water

**Figure 6** Reactive silicate concentrations in Shenzhen Bay, at the sites shown in Figure 1. The broken line represents bottom water, the solid line surface water. [From Qi, Y. Z., Zhang, J. P., Wu, K. D., Li, J. R., and Qian, H. L., *J. Jinan Univ. (Red Tide Res. Vol. 10)*, 1989 (in Chinese). With permission.]

## VI. SEDIMENT PROPERTIES

Sediment features were studied in the May 1986 investigation. The slope of the sea bed of the Bay is gentle. The sediment is mainly composed of sandy clay. The sediments were sand and clayey sand in sites 5 and 6, and sandy clay in sites 4 and 8. Organic matter was almost evenly distributed in the sediments of the Bay with a typical organic content of about 1.5% and only slight variations were apparent. The oil content of the sediment in all samples was about 100 mg kg$^{-1}$, except 615.5 mg kg$^{-1}$, at site 4 and 553 mg kg$^{-1}$, at site 8. Sulfite content was maximal, 284 mg kg$^{-1}$, at site 4. Total mercury levels were higher in May than October 1986, with a maximum of 0.23 mg kg$^{-1}$, at site 8. The average concentrations of Cu, Pb, Cd, Zn in the sediments were 18.0, 55.5, 2.8 and 63.1 mg kg$^{-1}$, respectively.

## VII. SYMPTOMS OF NUTRIENT ENRICHMENT

### A. SPECIES COMPONENT

Phytoplankton blooms are the major sign of eutrophication in Shenzhen Bay, particularly the red tide species *Noctiluca scintillans*, *Gymnodinium* spp, and *Skeletonema costatum*. Phytoplankton in Shenzhen Bay are mostly coastal species. The major genera are *Coscinodiscus*, *Chaetoceros*, *Rhizosolenia*, *Skeletonema*, *Ditylum*, *Biddulphia*, *Planktoniella*, *Ceratium* and *Noctiluca*. *Skeletonema costatum* and *Noctiluca scintillans* can be very abundant and were the dominant species, respectively, in the investigations of October 1987 and April 1988 (Table 1).

Planktonic diatoms are the main components of the phytoplankton. Among them, *Chaetoceros decipiens*, *Chaet. lorenzianus*, *Coscinodiscus centralis*, *Cos. excentricus*, *Cos. oculus-iridis*, *Ditylum brightwellii*, *Melosira sulcata* and *Planktoniella sol* often became dominant in the phytoplankton communities. *Coscinodiscus* species, for example, made up 76–93% of the phytoplankton in sites 4–10 in

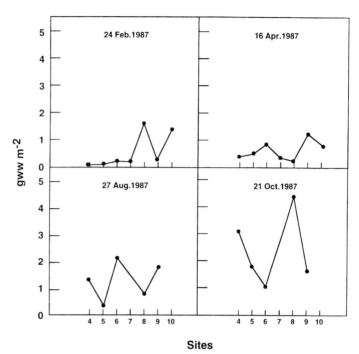

**Figure 7** Phytoplankton biomass in Shenzhen Bay, at the sites shown in Figure 1. Samples were collected in 1987 on (a) February 24th, (b) April 16th, (c) August 27th and (d) October 27th. Phytoplankton were collected using a phytoplankton sampler, 37 cm diameter, and 140 cm long, and results are expressed as g (wet weight)/m².

May 1986. *Cos. oculus-iridis* comprised 46% of the total phytoplankton biomass in March 1989, but dropped to only 1% in May and June of the same year, while *Cos. centralis* was only 5.2% in March 1989, increased to 20.8% in May and reached 80.5% in June 1989, showing a dramatic seasonal species succession.

## B. PHYTOPLANKTON BIOMASS

Phytoplankton biomass in Shenzhen Bay fluctuated greatly; it was usually below 10 gww m$^{-2}$, but sometimes reached over 100 gww m$^{-2}$, as recorded in October 1987 (Figure 7, Table 2).

The reproduction, growth and senescence of phytoplankton have close and complicated relations with environmental factors, which are not always revealed by statistical analyses. Linear regressions were carried out on study data and indicated that phytoplankton biomass was negatively correlated with DIN and reactive Si(OH)$_4$, positively correlated with zooplankton biomass, while no correlation was found with DIP. Multivariate regression using the same data indicated that phytoplankton biomass was statistically negatively correlated with DIN, DIP and reactive Si(OH)$_4$, and positively correlated with COD, DO, salinity and temperature.

**Table 1** Numbers and Percentages of Red Tide Organisms

| Sites | *Skeletonema costatum* (October 21, 1987) Biomass | | | | *Noctiluca scintillans* (April 21, 1988) Biomass | | | |
|---|---|---|---|---|---|---|---|---|
| | gww m$^{-2}$ | % | 10$^4$ cell L$^{-1}$ | % | gww m$^{-2}$ | % | 10$^4$ cell L$^{-1}$ | % |
| 4 | 43.8 | 35.1 | 92.6 | 63.0 | 11.7 | 75.6 | 122.9 | 7.0 |
| 5 | 11.6 | 16.1 | 26.0 | 63.4 | 7.8 | 84.5 | 97.7 | 8.2 |
| 6 | 10.0 | 22.1 | 9.1 | 63.3 | 60.3 | 94.1 | 274.0 | 12.5 |
| 7 | — | — | — | — | 7.0 | 71.8 | 45.7 | 4.8 |
| 8 | 47.8 | 27.6 | 39.9 | 81.0 | 1.0 | 29.5 | 4.2 | 0.5 |
| 9 | 16.5 | 24.8 | 13.8 | 76.3 | 24.3 | 94.9 | 347.0 | 10.5 |
| 10 | — | — | — | — | 4.5 | 65.0 | 40.6 | 1.2 |

**Table 2** Phytoplankton Biomasses (gww m⁻²) in Different Cruises in Shenzhen Bay

| Sites | 05.15 (1986) | 02.24 (1987) | 04.16 (1987) | 04.19 (1987) | 07.27 (1987) | 08.27 (1987) | 10.27 (1987) | 04.27 (1988) |
|-------|-------|-------|-------|-------|-------|-------|-------|-------|
| 4 | 0.21 | 0.03 | 0.34 | 0.08 | 4.98 | 1.37 | 124.02 | 11.47 |
| 5 | 0.44 | 0.04 | 0.44 | 0.59 | 5.15 | 0.42 | 72.20 | 9.25 |
| 6 | 0.48 | 0.20 | 0.79 | 1.58 | 5.13 | 2.20 | 45.12 | 64.10 |
| 7 | 0.27 | 0.17 | 0.28 | 1.33 | — | — | — | 9.80 |
| 8 | 0.80 | 0.27 | 1.13 | 10.47 | 3.89 | 1.82 | 66.60 | 25.60 |
| 9 | 0.35 | 1.51 | 0.17 | 0.02 | 8.13 | 0.82 | 173.05 | 3.40 |
| 10 | 0.18 | 1.26 | 0.71 | 0.23 | — | — | — | 6.87 |
| Average | 0.39 | 0.50 | 0.55 | 0.61 | 5.44 | 1.33 | 96.18 | 18.64 |

The positive correlation between phytoplankton and zooplankton biomass was expected. Zooplankton in the Bay are mainly copepods with *Acartia spini* dominant.

The lack of a significant positive correlation between phytoplankton biomass and DIN or DIP may indicate that the concentrations of these nutrients are not a major limiting factor for phytoplankton growth. The negative correlation between phytoplankton biomass and DIN is attributable to the high DIN concentrations during the wet season when there was little phytoplankton growth (Figure 7), with increased phytoplankton levels in the dry season when conditions (salinity, temperature and light) were more favorable, yet DIN concentrations were comparatively lower. Based on N/P atomic ratios, nitrogen appeared to be more limiting to phytoplankton growth than phosphorus.

The seasonal dynamics of active $Si(OH)_4$ concentrations were similar to those of DIN; however, the dramatic drop in reactive silicon concentrations during the October 1987 bloom of *Skeletonema costatum* (Figure 6) indicated that silicon is potentially a limiting factor for diatom growth in this year-round eutrophic bay. During this bloom, calculations revealed that the amount of silicon bound in *Skeletonema* was over 30 times higher than the amount in the water column; the amount of nitrogen bound in the alga was double that in the water column, the amount of phosphorus was approximately the same.

## VIII. CONCLUSION

In conclusion, Shenzhen Bay is a small, shallow and eutrophic bay. DIN and DIP in the Bay were over the criteria of eutrophication. The pollution is not yet severe as the COD values were mostly about 1 mg $L^{-1}$. From the point of algal ecology and the distribution of year-round N/P atomic ratios, DIN was the more limiting element, compared with DIP, and reactive $Si(OH)_4$ was a potential limiting element for the growth of diatoms. The seasonal change and distribution of nutrients were influenced by the interaction between oceanic exchange and the freshwater from the two main rivers and minor catchment runoff.

The increased frequency of red tides in recent years may indicate that water quality is declining.

## ACKNOWLEDGMENTS

This research was supported by the National Natural Science Foundation of China.

The authors would like to thank Ms. Li Jingrong and Mr. Qian Hongling from the South China Sea Branch, State Administration of Oceanography, Guangzhou; and Wu Kundong and Chen Wei from the Institute of Hydrobiology, Jinan University, for their participation in the assessment of nutrient data analyses and fruitful discussion.

## REFERENCES

1. **Zhang, J. P., Lu, S. H., and Qi, Y. Z.,** Studies on phytoplankton community structures in Shenzhen Bay and their relations with red tides, *J. Sci. Med., Jinan Univ.*, 3, 97, 1987 (in Chinese).
2. **Qi, Y. Z. and Chang, W. Y. B.,** The distribution, occurring mechanism and influences of red tides, *Environ. Prot.*, 19, 1988 (in Chinese).
3. **Qi, Y. Z., Zhang, J. P., Wu, K. D., Li, J. R., and Qian, H. L.,** Red tides in coastal China — with a special case study in South China Sea, *J. Jinan Univ. (Red Tide Res. Vol. 10)*, 1989 (in Chinese).

4. **Pong, K. L. and Li, J. R.,** A preliminary report of a red tide blooming in Chiwan Bay, Shenzhen Special Economic Zone, *J. Jinan Univ. (Red Tide Res. Vol. 86),* 1989 (in Chinese).

5. **Steward, W. D.,** *Algal Physiology and Biochemistry,* University of California Press, Berkeley, 610–614, 1974.

6. **Qi, Y. Z. and Wu, K. D.,** Community structure of phytoplankton in Shenzhen Bay and the fuzzy cluster analyses of environmental factors, research report, 1989 (unpublished).

7. **Rhee, G.-Y.,** Effects of N:P atomic ratios and nitrate limitation on algal growth, cell composition and nitrate uptake, *Limnol. Oceanogr.,* 23, 10–25, 1978.

8. **Zou, J. Z. and Dong, L. P.,** Preliminary study of eutrophication and red tides in Behai Bay, *Environ. Sci. Oceanogr.,* 2, 41, 1983.

9. **Werner, D.,** Silicon metabolism, in *Biology of Diatoms,* lst ed., Werner, D., Ed., Blackwell Scientific Publications, Oxford, 1977, 110.

10. **Pratt, D. M.,** The winter-spring diatom flowering in Narragansett Bay, *Limnol. Oceanogr.,* 10, 173, 1965.

11. **Morris, I.,** *The Physiological Ecology of Phytoplankton,* lst ed., Blackwell Scientific Publications, Oxford, England, 1980, 275.

12. **Mechanical Industry Association of Japan,** *Eutrophication of Waters and Protection Strategies,* Yang, Z. K. and Hu, B. L., Eds., transl., Environmental Science Publishing House, Beijing, 1981, 73 (in Chinese).

*Chapter 5*

# A Case Study of Tolo Harbour, Hong Kong

*I. J. Hodgkiss and W. W.-S. Yim*

## CONTENTS

I. Introduction ...................................................................................................................41
II. Physical Properties .......................................................................................................43
III. Catchment ....................................................................................................................45
IV. Water Movement and Exchange ..................................................................................46
V. Nutrient Concentrations ...............................................................................................47
VI. Sediment Properties......................................................................................................49
VII. Symptoms of Nutrient Enrichment .............................................................................51
VIII. Consequences of Nutrient Enrichment ........................................................................52
IX. Discussion....................................................................................................................53
References ................................................................................................................................55

## I. INTRODUCTION

Tolo Harbour, located in the northeastern part of Hong Kong's New Territories (Figure 1) near the northern limit of the South China Sea, is a subtropical estuary without major rivers draining into it. It is a system under stress due to a combination of natural and anthropogenic factors. Because of the bottle-necked configuration of the coastline and a small tidal range (below 3 m), the harbour is naturally eutrophic. It is poorly flushed by tidal currents under normal conditions. Human interference in the coastal environment, mainly through coastal land reclamation and the discharge of effluents, has accelerated the rate of eutrophication. The capacity of the system to assimilate agricultural, domestic and industrial effluents has been grossly exceeded.[1-2]

The Tolo Harbour system (Figure 2) is of considerable importance because it has currently a total population of over 1 million, mostly at the two new towns Sha Tin and Tai Po. The catchment area is also important in agriculture, with an estimated 10,000 pigs and 230,000 chickens.[3] Although no records of production are available, there is also a traditional inshore fishing industry. Since the 1980s, fish farming has also been introduced into the area, adding significantly to the income derived from fisheries. In addition, the harbour is a popular water recreation area.

A number of historical events have been important in influencing water and land use of the Tolo Harbour system. At the end of the 19th century, Tolo Harbour supported only a sparse population with Tai Po already established as a market town. Two storm surges during 1906 and 1937 led to extensive coastal flooding within the harbour and estimated death tolls of 10,000 and 11,000 for the whole of Hong Kong. The bulk of the casualties during the 1937 event was from within Tolo Harbour where the maximum sea level was recorded.

It was not until after the Second World War that the area became important in market gardening to supply the urban areas of Hong Kong. In 1964, the inlet of Plover Cove was dammed off to form a part of the Plover Cove Water Scheme. Further reduction of the Tolo Harbour catchment took place in 1979 with the construction of the High Island Water Scheme. At the same time, the two market towns, Sha Tin and Tai Po, with populations in 1973 of about 30,000 and 40,000, respectively, began their expansion into new towns of over 750,000 and 250,000, respectively. In order to provide land for urbanization, major coastal land reclamation schemes were undertaken within the inner harbour.

Since 1964, these changes have resulted in a reduction to the catchment area of some 68% to 5000 ha, a reduction in water area by some 28%, a reduction in the length of the coastline of some 22% to 109 km and, most significantly in terms of its value as a nutrient sink, a 42% reduction in the mangrove coastline.[4]

42

**Figure 1** Oblique air photograph of Tolo Harbour taken during January 1987, showing the Plover Cove Reservoir, the two new towns — Sha Tin (right) and Tai Po (left), and the Ma On Shan extension (middle). The two main fish culture areas at Three Fathoms Cove and Yim Tin Tsai are also shown. Aerial photograph reproduced with permission of The Director of Lands, ©Hong Kong Government.

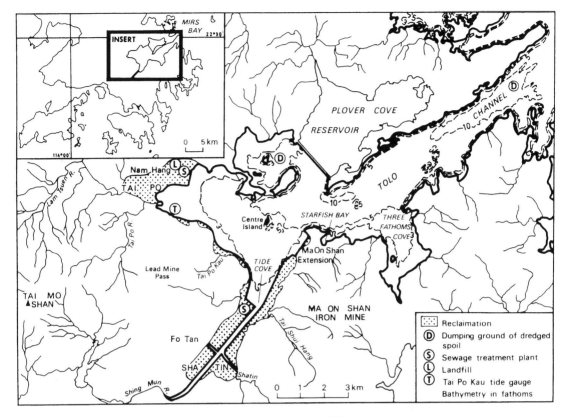

**Figure 2** Location map of the Tolo Harbour system.

## II. PHYSICAL PROPERTIES

Tolo Harbour lies just south of the Tropic of Cancer to the east of the Pearl River Delta. Because of the coastal configuration of Hong Kong, discharges from the Delta have only a minimal influence. The present day coastline is in part the product of a highland area drowned by the Holocene transgression. The sea level was close to that of the present day during interglacial periods in the past,[5] and it is probable that at least some of the coastal erosional features were formed as the result of Pleistocene inheritance.[6]

The overall surface area of Tolo Harbour is about 52 km² and the length from the tidal limit at Tide Cove to Mirs Bay is 23 km. Figure 2 shows the bathymetry of the harbour. Much of the inner harbour is shallow, with a water depth of less than 10 m or about 5 fathoms. The deepest water, at about 20 m, is in Tolo Channel, giving an average depth of about 12 m. Near the exit into Mirs Bay, the maximum width of Tolo Channel is 1.3 km. Because of the narrow entrance of the harbour, tidal circulation is poor and water residence time is long.

The present day climate is tropical monsoonal. Summers are hot and wet with predominantly southwesterly winds, while winters are cool and dry with predominantly northeasterly winds. Sea-surface temperatures fall to between 14 and 16°C in January and rise to between 30 and 32°C in September.[7] Mean annual rainfall is about 2134 mm, a large portion of which is associated with the tropical cyclone or typhoon season between July and September.

Mean sea-surface temperatures (high of 30.5°C in July and low of 15°C in February) recorded by Hodgkiss and Chan[1] in 1978 were significantly different from these noted 6 years earlier by Trott and Fung[8] — 32.3 and 11.6°C. Since the mean air temperatures recorded during both surveys did not differ significantly, it is unlikely that the decreased range is related to differences in air temperature. It is more likely that this narrower range reflects a stenothermal state in Tolo Harbour, possibly related to the changes in configuration of the harbour as a result of the extensive reclamation referred to above, and also the increases in turbidity related both to this reclamation and an increased pollution load resulting from the very large increase in population.

Thermal stratification is not marked in Tolo Harbour except following monsoon rains when thermoclines are generated by the inflow of freshwater. Thus, average surface and bottom water temperatures only differ by some 0.35°C in the inner harbour and some 1.33°C in the channel.[1]

Higher turbidity is common in summer as a result of turbid freshwater inputs and high winds keeping the water well-stirred. Secchi-disc depth readings showed a marked increase from the inner harbour to the channel (averages 2.17 m and 5.67 m, respectively).

The relative elevation of datums and sea levels in Tolo Harbour is summarized in Figure 3. The highest sea level predicted from astronomical tide tables in Hong Kong is 2.7 m above principal datum (P.D. — approximately 1.15 m above mean sea level). This level is frequently exceeded during storm surges associated with the passage of typhoons. Based on records of the Royal Observatory, during an unnamed typhoon in 1937 and typhoon Wanda in 1962, maximum sea levels were 6.10 and 4.88 m P.D., respectively.[9] Typhoons with a track close to Hong Kong, traveling at speed and associated with heavy rainfall within the Tolo Harbour catchment, are expected to reduce the water residence time. This aspect will be examined in greater detail in a later section.

The average diurnal tidal difference is 0.97 m, mean high tide is 1.75 m and mean low tide 0.78 m.

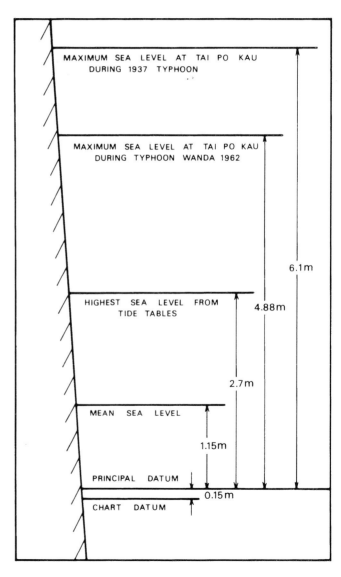

**Figure 3** Relative elevation of datums and sea levels in Tolo Harbour. Redrawn from Yim.[5]

## III. CATCHMENT

As mentioned previously, the Tolo Harbour catchment is some 5000 h and includes agricultural land, industrial areas and several large urban developments in addition to numerous small rural hamlets and a few fishing villages. The population has increased from about 75,000 in 1975 to over 1 million by 1990. Most of the urban growth is in high-rise blocks built on land reclaimed from mangrove swamps and other coastal areas and by cutting platforms on the hillsides. The three major developments are at Sha Tin, Tai Po and Ma On Shan.

A decline in arable land following rapid urbanization has brought a change in agricultural land use and the industry began to concentrate on the rearing of pigs and poultry. However, with enactment of recent livestock control legislation the numbers of livestock have declined significantly.

Effluent-producing industries in Tolo Harbour fall into seven main categories: textiles, food manufacturing, beverages, fabricated metal products, basic metal, chemicals and nonmetallic products. These are located mainly at the two industrial estates, Fo Tan and Nam Hang of the Sha Tin and Tai Po new towns, respectively (Figure 2). Before 1980, liquid waste was mostly discharged into sewers untreated. From 1980 onwards, disposal methods used include controlled tipping, recycling (particularly of heavy metals), and physical and chemical treatment.

Very little is known about the geological input of nutrients within Tolo Harbour and its catchment. Recent studies[10-12] have revealed the presence of Permo-Carboniferous rocks including coal seams on Centre Island and beneath the Ma On Shan Extension (Figure 2). Furthermore, organic-rich black shales outcrop along the shoreline at several localities in the middle harbour.[13] Although much of the sea floor of the inner part of Tolo Harbour has not been drilled, it is probable that Permo-Carboniferous rocks containing coal seams are also present in the area. Because of this, the organic C occurring in sea-floor sediments[14] may be accounted for at least in part by a bedrock derivation.

Of the rivers and streams entering Tolo Harbour, only four are of any significance in terms of size and volume of flow. These are the Tai Shui Hang Stream, Shing Mun River, Lam Tsuen River and Tai Po River. Their average annual discharge rates are 4.01, 6.88, 2.49 and 5.33 million $m^3$ per annum (i.e., 18.7 million $m^3$ in total).[15] River flows reflect the monsoonal seasonality of rainfall, river courses often being dry during the "dry" season (November to March).

During heavy rains in summer, turbidity values are very high in these streams/rivers since they carry a high load. For example, in 1978/79 the mean extinction coefficient k had a value of 4.83 in the Shing Mun River and 4.82 in the Tai Po River, whereas in the Tai Shiu Hong Stream and Tai Po River, values were much lower at 1.31 and 1.53, respectively.[1]

Since all four rivers/streams arise in granitic and volanic rock formations, the water is soft and could be expected to have an acidic pH. The acidic nature of the water near the mouth of all four of these streams/rivers at certain times of the year probably results from the organic enrichment to which all upper stream courses are subjected.

By analyzing the nutrient concentrations in the four streams, it was possible to calculate their nutrient input to Tolo Harbour,[1,15] as shown in Table 1. Thus, from these four major catchment streams/rivers, 58,450 kg of total inorganic nitrogen and 6,281 kg of orthophosphate phosphorus enter Tolo Harbour annually.

According to the eutrophication criteria of Reid and Wood,[16] Vollenweider,[17] Golterman,[18] and Butler and Tibbitts,[19] all four streams can be considered polluted, with the possible exception of the Tai Shui Hang River. The nutrient loads from the four streams, however, are almost insignificant (particularly from a management point of view) when compared to the input from sewage outfalls, drainage channels, sewage treatment plant outfalls and a potable water processing plant outfall. The estimated nitrate and phosphate loading from these other sources is 32,460 and 3,190 kg day$^{-1}$, respectively, compared with the mean annual discharge from the four streams of 14,291 and 6,281 kg, respectively. Thus, they represent some 830 and 185 times the stream loadings, respectively.

Fecal coliform to fecal streptococcal ratios (FC:FS)[20] for the four streams are shown in Table 2. Over 60% of all records indicate that pollution was derived from human wastes and a further 14% from predominantly human wastes, whereas only 7.1% indicate livestock/poultry waste pollution of this watershed.

Phytoplankton standing crops vary at these four streams/rivers but, using chlorophyll $a$ content as a measure of productivity and using a criterion of 0.5 mg chlorophyll $a$ per $m^3$ as indicative of a reasonably productive water, all four water courses are very fertile (Table 3).

**Table 1** Annual Discharge Rates, Average Nutrient Concentrations and Annual Nutrient Input Values of the 4 Major Streams/Rivers Entering Tolo Harbour

| | Tai Shui Hang Stream | Shing Mun River | Lam Tsuen River | Tai Po River | Totals |
|---|---|---|---|---|---|
| Annual discharge rate ($m^3\ yr^{-1}$) | $4.01 \times 10^6$ | $6.88 \times 10^6$ | $2.49 \times 10^6$ | $5.33 \times 10^6$ | $18.71 \times 10^6$ |
| Average nutrient concentration ($mg\ dm^{-3}$) | | | | | |
| $NH_3$-N | 0.06 | 3.09 | 1.84 | 2.26 | |
| $NO_2$-N | 0.03 | 0.40 | 0.44 | 0.40 | |
| $NO_3$-N | 0.21 | 0.80 | 1.47 | 0.77 | |
| $PO_4$-P | 0.03 | 0.31 | 0.53 | 0.51 | |
| Annual nutrient load ($kg\ yr^{-1}$) | | | | | |
| $NH_3$-N | 224 | 21,270 | 4,583 | 12,055 | 38,132 |
| $NO_2$-N | 106 | 2,682 | 1,104 | 2,135 | 6,027 |
| $NO_3$-N | 855 | 5,481 | 3,661 | 4,294 | 14,291 |
| Total inorganic | | | | | |
| N | 1,185 | 29,433 | 9,348 | 18,484 | 58,450 |
| $PO_4$-P | 106 | 2,154 | 1,315 | 2,706 | 6,281 |

Modified from Hodgkiss, I. J. and Chan, B. B. S., *Mar. Environ. Res.,* 10, 1, 1983.

Studies during 1972 and 1973 revealed that 40% of all water courses in the New Territories of Hong Kong were polluted and, further, that in the lowlands the water courses were with few exceptions either polluted or grossly polluted.[21] Overall, it has been calculated that the pollution load was 68% agricultural, 24% domestic and 8% industrial.[22] However, this ratio obviously varies from place to place in the New Territories, and given the very large townships (over 1 million people) discharging their wastes into Tolo Harbour, it is likely that a considerable portion of the load there is domestic in origin.

## IV. WATER MOVEMENT AND EXCHANGE

Estimated water residence times in the inner harbour range from 16 to 42 days,[23,24] except during extreme events such as storm surges and/or periods of abnormally heavy rainfall. The analysis of past storm surge records revealed that the highest maximum sea level in Hong Kong was found to occur within Tolo Harbour.[25] This is explained by the coastal configuration, which, because of its easterly exit into Mirs Bay, is influenced by extreme seiching. Maximum sea levels in the past were associated with typhoons with straight west-northwesterly tracks over the Luzon Strait with the eye passing over the southeastern part of Hong Kong.[26] Therefore, it is expected that low frequency and high magnitude typhoons passing close to Hong Kong are likely to have the greatest impact on lowering the water residence time. Furthermore, during periods with heavy rainfall associated with the summer monsoon, thermoclines generated by the inflow of freshwater are well developed. Their role in displacing water within the harbour and the mixing effect caused by the passage of a typhoon is however poorly understood at present. Nevertheless, since Hong Kong is affected by at least a few typhoons passing within a distance

**Table 2** Percentage Frequency of the Various Fecal Coliform to Fecal Streptococcal Ratios for the Total Input from the 4 Major Streams/Rivers Entering Tolo Harbour

| FC:FS ratio | % Frequency | Significance |
|---|---|---|
| > 4 | 60.7 | Human waste |
| 2–4 | 14.3 | Predominance of human waste in mixed pollution |
| 1–2 | 16.1 | Uncertain |
| 0.7–1 | 1.8 | Predominance of livestock and poultry wastes in mixed pollution |
| < 0.7 | 7.1 | Livestock/poultry wastes |

Modified from Hodgkiss, I. J. and Chan, B. B. S., *Mar. Environ. Res.,* 10, 1, 1983.

**Table 3**  Phytoplankton Standing Crop and Chlorophyll *a* Content in the Major Streams/Rivers Entering Tolo Harbour

| Stream/river | Phytoplankton standing crop (cells cm$^{-3}$) | Chlorophyll *a* content (mg m$^{-3}$) |
|---|---|---|
| Tai Shui Hang Stream | 4,636 | 30.14 |
| Shing Mun River | 3,954 | 51.18 |
| Lam Tsuen River | 954 | 38.51 |
| Tai Po River | 3,345 | 30.44 |

Modified from Hodgkiss, I. J. and Chan, B. B. S., *Mar. Environ. Res.*, 10, 1, 1983.

of 150 km each year, the water residence time of 16 to 42 days is likely to be applicable under normal conditions only. Typhoons with a return period of 10 or more years are likely to shorten the water residence time considerably, possibly to a few days only.

Human interferences have been partly responsible for changing the pattern of water movement and exchange in Tolo Harbour since the 1950s. These include 20 years of discharging iron-mine tailings,[27,28] the isolation of Plover Cove in 1967 with the construction of the Plover Cove Reservoir, large scale coastal land reclamations associated with the construction of the towns of Sha Tin and Tai Po, the redistribution of contaminated spoils within the harbour,[29] sewage discharge and the establishment of two fish farming areas near the Plover Cove Dam and Three Fathoms Cove (see Figure 2).

Although the overall impact of the above factors on water movement and exchange pattern is not well understood, numerous workers have concluded that coastal land reclamation in the past had the effect of reducing tidal circulation.[3,14,30] In spite of efforts made in water quality modeling by mathematical models,[31] the gazetting of Tolo Harbour as the first water control zone by the Hong Kong Government in 1982 and the installation of secondary sewage treatment plants in the 1980s, accelerated eutrophication of the waters of Tolo Harbour continued.

The hydrology of Hong Kong is affected by two major factors, first, the heavy rainfall (on average 2240 mm) which dilutes the sea water from May to September each year, and second, the large spate of freshwater issuing from the Pearl River in summer, which affects mainly the coastal waters.[32]

Localized drainage of freshwater into Tolo Harbour also has an effect on its salinity because of its virtually land-locked nature. As a result of such dilution the salinity of Tolo Harbour surface water ranges from 24 to 33‰ and bottom water from 29 to 34‰, with the lower values occurring during the hot wet summer months.[1,33]

However, there is also an obvious and more persistent impact of freshwater on the salinity of Tolo Harbour in that the surface waters tend to have a consistently lower value. Thus, in the study of Hodgkiss and Chan,[1] the mean vertical gradient varied between 0.94 and 1.23‰ at the four stations investigated in Tolo Harbour. This effect was magnified during monsoon rains when dramatic decreases in surface water salinity resulted, whereas the bottom waters remained virtually unaffected. The mean salinity levels for four stations (two in the inner harbour, one in mid harbour and the last in the channel leading to Mirs Bay) varied only between 30.87‰ at the innermost station and 31.58‰ in the channel for the surface waters and between 32.10 and 32.70‰ at the same stations for the bottom waters. Thus, the gradient from the inner harbour to the station nearest the open sea was very small.[1]

In fact, as mentioned earlier, under normal conditions, Tolo Harbour is a confined body of water with a severely restricted rate of tidal exchange[34] and only the very surface waters tend to be moved by the prevailing northeasterly winds which move the near surface water landward and thus prevent rapid removal of less saline and thus less dense terrestrial runoff.[35]

## V. NUTRIENT CONCENTRATIONS

The mean values for nutrient concentrations (together with $O_2$ and pH values) during the studies by Hodgkiss and Chan[1] are shown in Table 4.

pH values were always alkaline throughout the investigation, with no obvious seasonal pattern, although there was a vertical pattern in that surface waters were slightly more alkaline than bottom waters, possibly indicating the accumulation of organic material at the bottom.

**Table 4**  Mean Values for Various Nutrient Concentrations, Dissolved Oxygen Levels and pH in Tolo Harbour

|  | Inner Harbour | | Channel | |
|---|---|---|---|---|
|  | Surface | Bottom | Surface | Bottom |
| Dissolved $O_2$ (mg dm$^{-3}$) | 5.97 | 4.39 | 5.24 | 4.21 |
| pH | 8.40 | 8.25 | 8.36 | 8.16 |
| NH$_3$-N (mg dm$^{-3}$) | 0.1256 | 0.0850 | 0.0151 | 0.0339 |
| NO$_2$-N (mg dm$^{-3}$) | 0.0052 | 0.0067 | 0.0033 | 0.0077 |
| NO$_3$-N (mg dm$^{-3}$) | 0.0150 | 0.0160 | 0.0097 | 0.0082 |
| PO$_4$-P (mg dm$^{-3}$) | 0.0080 | 0.0119 | 0.0060 | 0.0230 |
| Reactive SiO$_3$-Si (mg dm$^{-3}$) | 0.4531 | 0.5218 | 0.3350 | 0.6198 |

Modified from Hodgkiss, I. J. and Chan, B. B. S., *Mar. Environ. Res.*, 10, 1, 1983.

The mean dissolved oxygen values varied between 5.24 and 5.97 mg dm$^{-3}$ for surface waters and were thus above the typical level reported for tropical water (1.4–2.9 mg dm$^{-3}$).[16] They were, however, at the lowest end of the range for Hong Kong of 5.43–9.3 mg dm$^{-3}$.[36]

There was a seasonal variation in dissolved oxygen content such that, as expected, when temperatures were high in summer, oxygen values were low and vice versa. A slight decrease in oxygen concentration occurred from the innermost harbour to the open sea. The mean difference between surface and bottom water was only +1.31 mg dm$^{-3}$.

There was little evidence of any seasonality in NO$_2$-N values in the surface waters, but the bottom waters exhibited their highest levels between July and October.[1] Bottom water means were significantly higher than surface water means. NO$_3$-N concentrations were also generally higher during the summer. There was a significant difference between mean values of NO$_3$-N at the inner and outer stations, the latter being much lower. Butler and Tibbitts[19] noted a similar trend of decreasing NO$_3$-N content with increasing salinity. Whereas bottom water means were higher than surface water means at the inner stations, they were lower at the outermost station.

PO$_4$-P concentrations fluctuated widely and no seasonal trend was observed. Bottom waters had higher mean concentrations than surface waters. Stirling and Wormald[37] found values high as 37.7 µg dm$^{-3}$ in the innermost harbour and between 30 and 45 µg dm$^{-3}$ further out. During the survey by Hodgkiss and Chan[1] similar values were found in the inner harbour, but in mid-harbour, values as high as 204 µg dm$^{-3}$ were observed and in the channel values as high as 326 µg dm$^{-3}$ were observed (i.e., the values were some 5 to 8 times higher in the later survey). These high PO$_4$-P concentrations resulted in consistently low NO$_3$:PO$_4$ ratios (never greater than 6.1:1), emphasizing the excessive phosphate loading on the harbour.

Bottom waters had a higher reactive silicate silicon concentration and values also increased from the innermost station to the channel station. This is probably a reflection of pollution levels in the inner harbour, since Mitchell[38] has shown that low silica concentrations are common in polluted sea water.

Nutrient concentrations in Tolo Harbour have been studied by a number of workers between 1972 and 1985.[39-41] These records show that dissolved inorganic phosphorus (DIP) and dissolved inorganic nitrogen (DIN) levels have increased alarmingly (e.g., there has been a 10-fold increase in DIP and a 5-fold increase in DIN between 1978 and 1985[41]).

This is supported by the Hong Kong Government's own data[42] for the period October 1973 and September 1984, during which DIN levels rose from 0.08 to 0.49 mg L$^{-1}$; total coliform counts from 3 to 350 per 100 mL; BOD$_5$ values from 1.3 to 2.7 mg L$^{-1}$ and chlorophyll *a* values from 3.1 to 17 mg m$^{-3}$ in inner Tolo Harbour.

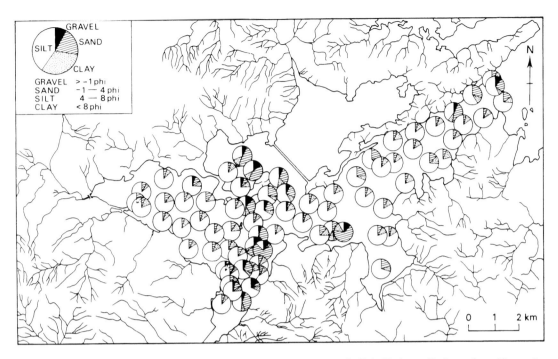

**Figure 4**   Sediment composition distribution of sea-floor sediments in Tolo Harbour. Redrawn from Yim and Leung.[14]

## VI. SEDIMENT PROPERTIES

Sediment properties in Tolo Harbour are comparatively well documented, with recent data on particle size, heavy metals, calcium, organic carbon and phosphorus levels available.[14,43,44]

A sediment composition map of Tolo Harbour is shown in Figure 4 and medium silt is seen to predominate. In the inner harbour and Tolo Channel, all samples show low sand and gravel fractions, which is in sharp contrast to the middle harbour near Centre Island where significant amounts are present: high sand and gravel fractions in this area are considered to represent reworked sediment scoured by relatively strong bottom currents on the basis of the foraminiferal assemblages found.[45] Near the entrance of Tolo Channel, the samples with significant amounts of sand size fractions may be accounted for by the dumping of dredged spoils containing constructional wastes.

All samples are either poorly sorted (80.3%) or very poorly sorted (19.7%) with a tendency for samples with a coarse mean size to be more poorly sorted than those with a finer mean size.[14] Based on Stokes' law, the dominant sediment type, medium silt, is predicted to fall below a bottom current velocity of 2 cm s$^{-1}$. This is suggestive of a low energy eutrophic environment.

The distribution of organic C and Ca in the −170 micron fraction are shown in Figures 5 and 6 respectively. Because of the moderately strong negative correlation found between them, the source of organic C is unlikely to be related to the presence of shells.[14] The highest organic C concentrations are found in the inner and middle harbour at a water depth exceeding 6 m or about 3 fathoms. The most likely explanation is that the bedrock in the vicinity is enriched in organic C as discussed previously. This is supported by the recognition of coal seams on Centre Island[46] and beneath the Ma On Shan Extension,[11] and the presence of black shales with plant fossils on Ma Shi Chau.[47] Tolo Harbour is therefore likely to be a naturally eutrophic system. Calcium content was attributed to the presence of shell remains.[14]

In the distribution of heavy metals, speciation was found to be an important controlling factor.[14] Based on a comparison of correlation coefficients between the Tolo and Victoria harbours,[48] the positive correlation found between Cr, Cu, Fe, Pb and Zn, and organic C in Tolo Harbour was explained by the natural input of metals including Cu, Pb and Zn, from the mineralized bedrock on Tai Mo Shan and Lead Mine Pass (Figure 2). This is supported by the absence of large scale industrial input of effluents before the date of sample collection. On the other hand, the dumping of dredged spoils originating from the inner harbour can be traced particularly well by the Pb and Zn contents.

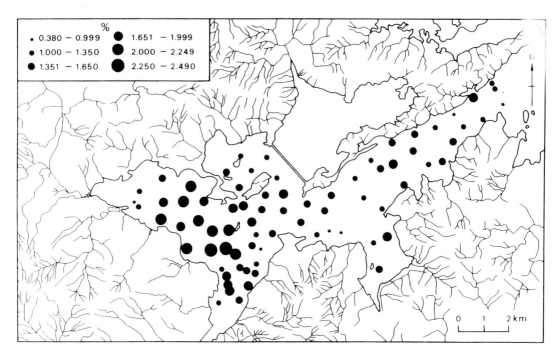

**Figure 5** Distribution of organic C in the −170 micron fraction of sea-floor sediments in Tolo Harbour. Redrawn from Yim and Leung.[14]

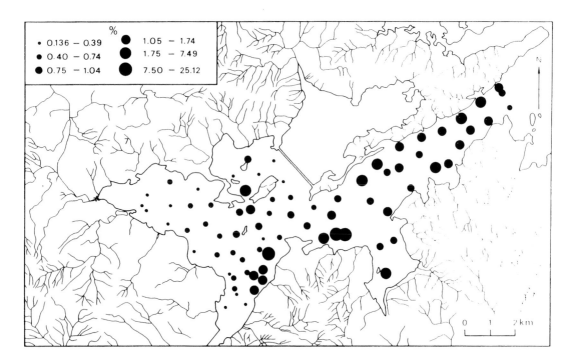

**Figure 6** Distribution of Ca in the −170 micron fraction of sea-floor sediments in Tolo Harbour. Redrawn from Yim and Leung.[14]

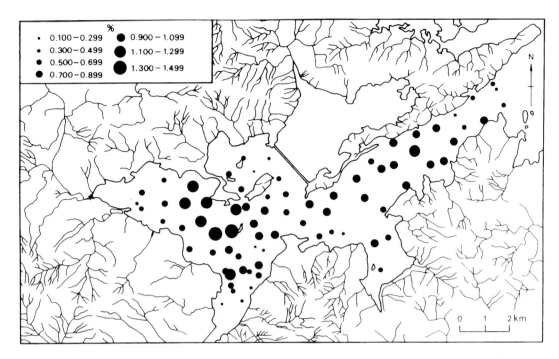

**Figure 7**  Distribution of total P in sea-floor sediments of Tolo Harbour. Redrawn from Thompson and Yeung.[43]

The P adsorption characteristics of sediments in Tolo Harbour have been studied in the laboratory and in the field.[37,43] The distribution of total P in sea-floor sediments (Figure 7) and organic C were found to show a good correlation although the degree of correlation was poorer in areas under the influence of coastal reclamation.[43] On the other hand, reclamation sediments were found to be particularly efficient in absorbing phosphates, with areas affected by reclamation showing the lowest P content.[37] The high P content in the inner part of Tolo Harbour at a water depth of below 6 m is best explained by the presence of a reducing environment and/or a local source of supply.

Since 1988, a number of other sediment characteristics with relevance to eutrophication have been monitored routinely at five stations in Tolo Harbour.[44] Results of relevance include the electrochemical potential and the total N content. Values for the former are within the range of −300 to −400 MV which is indicative of severe oxygen depletion. For the latter, the total N content was in the range 2000 to over 3000 ppm and was considered to be unsatisfactory.

Removal of waste materials from Tolo Harbour is severely impeded as a result of the long flushing period, poor tidal exchange, and the slow rate at which the terrestrial run off falls away.[23,24,34] Sedimentation and the absence of any thorough mixing result in much of this nutrient accumulation taking place at the bottom of the harbour. Even as long ago as 1975, Preston[24] noted phosphate accumulation and the development of anoxic conditions. Stirling and Wormald[37] showed that Hong Kong sediments have a high capacity to absorb dissolved phosphate from sea water and "lock it up" by a process which is virtually irreversible under polluted conditions such as those which exist in Tolo Harbour. Hodgkiss and Chan[1] showed that over a 10-month period the $PO_4$-P concentrations in the sediments showed a 60% increase. Thus, the sediments are reflecting the eutrophic nature of the overlying waters.

## VII. SYMPTOMS OF NUTRIENT ENRICHMENT

The deteriorating water quality in Tolo Harbour has been noted by numerous authors since 1971.[1,3,4,7,8,15,23,24,33-35,37,39,40,43,49-58] In an editorial in *Marine Pollution Bulletin,* Morton[56] referred to Tolo Harbour as "Hong Kong's First Marine Ecological Disaster" and described how surveys of the harbour in 1977, 1980, 1983 and 1986 demonstrated a steady decline not only of harbour water quality but also of the channel fauna towards the open sea. As a result, by 1986 inner Tolo Harbour was effectively dead, and coral destruction up the channel had proceeded at a rate of about 1 km per year.

The Tolo Harbour phytoplankton has been studied since 1974, and the documented changes in species dominance and biomass are symptomatic of progressive eutrophication. The dominance of diatoms has decreased from 80–90% of the population to 53% in 1982–1985,[55] while levels of dinoflagellates have increased, as attested by an increase in the number of red tides from 2 in 1977/78 to 17 in 1984.[3,41] In the same time period, chlorophyll $a$ levels in the inner harbour and channel have also at least trebled to 17.1 mg m$^{-3}$ and 4.1 mg m$^{-3}$, respectively.[41] Chan and Hodgkiss[59] calculated recent phytoplankton productivity indices of 38.05 for surface waters and 46.66 for bottom waters in the inner harbour and 25.36 and 30.24, respectively, for the channel. Since indices of 9–14 are considered typical of eutrophic waters,[60] it is clear that Tolo Harbour is in an advanced state of eutrophication.

The changes in the phytoplankton community in Tolo Harbour can be related to developments in the catchment which have resulted in changes in the nutrient loading to the embayment. Progressive eutrophication has also caused decreased species richness and declining populations of bivalves and gastropods,[61-63] coral,[61,64,65] the epibenthic community[66-71] and fish.[58,72] Total coliform and fecal coliform levels in the water and shellfish have increased dramatically since 1972,[40,51,54,57,73] and on the basis of EEC[74] and WHO[75,76] standards, the water is potentially hazardous for swimming and poses a threat through the collection of shellfish contaminated with pathogenic organisms.

## VIII. CONSEQUENCES OF NUTRIENT ENRICHMENT

Red tides in Tolo Harbour have shown a progressive increase since 1977[3,41] and the numbers are shown in Table 5. The results of such red tides can be fish kills due to oxygen depletion or toxin production, and Table 5 also shows the number of fish kills recorded in Tolo Harbour. Holmes and Lam[53] report that only one of these fish kills in Tolo Harbour related to red tides could be considered as a toxic effect; the others were as a result of oxygen depletion.

Between 1980 and 1984, 11 fish kills in mariculture areas were attributed to red tides, with a loss of 86 tonnes of cultured fish (35% of total cultured fish weight) valued at HK$4.2 million.[77] Shellfish contamination by such red tide toxins can also lead to human health problems on consumption. There is no recorded evidence that the latter has occurred in Tolo Harbour though a number of shellfish intoxications have been noted in Tolo Harbour as well as elsewhere in Hong Kong. Thus, toxic forms such as *Gymnodinum* spp. and *Gonyaulax* spp. have been regularly found in inner Tolo Harbour (11.8% of samples contained them) in the period 1976–1981[77] and PSP levels in shellfish in the harbour have tripled between 1984 and 1987.[78]

Toxic red tides have also been reported to cause minor irritations to the eyes, skin and respiratory system of swimmers but this has not been recorded in Hong Kong.[77]

During the period 1979 to 1984, fisheries production in Tolo Harbour declined from 935 to 606 tonnes (a decrease of 35%) and in Tolo Channel from 1046 to 786 tonnes (a decrease of 24%). The majority of this decline has been attributed to pollution since production remained high in Mirs Bay outside Tolo Harbour.[72]

Nutrient enrichment, particularly of phosphates, is the main cause of anoxic conditions in this harbour.[1] The sources of these nutrients are largely from human wastes (both sewered and unsewered) and animal wastes flushed into catchment streams. (Other forms of pollution have also occurred, particularly heavy metals, as a result of industrial activity in the catchment, but that, as they say, is another story!)

Tolo Harbour is considered to be very eutrophic and this poses a serious threat to the ecosystem and to human health. Hodgkiss and Chan[1] concluded that Tolo Harbour was an environment already under stress, and based on a continuing and increasing nutrient loading, that gross pollution could be predicted. Thus, the harbour's limited capacity to assimilate wastes had been exceeded, so that continuing changes to its configuration together with vastly increased sewage inputs threaten to turn it into an environmental disaster area. The Government's Environmental Protection Department (or Agency as it then was) stated in 1984 that Tolo Harbour had reached a stage where nutrient loading exceeded its receptive capacity, and thus deteriorating water quality would continue until domestic and agricultural loads were reduced by collection and/or *in situ* treatment.[2]

They further noted that plans for sewage treatment facilities would only partially reduce the load, since (a) the two sewage treatment plants proposed would still result in significant levels of $NO_3$-N

**Table 5**  Red Tides and Associated Fish Kills in Tolo Harbour

| Year | Number of red tides | Number of related fish kills |
|------|---------------------|------------------------------|
| 1977 | 2 | 1 |
| 1978 | 1 | 0 |
| 1979 | 1 | 0 |
| 1980 | 4 | 2 |
| 1981 | 3 | 0 |
| 1982 | 3 | 0 |
| 1983 | 11 | 1 |
| 1984 | 16 | 2 |
| 1985 | 16 | 2 |
| 1986 | 26 | 4 |
| 1987 | 19 | 1 |
| 1988 | 43 | 2 |
| 1989 | 20 | 0 |
| 1990 | 17 | 0 |

*Note:*  1977–1986 data abstracted from Wu[3], 1987/88 from Environmental Protection Department[44] and 1989/90 from Mak.[81]

and $PO_4$-P in their effluents, and (b) farm wastes would also have to be prevented from entering the harbour.

By 1988, Tolo Harbour had reached its most critical stage[56] and in that year a Tolo Harbour Action Plan (THAP) was developed by the Government to counter this disastrous trend.[4]

## IX. DISCUSSION

Starting in 1967, reclamation of Tolo Harbour for Plover Cove Reservoir; for Tai Po, Shatin and Ma On Shan new towns; for a racecourse; for a sewage treatment plant; and for industrial development have all led to an even more reduced tidal exchange rate than that described by earlier workers.

The population of the catchment has increased from 70,000 in 1973 to 600,000 in 1988 and will eventually develop to more than 1 million. Two secondary sewage treatment plants handle the human waste plus some of the industrial load, but these plants themselves result in N and P enrichment of the harbour by their effluents. In addition there is an even greater unsewered human waste load.

Total BOD loads increased gradually in 1976–1981 and then peaked dramatically in 1982; in 1983 and 1984 they were lower as the Sha Tin and Tai Po Sewage Treatment Plants began to have an effect. However, N and P loads continued to increase from 1976 to 1984 as a result of high nutrient values in sewage plant effluent and also increased agricultural loads.

New legislation to control animal wastes has subsequently led to a decrease in the pollution load from this source, but this has not resulted in any significant improvement in Tolo Harbour because of the enormous sewered and unsewered human organic waste load. Industrial wastes also add to the load. The relative contributions of these various sources[4] to the total load are given in Table 6. For comparison, urban runoff has been estimated to contribute an insignificant 2% to the total BOD load.

In February 1982, Tolo Harbour was gazetted as Hong Kong's first water control zone under the 1980 Water Pollution Control Ordinance. Controls were enforced under this Ordinance in 1988. Unfortunately, a number of the water quality objectives set for the zone have still not been attained.

Since there is no simple way of increasing flushing rates in Tolo Harbour to alleviate its receptive capacity, pollution loads have to be reduced still further in order to meet these water quality objectives. Between 1986 and 1988, the Tolo Harbour Action Plan (THAP) was developed by the Hong Kong Government in order to attempt to restore the water quality to the level specified in the water quality objectives established in 1982. (The three major objectives are shown in Table 7 for Tolo Harbour and Channel.) This was because in 1986 compliance was poor in terms of bottom DO and surface chlorophyll *a* values, both in the inner harbour and in the channel.[4,79]

**Table 6** The Relative Contributions of Various
Sources to the Total Waste Load on Tolo Harbour

| Source | % Contribution | | |
|---|---|---|---|
| | BOD₅ | Nitrogen | Phosphorus |
| Unsewered population | 29 | 12 | 12 |
| Sewered population | 8 | 69 | 56 |
| Poultry | 20 | 8 | 24 |
| Pigs | 20 | 11 | 8 |
| Industry | 23 | — | — |

Modified from Holmes, P. R., *J. Inst. Water Environ. Manag.*, 2, 171, 1988.

**Table 7** The 3 Major Water Quality Objectives Gazetted
by the Hong Kong Government for Tolo Harbour

| | Harbour | Channel |
|---|---|---|
| Chlorophyll *a* (mg m⁻³) | <20 | <6 |
| DO — surface to 2 m above bottom (mg L⁻¹) | >4 | >4 |
| DO — below 2 m from bed (mg L⁻¹) | >2 | >4 |

In order to reduce total pollution loads, the plan involves (a) upgrading effluent treatment at Sha Tin and Tai Po sewage treatment plants; (b) exporting effluents from the catchment; (c) increasing the sewered population by providing more connections, and (d) decreasing agricultural loads.

Sha Tin sewage treatment plant currently contributes about 44% of the total N load and 51% of the total P load on Tolo Harbour and Tai Po sewage treatment plant 18% and 8%, respectively (but they both contribute only 8% of the BOD load). Further reduction of N and P loads will only be achieved by the introduction of advanced sewage treatment involving the chemical removal of nutrients or by re-using the effluents at, for example, Sha Tin racecourse and the nearby Jubilee sports center for irrigation purposes or by re-using the effluent for domestic flushing. The daily irrigation demand would only account for about 7% of the effluent flow but flushing would account for another 25% of the flow. The Government of Hong Kong has, however, already decided that export of the effluent is a better proposition!

Thus, starting in 1994, the sewage effluent (as well as waterworks sludge from the nearby Shatin water treatment facility) is to be diverted into Victoria Harbour (the area of water between Hong Kong Island and the mainland) via tunnels, on the assumption that the latter harbour has a greater receptive capacity. There are doubts concerning the effects of this diversion, since Victoria Harbour is already severely polluted.

Interceptor sewers are planned under Hong Kong's long term sewerage strategy and new developments are already being sewered automatically. This stategy also involves the collection of the sewage for disposal along a submarine outfall well to the south of Hong Kong.

As stated earlier, Hong Kong's animal waste control legislation has begun to have an effect and the contribution of agricultural wastes to the total N and P load of Tolo Harbour decreased from 66% and 80%, respectively, in 1976 to 26% and 35%, respectively, in 1986.[80] It is likely that these values have since been reduced by even more, but no post-1986 data are available concerning these loads. Further livestock control measures are planned.

In addition, industrial effluents are to be more closely controlled, under a licensing system introduced in 1987, first, by removing exemption clauses in the legislation which permit the discharge of effluents in existence before enforcement of the legislation to continue, and second, by removing the so-called "30% clause" which allowed existing effluents to be increased by up to 30% without penalty.

Eutrophication is well advanced and there has so far been little capital investment to stop it. It can only be hoped that THAP will take effect before destruction becomes irreversible.

# REFERENCES

1. **Hodgkiss, I. J. and Chan, B. S. S.,** Pollution studies on Tolo Harbour, Hong Kong, *Mar. Environ. Res.,* 10, 1, 1983.
2. **Environmental Protection Agency,** Environmental Protection in Hong Kong 1983–84, Government Printer, Hong Kong, 1984.
3. **Wu, R. S. S.,** Marine pollution in Hong Kong: a review, *Asian Mar. Biol.,* 5, 1, 1988.
4. **Holmes, P. R.,** Tolo Harbour — the case for integrated water quality management in a coastal environment, *J. Inst. Water Environ. Manag.,* 2, 171, 1988.
5. **Yim, W. W.-S.,** unpublished data, 1988.
6. **Hopley, D.,** Geomorphological development of modern coastlines: a review, in *Geomorphology: Themes and Trends,* Pitty, A. F., Ed., B and N Publishers, Croom Helm, 1985, 56.
7. **Morton, B.,** An introduction to Hong Kong's marine environment with special reference to the north-eastern New Territories, in *Proc. 1st Int. Marine Biological Workshop: The Marine Flora and Fauna of Hong Kong and Southern China, 1980,* Morton, B. S. and Tseng, C. K., Eds., Hong Kong University Press, Hong Kong, 1982, 25.
8. **Trott, L. B. and Fung, A. Y. C.,** Marine pollution in Hong Kong, *Mar. Pollut. Bull.,* 4, 13, 1973.
9. **Chan, Y. K.,** Statistics of extreme sea-levels in Hong Kong, *Royal Observatory Hong Kong Technical Note,* No. 35, 1983.
10. **Yim, W. W.-S., Nau, P. S., and Rosen, B. R.,** Permian corals in the Tolo Harbour Formation, Ma Shi Chau, Hong Kong, *J. Paleontol.,* 55, 1298, 1981.
11. **Enpack Consultants (Asia) Limited,** Personal communication, 1990.
12. **Addison, R.,** The Geology of Sha Tin, Geotechnical Control Office, Civil Engineering Services Department, Hong Kong Government, 1986.
13. **Yim, W. W.-S. and Ng, C. Y.,** Distribution of metals in some blackish shales of Hong Kong, *Geol. Soc. Hong Kong Newsl.,* 4/1, 13, 1986.
14. **Yim, W. W.-S. and Leung, W. C.,** Sedimentology and geochemistry of sea-floor sediments in Tolo Harbour, Hong Kong — implications for urban development, *Geol. Soc. Hong Kong Bull.,* 3, 493, 1987.
15. **Hodgkiss, I. J. and Chan, B. S. S.,** Studies on four streams entering Tolo Harbour, Hong Kong in relation to their impact on marine water quality, *Arch. Hydrobiol.,* 108, 185, 1986.
16. **Reid, G. K. and Wood, R. D.,** *Ecology of Inland Waters and Estuaries,* Van Nostrand Company, New York, 1976.
17. **Vollenweider, R. A.,** *A Manual on Methods for Measuring Primary Production in Aquatic Environments,* IBP Handbook No. 12, Blackwell Scientific Publications, Oxford, 1974.
18. **Golterman, H. L.,** Natural phosphate sources in relation to phosphate budgets, *Water Res.,* 7, 3, 1973.
19. **Butler, E. I. and Tibbitts, S.,** Chemical survey of the Tamar Estuary. I. Properties of the water, *J. Mar. Biol. Assoc. U.K.,* 52, 681, 1972.
20. **Geldreich, E. E. and Kenner, B. A.,** Concepts of faecal streptococci in stream pollution, *J. Water. Poll. Contr. Fed.,* 41, 336, 1969.
21. **Government of Hong Kong,** New Territories Stream Pollution Study, Government Printer, Hong Kong, 1974.
22. **Environmental Resources Ltd.,** *Control of the Environment in Hong Kong: Final Report,* Environmental Resources Limited, Hong Kong, 1977.
23. **Oakley, H. R. and Cripps, T.,** Marine pollution studies at Hong Kong and Singapore, in *Marine Pollution and Sea Life,* Ruivo, M., Ed., Fishing News (Books) Limited, London, 1972, 83.
24. **Preston, T. R.,** An account of investigation carried out into marine pollution control needs in Hong Kong with special reference to the existing and future urban centres about Victoria and Tolo Harbour, in *Proc. Pacific Science Association Spec. Symp. on Marine Science,* Morton, B., Ed., Government Printer, Hong Kong, 1975, 91.
25. **Yim, W. W.-S.,** An analysis of tide gauge and storm surge data in Hong Kong, in *Future Sea-Level Rise and Coastal Development,* Yim, W.W.-S., Ed., Geological Society of Hong Kong and University of Hong Kong, abstracts 5, 18, 1988.
26. **Petersen, P.,** Storm surge statistics, *Royal Observatory Hong Kong Technical Note,* No. 20, 1975.
27. **Wan, C. M.,** Geomorphological Impact of the Ma On Shan Iron Mine Tailings on the Coastal Environment, B.Sc. thesis, University of Hong Kong, Hong Kong, 1981.
28. **Wong, M. H., Chan, K. C., and Choy, C. K.,** The effect of the iron mine tailings on the coastal environment of Tolo Harbour, Hong Kong, *Environ. Res.,* 15, 342, 1978.
29. **Engineering Development Department,** Hong Kong Government, personal communication, 1982.
30. **Morton, B.,** Pollution of the coastal waters of Hong Kong, *Mar. Pollut. Bull.,* 20, 310, 1989.
31. **Bowler, R. A.,** The Tolo water quality model project, *Hong Kong Engineer,* 13/11, 13, 1985.
32. **Morton, B. S. and Wu, S. S.,** The hydrology of the coastal waters of Hong Kong, *Environ. Res.,* 10, 319, 1975.
33. **Trott, L. B.,** Preliminary hydrographic studies of Tolo Harbour, Hong Kong, *J. Chinese University Hong Kong,* 1, 255, 1973.
34. **Gordon, M. S.,** Water pollution control in Hong Kong: What can be done? *Pacific Science Association Special Symposium on Marine Sciences,* Hong Kong, 1973, 100, 1975.

35. **Watson, J. D. and Watson, D. M.,** Marine Investigation into Sewage Discharges: Report and Technical Appendices, Government Printer, Hong Kong, 1971.
36. **Chau, Y. K. and Abesser, R.,** A preliminary study of the hydrology of Hong Kong territorial waters, *Hong Kong Univ. Fish. J.,* 3, 1, 1958.
37. **Stirling, H. P. and Wormald, A. P.,** Phosphate/sediment interaction in Tolo and Long Harbours, Hong Kong and its role in estuarine phosphate availability, *Estuarine Coastal Mar. Sci.,* 5, 631, 1977.
38. **Mitchell, R.,** *Water Pollution Microbiology,* Vol. 1, Wiley Interscience, New York, 1972.
39. **Fung, A. Y. C.,** A study of Selected Pollution Problems in Tolo Harbour, M. Sc. thesis, Chinese University of Hong Kong, Hong Kong, 1972.
40. **Kueh, C. S. W.,** An investigation on the nutrients, coliform bacteria and other indicators of marine pollution in Tolo Harbour, Hong Kong, *Hong Kong Fish. Bull.,* 4, 115, 1974.
41. **Hodgkiss, I. J. and Chan, B. S. S.,** Phytoplankton dynamics in Tolo Harbour, *Asian Mar. Biol.,* 4, 103, 1987.
42. **Engineering Development Department,** *Data Report No. 10. Monitoring of Local Waters and Sewage Characteristics,* Pollution Control (Liquid and Solid Waste) Division, Hong Kong Government, Hong Kong, 1986.
43. **Thompson, G. B. and Yeung, S. K.,** Phosphorus and organic carbon in the sediments of a polluted subtropical estuary, and the influence of coastal reclamation, *Mar. Pollut. Bull.,* 13, 354, 1982.
44. **Environmental Protection Department,** Marine Water Quality in Hong Kong, Hong Kong Government, Hong Kong, 1990.
45. **Li, Q. Y. and Yim, W. W.-S.,** Foraminiferal thanatocoenoses in surficial sea-floor sediments of Hong Kong, *Acta Micropaleontologica Sinica,* 5, 221, 1988 (in Chinese).
46. **Lee, C. M.,** personal communication, 1987.
47. **Nau, P. S.,** personal communication, 1986.
48. **Yim, W. W.-S.,** Geochemical mapping of bottom sediments as an aid to marine waste disposal in Hong Kong, *Conservation and Recycling,* 7, 309, 1984.
49. **Trott, L. B.,** Marine ecology in Tolo Harbour, Hong Kong, *Chung Chi J.,* 11, 26, 1972.
50. **Wong, M. H., Ho, K. C., and Kwok, T. T.,** Degree of pollution of several streams entering Tolo Harbour, Hong Kong, *Mar. Pollut. Bull.,* 11, 36, 1980.
51. **Wong, M. H., Yeung, Y. F., Wong, K. S., and Leung, K. N.,** A preliminary survey of organic pollution of shellfish in Tolo Harbour, Hong Kong, *Hydrobiologia,* 54, 141, 1977.
52. **Wear, R. G., Thompson, G. B., and Stirling, H. P.,** Hydrography, nutrients and plankton in Tolo Harbour, Hong Kong, *Asian Mar. Biol.,* 1, 59, 1984.
53. **Holmes, P. R. and Lam, C. W. Y.,** Red tides in Hong Kong waters — response to a growing problem, *Asian Mar. Biol.,* 2, 1, 1985.
54. **Hodgkiss, I. J.,** Bacteriological monitoring of Hong Kong marine water quality, *Environ. Int.,* 14, 495, 1988.
55. **Lam, C. W. Y. and Ho, K. C.,** Phytoplankton characteristics of Tolo Harbour, *Asian Marine Biology,* 6, 5, 1989.
56. **Morton, B.,** Hong Kong's first marine ecological disaster, *Mar. Pollut. Bull.,* 19, 299, 1988.
57. **Kueh, C. S. W. and Trott, L. B.,** A preliminary bacteriological examination of water and shellfish in Shatin Hoi, Hong Kong, *Chung Chi J.,* 12, 79, 1974.
58. **Morton, B.,** Pollution and the subtropical inshore hydrographic environment of Hong Kong, in *Proc. 2nd Int. Mar. Biol. Workshop: The Marine Flora and Fauna of H.K. and Southern China, 1986,* Morton, B., Ed., Hong Kong University Press, Hong Kong, 1990, 3.
59. **Chan, B. S. S. and Hodgkiss, I. J.,** Phytoplankton productivity in Tolo Harbour, *Asian Marine Biology,* 4, 79, 1987.
60. **Takahashi, M. and Bienfang, P. K.,** Size structure of phytoplankton biomass and photosynthesis in subtropical Hawaian water, *Mar. Biol.,* 76, 203, 1983.
61. **Dudgeon, D. and Morton, B.,** The coral associated mollusca of Tolo Harbour and Channel, Hong Kong, in *Proc. 1st. Int. Mar. Biol. Workshop: The Mar. Flora and Fauna of H.K. and Southern China, 1980,* Morton, B. and Tseng, C. K., Eds., Hong Kong University Press, Hong Kong, 1982, 627.
62. **Lee, S. Y. and Morton, B.,** The Hong Kong Mytilidae, in *Proc. 2nd Int. Workshop on the Malacofauna of Hong Kong and Southern China, 1983,* Morton, B. and Dudgeon, D., Eds., Hong Kong University Press, Hong Kong, 1985, 49.
63. **Lee, S. Y.,** Growth and reproduction of the green mussel *Perna viridis* (L.) (Bivalvia, Mytilacea) in constrasting environments in Hong Kong, *Asian Mar. Biol.,* 3, 111, 1986.
64. **Scott, P. J. B. and Cope, M.,** Tolo revisited: A resurvey of the corals in Tolo Harbour and Channel 6 years and a million people later, in *Proc. 2nd Int. Mar. Biol. Workshop: The Marine Flora and Fauna of H.K. and Southern China II, 1986,* Morton, B. Ed., Hong Kong University Press, Hong Kong, 1990, 1263.
65. **Scott, P. J. B. and Cope, M.,** The distribution of scleractinian corals at six sites within Tolo Harbour and Channel, in *Proc. 1st Int. Mar. Biol. Workshop: The Marine Flora and Fauna of Hong Kong and Southern China, 1980,* Morton, B. and Tseng, C. K., Eds., Hong Kong University Press, Hong Kong, 1982, 575.
66. **Wu, R. S. S.,** Periodic defaunation and recovery in a sub-tropical epibenthic community, in relation to organic pollution, *J. Exp. Mar. Biol. Ecol.,* 64, 253, 1982.

67. **Shin, P. K. S.,** A trawl survey of the subtidal Mollusca of Tolo Harbour and Mirs Bay, Hong Kong, in *Proc. 2nd Int. Workshop on the Malacofauna of Hong Kong and Southern China, 1983,* Morton, B. and Dudgeon, D., Eds., Hong Kong University Press, Hong Kong, 1985, 439.

68. **Shin, P. K. S.,** Benthic invertebrate communities in Tolo Harbour and Mirs Bay: a review, in *Proc. 2nd Int. Mar. Biol. Workshop: The Marine Flora and Fauna of H.K. and Southern China II, 1986,* Morton, B., Ed., Hong Kong University Press, Hong Kong, 1990, 883.

69. **Horikoshi, M. and Thompson, G.,** Distribution of subtidal molluscs collected by trawling in Tolo Harbour and Tolo Channel, Hong Kong, in *Proc. 1st Int. Workshop on the Malacofauna of Hong Kong and Southern China, 1977,* Morton, B., Ed., Hong Kong University Press, Hong Kong, 1980, 149.

70. **Ansell, A. D. and Morton, B.,** Aspects of natacid predation in Hong Kong with special reference to the defensive adaptations of *Bassina calophylla (Bivalvia),* in *Proc. 2nd Int. Workshop on the Malacofauna of Hong Kong and Southern China, 1983,* Morton, B. and Dudgeon, D., Eds., Hong Kong University Press, Hong Kong, 1985, 635.

71. **Wu, R. S. S. and Richards, J.,** Mass mortality of benthos in Tolo Harbour, *Hong Kong Fisheries Occasional Paper,* No. 2, Agriculture and Fisheries Department, Hong Kong, 1979.

72. **Richards, J.,** personal communication, 1985.

73. **Ni, C. Z. and Huang, Z. G.,** Faecal coliform contamination of intertidal bivalves from Hong Kong, in *Proc. 2nd Int. Workshop on the Malacofauna of Hong Kong and Southern China, 1983,* Morton, B. and Dudgeon, D., Eds., Hong Kong University Press, Hong Kong, 1985, 473.

74. **Council of the European Economic Community,** Council directive concerning the quality of bathing water (76/160/ EEC), *Off. J. Eur. Community,* L31, 1, 1975.

75. **World Health Organization,** *Control of Water Pollution,* W.H.O., Geneva, 1967.

76. **World Health Organization,** *Guidelines for Health Related Monitoring of Coastal Water Quality,* U.N.E.P. and W.H.O., Copenhagen, 1977.

77. **Wong, P. S. and Wu, R. S. S.,** Red tides in Hong Kong: problems and management strategy with special reference to the mariculture industry, *J. Shoreline Manag.,* 3, 1, 1987.

78. **Environmental Protection Department,** Environment Hong Kong 1988: A Review of 1987, Environmental Protection Department, Hong Kong, 1988.

79. **Reed, B. S.,** Sewage and a city by the sea — Hong Kong: a case in point, *J. Inst. Water Environ. Manag.,* 2, 13, 1988.

80. **Phillips, D. J. H.,** personal communication, 1986.

81. **Mak, P. M. S.,** personal communication, 1991.

# Chapter 6

# Eutrophication of the Lagoon of Venice: Nutrient Loads and Exchanges

*A. Marcomini, A. Sfriso, B. Pavoni, and A. A. Orio*

## CONTENTS

I.  Historical Evolution .................................................................................................... 59
II.  Physical Properties ................................................................................................... 61
III.  The Catchment .......................................................................................................... 61
   A.  Drainage Area of Rivers Feeding the Lagoon ................................................. 61
   B.  Catchment Area of Direct Inputs ...................................................................... 62
   C.  Estimated and Measured Nutrient Loads ........................................................ 62
IV.  Water Movement and Exchange ............................................................................. 63
V.  Sediment Properties ................................................................................................. 63
VI.  Biological Evolution and Biomass Distributions ................................................... 64
   A.  Central Lagoon ................................................................................................... 64
   B.  Northern and Southern Lagoons ...................................................................... 67
VII.  Nutrient Concentrations .......................................................................................... 69
   A.  Concentration Ranges and Spatial Distributions in Water ............................ 69
   B.  Concentration Ranges and Spatial Distribution in the Sediment .................. 70
   C.  Seasonal Trends and N:P Ratios in Water and Sediment ............................. 71
   D.  Nutrients Recycled by Macroalgae and Exchanged Between Sediment-Water
      and Lagoon-Sea ................................................................................................. 73
VIII.  Conclusions ............................................................................................................. 75
References ............................................................................................................................ 77

## I. HISTORICAL EVOLUTION

The lagoon of Venice is a shallow, semi-enclosed body of water lying parallel to the coastline and separated from the open sea by barrier islands (Figure 1). It is located along the Northern Adriatic Sea west coast, in the central Mediterranean Sea, where it originated some 12,000 years ago and evolved according to natural factors up to the 12th century.[1,2] At that time the transitional environmental conditions were progressively disappearing due to the dominant influence of the river-borne solid material over eustatic sea level rise and erosion. Starting with the 13th century, however, the natural trend was broken off by human activities.[3] The government of the city state of Venice, ruling the city and the surrounding hinterland up to 1797, diverted the major tributary rivers (Brenta, Piave, Sile) from the lagoon to the sea, and used stone dams to protect the littoral zone against the erosive action of the sea. Just at the end of its dominion, the Venetian city state government accurately fixed the perimeter of the lagoon, allowing the evaluation of the lagoon area (586 km$^2$) at that time.

From the beginning of the 19th century up to now, three main interventions have been performed: (1) 30% of the 586 km$^2$ were reclaimed to become agricultural and industrial land and diked fishing ponds; (2) the port entrances were deepened and twin jetties, extending 2–3 km seaward, were constructed; (3) the waterways Vittorio Emanuele (1926) and Malamocco-Marghera (1968) were dredged to connect the sea with the industrial district of Porto Marghera. The latter grew in the last 70 years, particularly after the Second World War. Most of the settled productive processes required high electricity use, exploitation of groundwater (stopped in 1975) and production of waste residues. This caused increasing air-water pollution and a severe subsidence (peak rate of 12–14 mm year$^{-1}$ recorded in 1968–69[4]). The latter, of both anthropogenic and natural origin, together with the eustatism, contributed to an overall ground lowering of 22.4 cm in the last century.[5] The hydrological and morphological changes favored the evolution of some areas, particularly those near the main canals, toward a marine environment.[6]

0-8493-6839-1/95/$0.00+$.50

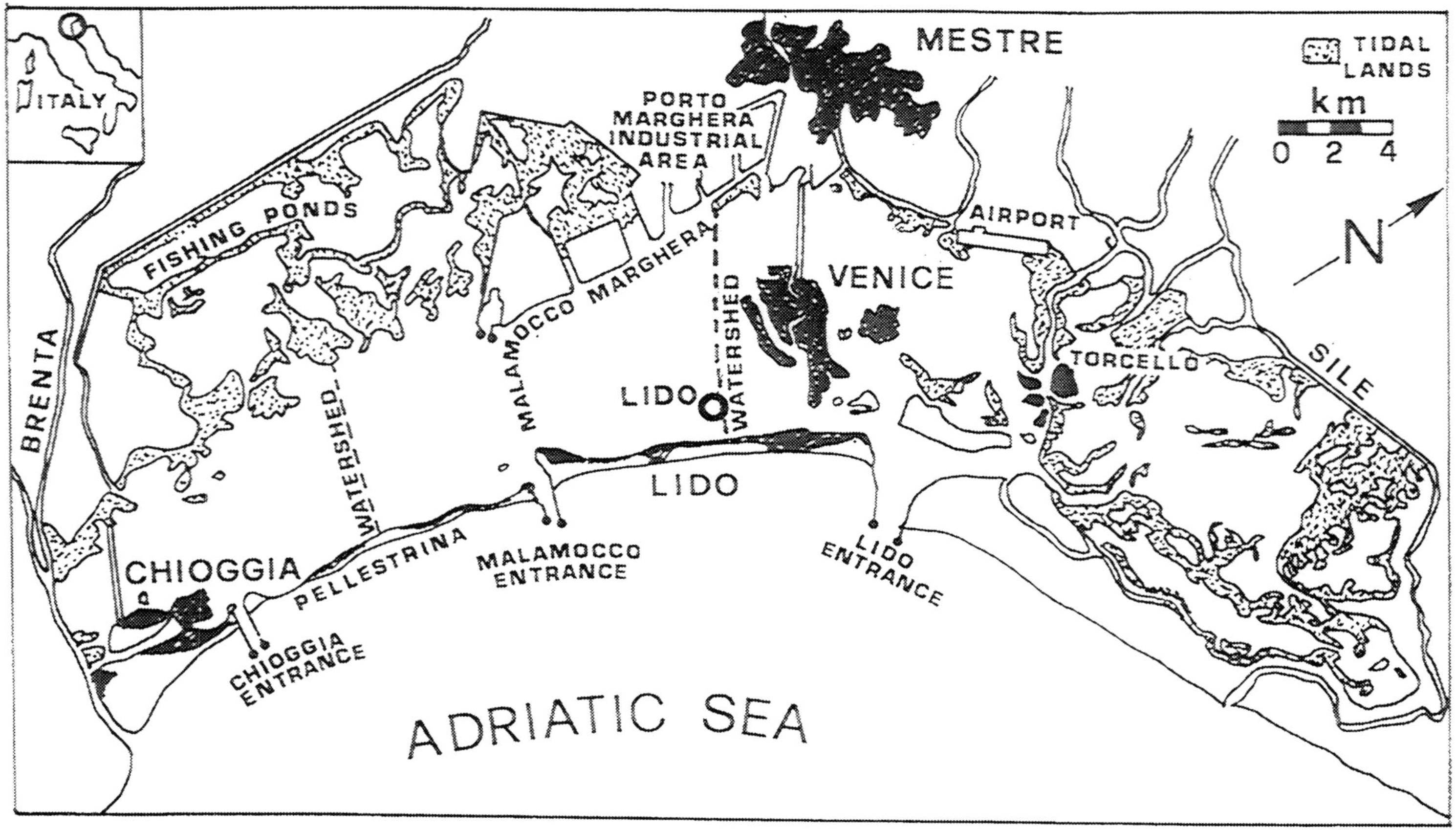

**Figure 1** The lagoon of Venice.

Tidal excursions in some quarters of the Venice historical center are higher than those in the open sea,[6] occasionally causing (5 to 10 times during the last autumn-winter periods) a "high water" phenomenon — exceptionally high tides with water levels easily reaching, in the lower parts of the city, 70–80 cm above the ground. Bottom erosion is becoming more intensive than in the past, threatening the basements of Venetian buildings; several sandbanks are disappearing.[6]

On the other hand, the massive proliferation of macroalgal biomass in the central lagoon during the last decade indicates a decreasing water renewal of some areas, favoring progressive silting up and nutrient recycling.[7] The increasingly frequent "high water" phenomenon and dystrophic conditions (massive biomass decomposition leading to the mortality of benthic organisms) are the worst consequences of the most recent evolution of the lagoon.

## II. PHYSICAL PROPERTIES

The lagoon of Venice, located in the Veneto region, is situated between the rivers Sile and Brenta-Bacchiglione, both of which formerly entered the lagoon but now flow into the sea. The lagoon, 52 km long and 8–14 km wide, is separated from the sea by the two littoral islands Lido and Pellestrina (10–12 km long, 0.5–1 km wide), and linked to the sea through the three port entrances of Lido, Malamocco and Chioggia.

The overall surface is 549 km$^2$, distributed among the three hydrological basins of Lido (276 km$^2$), Malamocco (162 km$^2$) and Chioggia (111 km$^2$). This surface includes the 19 diked fishing ponds, accounting for 18.3% (87 km$^2$) of the total surface.

Water occupies approximately 60% of the Lido basin; an additional 12% are tidal lands; the remaining surface (accounting for about 7% of the total lagoon) is occupied by emerged land, including the historical center of Venice and the most important lagoon islands (Lido, Murano, Burano and Torcello). The hydrological basin of Chioggia displays the largest surface percentage (75.5%) available to tidal expansion. Intermediate conditions, in terms of water-to-land ratio, are found in the Malamocco basin.[8]

The network of canals, approximately 800 km in length, represent the old alluvial plain system. The Malamocco-Marghera and Vittorio Emanuele canals are the most important, both approximately 100 m wide and 12–20 m deep, and allow the ships from the Malamocco and Lido port entrances, respectively, to reach the docks.

About 75% of the lagoon is less than 2 m deep; only the main canals (accounting for 5% of the total lagoon area) are more than 5 m deep. The average depth of the lagoon is approximately 1 m. Due to the shallow water, the water and air temperatures are similar (±2°C), except in the deeper canals during seasonal transitions, where differences between surface and bottom waters as high as 4–5°C can be found.[9] In winter the surface temperatures are lower than bottom temperatures and vice versa in summer.

Water temperature varies annually from 0–33°C, with minima in winter and maxima in summer. The greatest annual variation is encountered near the tidal lands in the inner part of the lagoon.

The salinity of lagoon waters varies with the mixing of freshwater and seawater. At the port entrances the salinity is about 35‰, the same as in the Northern Adriatic sea, and decreases to an average of 26‰ in the innermost areas. The seasonal extremes of salinity in the lagoon occur in winter and summertime, corresponding to maximum sea water (winter) and minimum freshwater (summer) inputs, respectively.[8]

During the warmer months the lagoon is subjected to a breeze regime consisting of NE winds between sunrise and 1 p.m. (land-based breeze)[10] and SE winds between 1 p.m. and sunset (sea-based breeze) of low intensity (up to 4 m sec$^{-1}$). The most intense winds (called "bora") have a NE direction and reach easily speeds up to 20 m sec$^{-1}$. The most frequent wind direction is NE (29%) and the average overall rate is between 2 and 4 m sec$^{-1}$ (44%). Several fog events, favored by absence of winds and low thermal inversions, occur in the late autumn-winter periods.[10]

In the last 10 years, the average rainfall was 840 mm (range: 758–1040 mm).[10]

## III. THE CATCHMENT

### A. DRAINAGE AREA OF RIVERS FEEDING THE LAGOON

The fresh surface waters flowing into the lagoon come from (1) a permanent drainage area, including the rivers and canals Zero, Dese, Marzenego, Osellino, Muson, Tergola, Trezze, as well as some streams (Lusore, Salso, etc.); and (2) an area partially drained by the rivers Sile, at the north, and Brenta-Bacchiglione, at the south.[11] The Sile River, diverted to the sea in 1683, communicates with the lagoon through the Silone canal, as well as through an artificial stream (located in Businello) carrying 7–10 m³ sec⁻¹ out of the approximately 54 m³ sec⁻¹ carried by Sile. Brenta and Bacchiglione are linked to the lagoon through the "Naviglio Brenta" running into the lagoon 7–10 m³ sec⁻¹ out of the 100 m³ sec⁻¹ carried on average by the Brenta and Bacchiglione rivers into the sea.

Average annual flow rate of the freshwaters entering the lagoon is between 31 and 45 m³ sec⁻¹;[12,13] under conditions of heavy rain, however, the flow reaches, and sometimes exceeds, 600 m³ sec⁻¹.

The drainage area covers a surface of 1839 km² with 1170 km² devoted to intensive agricultural activities: maize (40%); beet and soy-bean (18%); forage (14%); grape (12%); wheat (10%); others (6%).[14]

The overall population of the permanent drainage area is 650,000 inhabitants, of which only about 50% is connected with sewage treatment plants.[15]

## B. CATCHMENT AREA OF DIRECT INPUTS

Other significant contributions of freshwater are represented by (1) domestic sewage released by the four Venice districts (historical center, Mestre-Marghera, Lido, Cavallino), Chioggia and the municipalities located in the Mirese area and (2) by effluents from the industrial zone of Porto Marghera, as well as the cooling waters used by industry.

The population of Venice is 347,000, including an average daily presence of 20,000 tourists.[16] About 60% of this population is served by eight sewage treatment plants (STP), out of which the mechanical-biological ones in Fusina and Campalto are the largest, treating domestic sewage of about 200,000 equivalent inhabitants. Due to the lack of a sewerage system, the historical center of Venice is the major source (33%) of untreated wastewater discharged directly into the lagoon, followed by Mestre-Marghera (23%) and the Mirese area (22%), because of their incomplete connection with the sewerage system.

The Porto Marghera industrial area covers ca. 2000 ha of land.[17] It is divided into three areas, of which the third is undeveloped. It faces the central part of the Venetian lagoon, extends for ca. 5 km, and is surrounded by the residential areas of Mestre and Marghera, as well as by the lagoon. The industrial zone dates back to the 1920–1930s, and includes coal distillation, production of sulfuric acid, phosphate fertilizers, pesticides, and wood products, mechanical and ship building production, oil refining and the storage of mineral oils. After World War II, the industrial zone was expanded to include petrochemical and related activities. The maximum development occurred during the 1960s when employees numbered 30,000, but has since then declined to employ only 17,000 workers. Production activities indicate a continued major importance of petroleum and its derivatives, of phosphates and fertilizers, of chemical products and, to a lesser degree, of metals.[17]

Maritime and industrial traffic data over the 1974–1988 period confirm a relative stagnation in production during the 1980s (average annual total movements: 21,400 metric tons).[17] The industrial wastewaters equate to approximately 40,000 equivalent inhabitants and are conveyed, after *in situ* primary treatment, to the consortial STP in Fusina.

The maximum flow of industrial cooling water is 49 m³ sec⁻¹, out of which 28 m³ sec⁻¹ are discharged by the ENEL (Electric Energy National Corporation) thermoelectric power plant. The maximum temperature difference between withdrawn and discharged water is 10°C.[11]

## C. ESTIMATED AND MEASURED NUTRIENT LOADS

Several estimates of the potential nutrient loads flowing into the lagoon were conducted in the last decade.[12,18-22] Only the most recent results can be used, since an increased number of inhabitants served by sewage treatment plants, on the one hand, and regulatory measures against phosphorus in detergents, on the other hand, have occurred in the 1980s. The percentage of phosphorus in both domestic and industrial Italian detergent products decreased from 8 to 6.5% in 1982, to 5% in 1983, to 2.5% in 1986, to 2% in 1988 and to 1% in 1989. Moreover, in Venice and in its hinterland domestic detergent phosphorus was banned in 1989.

The sum of total direct and indirect nutrient loads leads to an overall estimate of 8,200–10,060 tons of nitrogen and 1,100–1,900 tons of phosphorus yearly delivered to the lagoon waters. The amounts of nitrogen and phosphorus carried by the rivers into the lagoon are 3,400 tons y⁻¹ and 790 tons y⁻¹, respectively.[21,22] These loads were, however, based on measurements carried out in periods of low rainfall, may underestimate the agriculture runoff.

The average nutrient concentrations measured in the influent rivers were 22–34 mmol m⁻³ for total inorganic nitrogen and 4–17 mmol m⁻³ for orthophosphate, with a N:P ratio range of 4–12.[12] The atmospheric fallout of nutrients has been estimated to be 870 tons y[-1] for nitrogen and 66 tons y[-1] for phosphorus.[20] The contribution of harbor activities, particularly through the leakage of fertilizers, accounts for 94–187 tons y[-1] of nitrogen and 42–85 tons y[-1] of phosphorus.[21]

## IV. WATER MOVEMENT AND EXCHANGE

Tidal excursions parallel sea level variations in the Northern Adriatic Sea. The tide wave propagates up to the inland border through the canals and, to a lesser extent, the water sheets between them.

The canals naturally shaped by the tides exhibit a cross section decreasing with increasing distance from the port entrances, and so the tidal current speed stays fairly constant. The man-made canals, however, have a constant cross section, and in them the tidal current speed is reduced with increasing distance from the sea current inlets.[23] Consequences even more important to the hydrological conditions of the lagoon began to occur in the second half of the past century, with the construction of the twin jetties at the port entrances, each consisting of parallel rock-fill dams, 500–900 m from each other and 2–3 km long. The exchanged flows (nowadays accounting for up to 8,000 m$^3$ sec$^{-1}$, corresponding to flow rates of 120–150 cm sec$^{-1}$) increased remarkably, in both ebb and flood, together with erosion leading to the disappearance of many sandbanks and tidal lands. The final effect of these man-induced modifications was an increasing flattening of the central lagoon bottoms. The completion (1846) of the railway bridge connecting the historical center with Mestre and its successive enlargement (1936) to allow motor-car traffic, also considerably restricted water movement across this part of the lagoon (Figure 1).

A few short-term studies have been carried out to estimate the water exchanged between lagoon and sea. Simultaneous flow measurements at the three port entrances were carried out on September 15–16, 1970.[24] The overall water volumes exchanged were: $350 \times 10^6$ and $287 \times 10^6$ m$^3$ in flood and ebb, respectively, with a tidal excursion of 115 cm in flood and 100 cm in ebb.

By integrating over 5 years (1979–1983) flow values obtained by applying a monodimensional model,[25] average annual volumes of water entering the lagoon were obtained: $56 \times 10^9$ m$^3$ y$^{-1}$ for the Lido basin; $56 \times 10^9$ m$^3$ y$^{-1}$ for the Malamocco basin; and $28 \times 10^9$ m$^3$ y$^{-1}$ for the Chioggia basin (overall volume: $140 \times 109$ m$^3$ y$^{-1}$). Based on satellite remote sensing[26] and salinity measurements,[27] it was found that a fraction of water leaving the lagoon in ebb tide returns to the lagoon in flood tide. This fraction was estimated to be 28% for the Lido basin and 16% for the Malamocco basin. No data are available for the Chioggia basin. However, the similarity of the transverse sections of the Chioggia and Malamocco mouths suggest the same value (16%) can be assigned for the Chioggia basin. Therefore, the estimated net lagoon-sea exchange of water is $111 \times 10^9$ m$^3$ y$^{-1}$, distributed: Lido, $40 \times 10^9$ m$^3$ y$^{-1}$; Malamocco, $47 \times 10^9$ m$^3$ y$^{-1}$; Chioggia, $24 \times 10^9$ m$^3$ y$^{-1}$.[19] The inflowing river freshwater is 1–2% of that exchanged with the sea.

In the central part of the lagoon, the tide arrives earlier than in the northern part, and this results in a transverse current, perpendicular to the translagoon bridge connecting the historical center with the mainland. Despite the bidimensional movement of the water, the current speed near the watershed between the Lido and Malamocco basins is quite reduced.

In the northern part, water movement is monodimensional and is controlled by the sections of the canals where the speeds are high. In the shallow areas the speeds show the same direction as in the canals in both flood and ebb tides. A bidimensional behavior was observed in semi-uniform bottoms such as those in front of the airport and S. Giuliano. In the southern basin the water movement is monodimensional with another watershed separating the basins of Malamocco and Chioggia.[8]

The propagation of the tide wave into the lagoon depends strongly on bottom morphology. Near the port entrances and in surrounding areas the length of the flood tide is higher than that of the ebb tide, which leads to bottom erosion, as confirmed by recent bathymetric data. In contrast, in the internal lagoon the lengths of ebb tides are higher, and favor sedimentation.

Information on the freshwater residence time in the different lagoon areas is quite poor. Based on salinity measurements,[27] the order of magnitude of the freshwater average residence time in the whole lagoon was estimated to be 15–30 days.

## V. SEDIMENT PROPERTIES

The sediment of the Venice lagoon is constituted, according to the lithological classification of Shepard, of sands, silty sands, sandy silts, clayey silts and silty clay.[28-30] Sand, silt and clay are defined by their granular size; 2.0–0.063 mm, 0.063–0.002 mm and <0.002 mm, respectively. The intermediate term (silty sands, sandy silts, etc.) indicate the dominance of one fraction over another in two types with sediments intermediate between the three main categories. The term "fine fraction" is used to indicate the sum of silt and clay, namely, the fraction <63 μm.

The grain size distribution of the lagoon surface sediment in the central part of the lagoon is reported in Figure 2. The fine fraction is the most abundant, and is attributable to the clayey silts and, to a less degree, the sandy silts.[29] The grain size distribution depends greatly on the hydrodynamic conditions of the lagoon. Where the tide currents are higher (i.e., at the port entrances), there is an accumulation of coarser material (sands and silty sands), whereas weak currents (i.e., in the innermost areas) favor the deposition of fine material (clayey silts and silty clays). The deposition of sandy silts and silts occurs under intermediate conditions. The fine sediments are also diffused in the area surrounding the watersheds, where the tidal excursion causes only vertical movements.[29]

Further information on the sediment characteristics is provided by the percentages of carbonates, mostly accounted for by calcite and dolomite[31] and organic material. The percentage of carbonates increases from the inland border seaward. The organic matter content, depending primarily on the sediment grain size, as well as on local hydrodynamic conditions, shows maximum values (5–8%) in sediment with high fine fraction percentages. The minimum values correspond to the sandy sediments at the port entrances and in areas influenced by canals with high current rates.[31]

The area most extensively studied for its geochemical properties is located between the historical city and the industrial zone.[32-34] Most sediments are clayey silt, typical of a lagoon depositional environment at very low energy. In this area, the average percentage of organic matter was $5.7 \pm 2.4$ on surface (0–2 cm) sediment and $4.8 \pm 2.1$ in the deeper (3–10 cm) sediment.[34]

Radiometrically dated ($^{210}$Pb and $^{137}$Cs) sediment cores collected in the central lagoon (at the mouth of the Silone canal and close to the industrial district of Porto Marghera) showed sedimentation rates of 0.59 cm y$^{-1}$ and 0.81 cm y$^{-1}$, respectively.[35,36]

## VI. BIOLOGICAL EVOLUTION AND BIOMASS DISTRIBUTIONS

Although hydrodynamic and some physicochemical properties would be better described by taking into account the individual basins, the pattern of eutrophication is such that most attention is given to the central part of the lagoon, located between the tidal lands of the Torcello Island and the Malamocco-Marghera Canal (Figure 1).

### A. CENTRAL LAGOON

Changes in the physicochemical characteristics of water and sediment (prompted by the increased availability of nutrients following the economic development of the catchment area, during the 1950–60s, particularly the industrial zone of Porto Marghera) and of the hydrodynamic conditions (see Section I) were characterized by a marked change in benthic population of the central part of the lagoon.[7,37-42] At the end of the 1960s, the species typical of sandy bottoms (*Melinna palmata* Grube, *Nucula nucleus* L., *Loripes lacteus* L.) were decreased in favor of those typical of mud bottoms (*Nereis diversicolor* O.F.M., *Marphysa sanquinea* Mont., *Cerastoderma edule glaucum* Brughiere). Concurrently, in the lagoon littorals both typology and distribution of macroalgal associations changed, nitrophile species (*Ulva rigida* C. Ag., *Enteromorpha* spp., *Cladophora* spp.) replacing both those most sensitive to eutrophic phenomena (particularly the sharp and prolonged fluctuations of oxygen) and the rhizophytes previously populating the bottoms around the historical center of Venice. Several species typical of oligotrophic environments, which were previously reported as dominant [*Cystoseira fimbriata* (Desf.) Bory, *Cystoseira barbata* J. Ag., *Dictyopteris membranacea* (Stackh.) Batt., *Cladostephus verticillatus* (Lightf.) Lyngb., *Taonia atomaria* (Woodw.) J. Ag., etc.], disappeared quickly. The phanerogamae *Zostera marina* L., *Cymodocea nodosa* Ascher. and *Zostera noltii* Hornem. remained only in the northern and southern parts of the lagoon, where dystrophic conditions have not occurred so far. Moreover, the number of macroalgal species decreased progressively from 141 in 1940,[38] to 104 in 1962[39] and 95 in 1987.[40]

The most recent classification of vegetation associations and dominant species in the lagoon is reported in Table 1. Drastic changes in the flora and vegetation associations of the macrophytes occurred in the central part of the lagoon, which is the most stressed one. *Ulva rigida* C. Ag., the dominant alga, has progressively replaced the other species, occupying close to 100% of the available space and often exhibiting a standing crop >10 kg m$^{-2}$ (fresh weight, fw).[43] Under these conditions, net photosynthetic

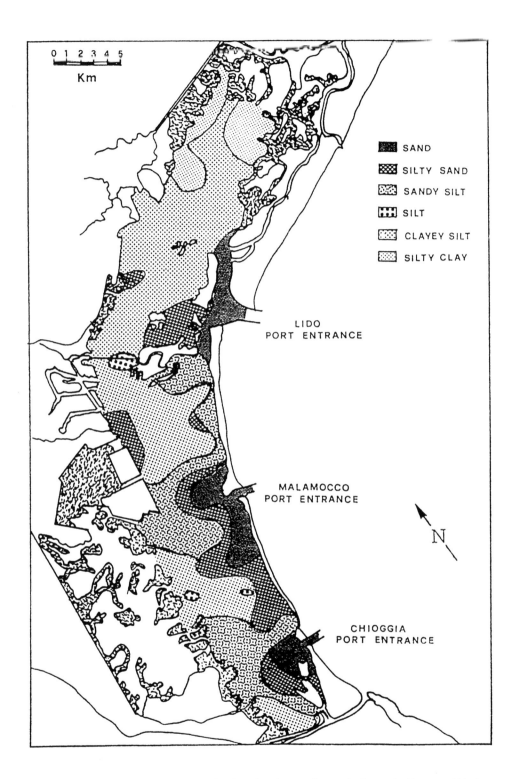

**Figure 2** Grain size distribution of surface sediments in the central lagoon of Venice.

rates calculated by macroalgal biomass variations were up to 30.5 g C m$^{-2}$ day$^{-1}$, which are much higher than those reported for phytoplankton (up to 186 mg C m$^{-3}$ h$^{-1}$ [44] and 580 mg C m$^{-3}$ h$^{-1}$ [45]; Table 2). An annual net macroalgal production of 646 ± 123 g C m$^{-2}$ was measured.[46]

Macroalgal biomass production was periodically interrupted by the occurrence of anoxia causing macroalgal decomposition and extensive mortality of macrofauna.[43] Under these conditions, the temporary

**Table 1** Vegetational Associations Monitored in the Venice Littorals in the 1980–85 Period

| Vegetational zones | | Sea associations | Port entrance associations | Lagoon associations |
|---|---|---|---|---|
| **Supralittoral** | | | | |
| Northern exp. | Perennating sp. | Blidingia minima | Blidingia minima | Blidingia minima |
| | Seasonal sp. | Bangia fuscopurpurea (w-spr) | — | Ulothrix pseudoflacca (w-spr) |
| Southern exp. | Perennating sp. | — | — | — |
| | Seasonal sp. | Bangia fuscopurpurea (w-spr) | — | Ulothrix pseudoflacca (w-spr) |
| **Upper Mediolittoral** | | | | |
| Northern exp. | Perennating sp. | Enteromorpha compressa<br>Gelidium spathulatum | Enteromorpha compressa<br>Gelidium spathulatum | Enteromorpha compressa<br>Enteromorpha prolifera<br>Petalonia fascia (w) |
| | Seasonal sp. | Scytosiphon lomentaria (spr)<br>Petalonia fascia (w) | Scytosiphon lomentaria (spr)<br>Petalonia fascia (w) | |
| Southern exp. | Perennating sp. | Enteromorpha compressa<br>Gymnogongrus griffithsiae | Enteromorpha compressa<br>Gymnogongrus griffithsiae | Enteromorpha compressa<br>Enteromorpha prolifera<br>Petalonia fascia (w)<br>Navicula spp. (sum) |
| | Seasonal sp. | Petalonia fascia (w)<br>Porphyra leucosticta (w) | Petalonia fascia (w)<br>Ulothrix implexa (w)<br>Navicula spp. (sum) | |
| **Lower Mediolittoral** | | | | |
| No differences have been observed for the two exposures | Perennating sp. | Ceramium ciliatum<br>Corallina elongata<br>Grateloupia filicina | Ceramium ciliatum<br>Polysiphonia breviarticolata<br>Grateloupia filicina | Bryopsis hypnoides<br>Polysiphonia sanguinea |
| | Seasonal sp. | Bryopsis disticha (spr)<br>Ulva fasciata (sum-a)<br>Lomentaria clavellosa (w) | Bryopsis disticha (spr)<br>Ulva fasciata (sum-a)<br>Lomentaria clavellosa (w) | —<br>Enteromorpha intestinalis (spr)<br>Ulva fasciata (sum-a)<br>Porphyra leucosticta (w)<br>Antithamnion cruciatum (w) |
| **Infralittoral Zone** | | | | |
| No differences have been observed for the two exposures | Perennating sp. | Ulva rigida<br>Bryopsis plumosa<br>Polysiphonia sanguinea | Ulva rigida<br>Codium fragile<br>Dictyota dichotoma<br>Rhodymenia ardissonei | Ulva rigida<br>Gracilaria verrucosa |
| | Seasonal sp. | Enteromorpha linza (sum)<br>Dasya pedicellata (sum-a)<br>Antithamnion cruciatum (w) | Enteromorpha linza (sum)<br>Halymenia floresia (sum-a)<br>Antithamnion cruciatum (w) | Punctaria latifolia (spr)<br>Cladophora vagabunda (sum-a)<br>Vaucheria dichotoma (w)<br>Ectocarpus siliculosus (w-spr) |

*Note:* exp.: exposure; sp: species; spp.: species plurimae; (spr): spring; (sum): summer; (a): autumn; (w): winter.

**Table 2** Estimation of Primary Production in the Lagoon of Venice

| Values | Range | Period | Location | Ref. |
|---|---|---|---|---|
| | | Phytoplankton | | |
| 0.6–16 mg | C m$^{-3}$ h$^{-1}$ | Winter | Industrial zone | Battaglia et al. (1983)[44] |
| 17.0–580 mg | C m$^{-3}$ h$^{-1}$ | Summer | | |
| 0.4–186 mg | C m$^{-3}$ h$^{-1}$ | | Industrial zone | Degobbis et al. (1986)[45] |
| 1.0–42 mg | C m$^{-3}$ h$^{-1}$ | Year | Sacca Sessola | |
| 0.8–20 mg | C m$^{-3}$ h$^{-1}$ | | Lido port entr. | |
| | | Macroalgae | | |
| up to 30.5 g | C m$^{-2}$ d$^{-1}$ | Spring | Lido watershed | Sfriso et al. (1988)[41] |
| 646 ± 123 g | C m$^{-2}$ d$^{-1}$ | Year | Lido watershed | |
| up to 34.2 g | C m$^{-2}$ d$^{-1}$ | Spring | Sacca Sessola | Sfriso et al. (1988)[77] |
| ~ 550 g | C m$^{-2}$ d$^{-1}$ | Year | Sacca Sessola | |

disappearance of zoobenthos and fish favored an abnormal increase of anaerobic species such as *Chironomus salinarius* Kieff., which invaded many areas of the lagoon.[47] In June 1987, when macroalgae showed their highest annual standing crop, the distributions of both macroalgal and phytoplankton biomass were obtained for the central part of the lagoon (Figure 3 A,B).[46] The highest macroalgal standing crops were located along the Lido watershed, south of Venice, and in the whole area north to Venice (in front of the airport). Low macroalgal biomass was measured south of the watershed and near the Lido and Malamocco port entrances. A negligible biomass was found in front of the Porto Marghera industrial zone, the lagoon area most chemically polluted. Here, phytoplankton accounted for average chlorophyll *a* (Chl. *a*) concentrations of 10–30 mg m$^{-3}$, with values easily exceeding 100 mg m$^{-3}$.

Relatively high concentrations of phytoplankton (Chl. *a* 5–10 mg m$^{-3}$) were also found near the central part of the Lido island,[46] following a sanitation intervention conducted by the Municipality. Macroalgae were removed by harvesting machines, in order to prevent the occurrence of anoxic conditions systematically observed over the last few years. During this intervention, biomass was lowered from 10–15 to 5–6 kg m$^{-2}$ (fw). The intervention decreased the level of macroalgal stratification but favored the release of nutrients from the disturbed sediment, which triggered phytoplankton blooms and faster macroalgal growth. The Chl. *a* fraction <20 μm, i.e., the nanoplankton, encountered in the central lagoon accounted for 82 ± 19% of the total Chl. *a*,[46] in good agreement with the result found in the Northern Adriatic Sea.[48]

The phytoplankton communities in the spring season were largely diatoms (87–97%).[49,50] The dominant species was *Skelenatum costatum* (Grev.), reaching bloom cell counts higher than 60,000 × 10$^3$ L$^{-1}$, and causing the water's brownish color. Based on its areal distribution, a preferred growth of this species was found in areas influenced by domestic sewages, north to the translagoon bridge.

The discharge of the industrial cooling waters did not appear to affect phytoplankton abundance and biomass, but its effect was possibly masked by other concurrent factors.[49]

## B. NORTHERN AND SOUTHERN LAGOONS

These parts of the lagoon suffered minor influence from human activities. Biological communities show higher diversity and lower biomasses than those recorded in the central lagoon,[51-55] extensive anoxia has not been detected and biological communities are well represented. The bottom is mostly populated by rhyzophites: *Cymodocea nodosa* prevails near the port entrances and in the areas where tidal currents are strong and sediment grain size is high (silty sand and sandy silt). *Zostera noltii* reaches the highest concentration in the innermost areas where currents are dampened. This species has a preference for clay sediments, and contributes to particulate material deposition. These spermatophytes allow the survival of a complicated trophic chain that is absent in areas populated by abnormal macroalgal biomass.[54]

The main species of macroalgae are *Chaetomorpha linum* (Mull.) Kütz., *Gracilaria bursa pastoris* (Gemel.) Silva and *Cladophora* spp. in the southern basin and *Valonia aegagrophyla* C. Ag. in the northern one. Their biomass is low (0–1 kg m$^{-2}$, fresh wt.) in the areas populated by rhyzophites but occasionally can exceed 20 kg m$^{-2}$ (fresh wt.) in restricted areas where rhizophyte biomass is negligible. The ribbon-like or ball-like morphology of these macroalgal species do not, however, pose the problems of *Ulva*, whose laminar fronds are particularly suited to fill up the water column, hamper water exchange

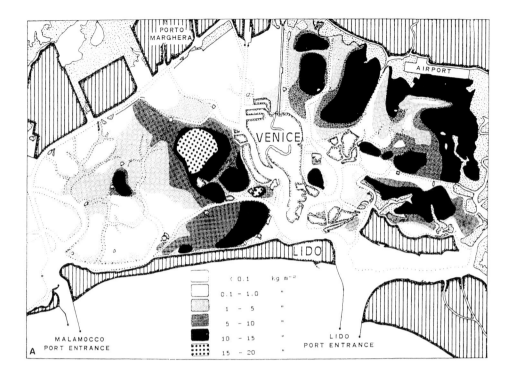

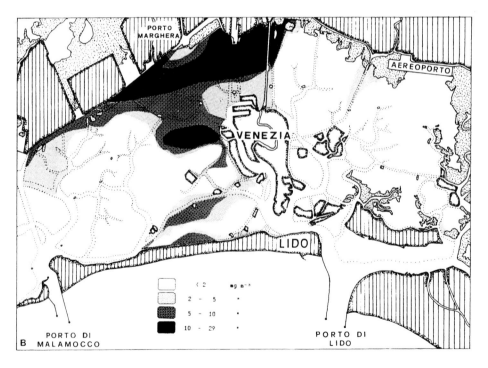

**Figure 3** Distributions of macroalgal biomass (A) and phytoplankton (B) in the central part of the lagoon of Venice.

and trigger biomass decay. In the last decade, however, *Ulva rigida* C. Ag. has increased significantly in some areas, particularly the Brenta pond, at the lagoon-inland border.

Many investigations were conducted on the spatial and temporal distributions of plankton in the northern and southern lagoon, with special emphasis on the Cona and Brenta ponds in the northern and southern parts, respectively. Phytoplankton associations typical of coastal, brackish and fresh waters were identified depending on hydrological conditions. The coastal assemblage is dominated by *Nitzschia delicatissima, Nitzschia longissima* f. *parva, Nitzschia longissima* var. *closterium* and *Ditiocha speculum.* The species *Gomphonema parvulum* and *Navicula* spp., as well as the naked flagellates, dominate the brackish group. *Nitzschia palea, Nitzschia* sp., *Cocconeis placentula, Fragilaria capucina, Fragilaria crotonensis* and *Fragilaria pinnata* were typical of the freshwater community. The effects of salinity showed the plankton diversity index decreasing significantly from marine to inner areas, and increasing only for phytoplankton with the water dilution.[51]

Maximum abundances of phytoplankton at the Malamocco port entrance were recorded in summer, with the predominance of microplanktonic diatoms.[52] In winter and in transition periods, the community is dominated by naked Flagellates. Some of the most representative species (*Nitzschia seriata, Rhizosolenia alata, Chaetoceros* spp., *Thalassionema nitzschioides, Thalassiothrix frauenfeldii*) occur in both summer and autumn and are neritic forms frequent in similar seasonal conditions in the Northern Adriatic Sea.

## VII. NUTRIENT CONCENTRATIONS

The concentrations of nutrients in both the water column and surface sediment depend primarily on the distance from point discharges of river waters, domestic and industrial wastes, and on the influence of algal blooms and subsequent sediment regeneration processes, as well as on the sea tide and climatic conditions. In contrast with industrial and municipal waste waters from the interlands (Porto Marghera, Mestre, Cavallino littoral and Lido Island) which were increasingly treated in the last 2 decades,[20] the sewage of the Venice historical center continues to be discharged into the lagoon after septic tank digestion.

### A. CONCENTRATION RANGES AND SPATIAL DISTRIBUTIONS IN WATER

Several monitoring programs were conducted in the last 30 years, particularly in the central lagoon waters.[56-70] Allowing for improved analytical methods, the comparison of historical sets of water nutrient concentrations shows a general decreasing trend in the last decade, particularly for ammonium. The latter was recorded in front of the industrial zone at concentrations of 360–2500 mmol m$^{-3}$ in the early seventies; 10–200 mmol m$^{-3}$ in the late seventies and <3–50 mmol m$^{-3}$ in the second half of the eighties. The oxidized inorganic forms of nitrogen, i.e., nitrate and nitrite, however, decreased to a negligible extent, which suggests an unaltered influence of agricultural runoff as the major source of input in the lagoon.

The results of a recent study[21] that includes the whole lagoon and a nearshore sea station, and the analysis of organic and inorganic nutrients in both water and sediment, are reported in Table 3. Based on the total P and Chl. *a* values proposed for the classification of the Adriatic Sea,[71] the lagoon waters are eutrophic to hypertrophic. The average nutrient concentrations inside the lagoon are at least one order of magnitude higher than those in the nearshore gulf of Venice.[21] The reported average concentration ranges, over a year, of nitrate + nitrite and ammonium in the whole lagoon were 10–50 mmol m$^{-3}$ and 3–5 mmol m$^{-3}$, respectively. The concurrent average concentration range for both orthophosphate and orthosilicate was 0.3–1.5 mmol m$^{-3}$.

**Table 3** Mean Values of Total Phosphorus ($P_{TOT}$), Total Inorganic Nitrogen (TIN) and Chl. *a* in the Whole Lagoon Over the Period July 1987–July 1988

| Parameter | No. of data | Mean values | Ranges | Standard deviations |
|---|---|---|---|---|
| PTOT (mmol m$^{-3}$) | 1,507 | 2.2 | 0.1–18 | 2.5 |
| TIN (mmol m$^{-3}$) | 1,507 | 24.5 | 0.2–264 | 21.6 |
| Chl. a (mg m$^{-3}$) | 1,509 | 3.7 | 0.0–131 | 6.9 |

Data from ENEA, *Final Report on the Actual Conditions of the Lagoon Ecosystem* (in Italian), Consorzio Venezia Nuova, Ed., Venzia, 1990.

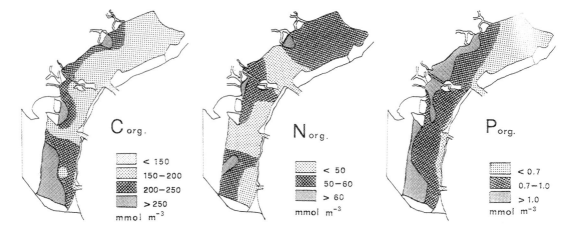

**Figure 4** Spatial distribution of annual average concentrations of organic carbon, nitrogen and phosphorus in the waters of the lagoon of Venice.

The average concentration ranges and distribution of organic nutrients,[21] reported in Figure 4, demonstrate the influence of the rivers as the main contributors to the non-point nutrient sources. The ratio of inorganic to organic forms is approximately 1 for phosphorus and at least 2 for nitrogen.

Typical results[70] for dissolved nutrients concentrations inside the Venice center and all around it are reported in Table 4: the results obtained for three representative areas (area A between the historical center and the Lido port entrance; area B around the historical center; area E between the historical center and the industrial zone) are presented.

Significant concentration differences of the nutrients in high and low tide were recorded in all areas, except in the historical center. Nitrogen and phosphorus concentrations in the major canals were negatively correlated with salinity, showing the influence of fresh water. In contrast, no significant correlation, with the exception of the area in front of the industrial zone, was found between salinity and the nutrient concentrations expected in the inflowing freshwaters on the basis of the nutrient concentrations found in the innermost lagoon waters.[72] This indicates that as soon the freshwaters enter the lagoon, biological and regeneration processes take place, overwhelming the dilution processes.

## B. CONCENTRATION RANGES AND SPATIAL DISTRIBUTION IN THE SEDIMENT

Only a few systematic investigations of sedimentary nutrients have been conducted (Table 5).[21,63,73] The data indicate a trend of decreasing nutrient concentrations in the surface sediment over the last 2 decades.

**Table 4** Mean Concentrations of TIN (mmol m$^{-3}$) and RP (mmol m$^{-3}$) in High and Low Tide Samplings in Three Areas of the Central Lagoon

|  | High tide | | Low tide | |
| --- | --- | --- | --- | --- |
|  | TIN | RP | TIN | RP |
| Area A |  |  |  |  |
| Mean | 11.2–28.5 | 0.3–0.5 | 10.3–78.9 | 0.9–1.5 |
| Std. Dev. | 1.5–8.1 | 0.1–0.2 | 2.7–13.2 | 0.3–0.5 |
| Area B |  |  |  |  |
| Mean | 9.2–47.7 | 0.2–0.8 | 6.1–68.4 | 0.5–1.4 |
| Std. Dev. | 0.9–4.7 | 0.0–0.2 | 1.5–3.5 | 0.1–0.2 |
| Area E |  |  |  |  |
| Mean | 13.7–48.8 | 0.3–1.7 | 17.4–79.1 | 0.4–3.5 |
| Std. Dev. | 0.6–6.5 | 0.0–0.5 | 2.7–10.7 | 0.1–0.7 |

Note: 4 seasonal samplings; Area A:6 stations; Area B:3 stations; Area E:3 stations.

Data from Pavoni, B., Donazzolo, R., Sfriso, A., and Orio, A. A., *Sci. Total Environ.*, 96, 235, 1990.

**Table 5** Concentration Ranges
(mg kg$^{-1}$, dry weight) of Sedimentary
Nitrogen and Phosphorus in the
Whole Lagoon

| Location | N$_{TOT}$ | P$_{TOT}$ |
|---|---|---|
| Northern lagoon | 990–3,800 | 298–383 |
| Central lagoon | 896–2,744 | 327–682 |
| Southern lagoon | 784–5,320 | 343–550 |

Data from Perin, G., Orio, A. A.,Pauoni, B.,
Donazzolo, R., Pastre, B., Carniel, A., Gabelli, A.,
and Pasquetto, A., *Acqua Aria,* 6, 623, 1983.

Concentrations of total P <4 10 mg kg$^{-1}$ (dry weight basis) were considered as background values, according to the vertical profiles of radiometrically dated sediment cores.[74] Values >650 mg kg$^{-1}$, typical of the innermost areas, indicated heavy contamination. Most of the total phosphorus (80%) was inorganic, the remaining being organic phosphorus. In contrast, most nitrogen (99%) was organic, with the inorganic one (1%) accounted for mostly by ammonium.[63]

## C. SEASONAL TRENDS AND N:P RATIOS IN WATER AND SEDIMENT

The central part of the lagoon is most severely threatened by macroalgal blooms during the spring-summer period. A thorough investigation was performed in an area populated by macroalgae, examining both water and sediment, as well as biomass fluctuations.[41,43,75] The variations of physicochemical parameters and nutrients (total inorganic nitrogen, TIN, was calculated as the sum of $NH_4^+$, $NO_2^-$ and $NO_3^-$, and reactive phosphate, RP, equals orthophosphate) in a dystrophic area of the central lagoon are presented in Figures 5 and 6. Dissolved oxygen percentage saturation (range:0–366%), pH and Eh showed similar profiles, with a flat trend in autumn and winter, when biological activity was negligible. Water and sediment pH showed almost the same values, around 7.5, in autumn and winter, and fluctuated between 6.8 and 9.0 in water, and between 6.8 and 8.0 in sediment, during the other seasons.[75] Water Eh was in the range 200–400 MV, except during anoxic conditions when values around 0 and –100 MV were measured. On the surface sediment, Eh varied between 0 and 300 mV under oxic conditions and was negative (0–200 MV) when decomposing macroalgae were sitting on the sediment. Inorganic nitrogen compounds contributed differently to the TIN concentration. Ammonium accounted for 40–97% of TIN from April to December, and decreased in winter when nitrates became predominant. Nitrites were below 3% from October to May, but accounted for up to 30% of the TIN between July and September. The highest contribution (up to 80%) of nitrates to TIN was observed between January and March, as well as during the anoxias in May through July. In contrast, reactive phosphorus exhibited the highest concentrations during the spring-summer period,[75] when the sediment was under anoxic conditions, and reducing conditions favored the release of sedimentary phosphorus.[76]

Sedimentary phosphorus and nitrogen, as well as carbon, are reported in Figure 7.[75] All these elements increased as a consequence of macroalgal decomposition and exhibited the minimum values in the cold season. Both organic carbon ($C_{org}$) and total nitrogen ($N_{tot}$) rose sharply during the acute anoxias occurring in June–July. The observed concentration trends of nutrients in both water and sediment are easily understandable by taking into account the biomass fluctuations occurring over a vegetative year. These fluctuations accounted for by macro- (scaled to kg m$^{-2}$, fw) and microalgal (Chl. *a* concentrations) biomass are shown in Figure 8. Macroalgae fluctuated mostly around 6–8 kg m$^{-2}$ (fw), before the sudden and fast decomposition under the April–May anaerobic conditions. Phytoplankton was systematically below 5 mg m$^{-3}$ (as Chl. *a*) until May 1985, when macroalgae quickly decomposed. The sudden availability of nutrients released in water triggered phytoplankton blooms (up to 100 mg m$^{-3}$). When macroalgae increased again, between June and July, phytoplankton decreased to 2.2 mg m$^{-3}$. After the complete decay of macroalgae in late July, phytoplankton fluctuated around 10 mg m$^{-3}$. Afterward, from September until May 1986, phytoplankton was systematically below 0.8 mg m$^{-3}$.

In Figure 9 the N:P atomic ratio in both water and sediment is shown.[75] The N:P values decreased from 78, monitored in water before the spring-summer algal growth, to 0.7 detected between April and August. On the contrary, N:P ratio in the surface sediment increased from values around 7 to more than 13 during

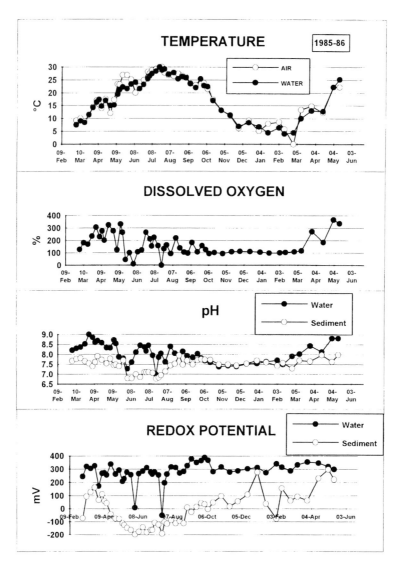

**Figure 5** Trends of water physicochemical parameters (pH, Eh, %DO) in a dystrophic area of the lagoon of Venice.

the same period. The sedimentary N:P ratio progressively approached the average value of 25 (range:22–30), measured in *Ulva*.[77] After August, when macroalgae biomass was negligible, this ratio returned to about the previous values in both water and sediment (N:P>20 in water, from 6 to 7 in sediment). Therefore, the constituent, if any, limiting the productivity of algae was nitrogen.

A recent investigation[21] on the whole lagoon confirmed these results showing the major role of sediment in controlling the overlying water composition. In this same study[21] the percentage occurrence frequency of the N:P ratio was calculated. The optimal values of the N:P ratio for algal growth (10–30) occurred mostly (occurrence frequency of 50–60%) in the northern and central part of the lagoon during the warm season. This indicates the predominant influence in such areas of *in situ* sedimentary nutrient regeneration over any external input fluctuations. The nutrient time trends in the northern and southern lagoon inner parts are more influenced by external input variations (river nutrient supply) as well as by biological activity and sedimentary regeneration (Table 5).[50,62,64]

An indirect confirmation of the dominant influence of macroalgal biomass is found in the behavior of orthosilicate, a major component of phytoplankton, which did not exhibit any significant seasonality on the whole lagoon.[21]

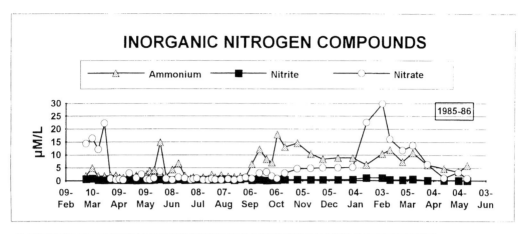

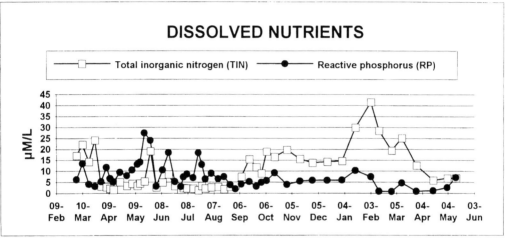

**Figure 6** Trends of ammonium, nitrite and nitrate, as well as total inorganic nitrogen (TIN), and reactive phosphorus (RP) in waters of a dystrophic area of the lagoon of Venice.

## D. NUTRIENTS RECYCLED BY MACROALGAE AND EXCHANGED BETWEEN SEDIMENT-WATER AND LAGOON-SEA

The total macroalgal standing crop measured in the central lagoon in 1985–86[46] was approximately 68,900 tons (dry weight), comprising approximately 17,680 tons of carbon, 1,900 tons of nitrogen and 170 tons of phosphorus. These values were used to estimate the amounts of nutrients assimilated annually by the biomass;[46] 28,000–37,000 tons of carbon, 3,000–4,000 tons of nitrogen and 270–360 tons of phosphorus. Thus, the net spring-summer biomass production of macroalgae in the central lagoon recycled amounts of nutrients quite similar to those carried annually in the whole lagoon by rivers (see Section III.C).

Based on a semiquantitative investigation,[78] macroalgal biomass production in the whole lagoon would account for as much as that found in the central lagoon. Moreover, the macrophytes (rhizophytes) in the southern and northern lagoon contribute presumably quite remarkably to the net primary production. More research is urgently needed on the overall lagoon primary production, with special emphasis on the comparison of areas with different species and standing crop. As the extent of nutrient exchange at the water-sediment interface depends on many factors (biomass stratification in the overlying water, redox conditions, physical and biological burial, etc.), benthic chambers were used to obtain an insight into the nutrient release by the surface sediment.[21,44,58,63] Average values of 0.29 mmol $m^{-2}$ day$^{-1}$ for phosphorus and 8.36 mmol $m^{-2}$ day$^{-1}$ for nitrogen[63] indicated a release of 8540 tons y$^{-1}$ of nitrogen and 650 tons y$^{-1}$ of phosphorus from the sediment to the overlying water. This estimate refers only to one-way (sediment-water) transfer; it is not a net contribution, since the concurrent deposition of particulate

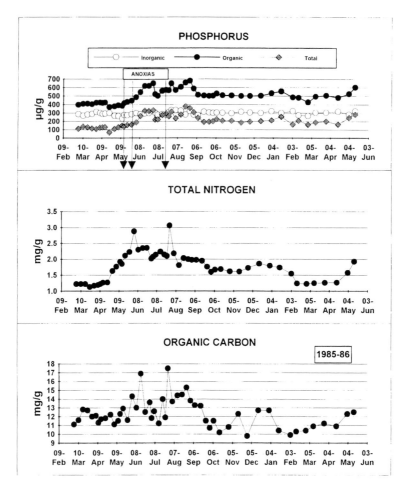

**Figure 7** Trends of sedimentary phosphorus, nitrogen and organic carbon in a dystrophic area of the lagoon of Venice.

forms is not included. Moreover, the nutrient fluxes from sediment vary strongly over the year, as most nutrients are recycled during a short time (June–August). An investigation[79] recently carried out in three stations differing markedly in biogeochemical features and trophic levels, showed that the amount of particulate matter deposited onto surface sediment ranged from 140 kg m$^{-2}$ y$^{-1}$ near the lagoon-inland border to 41 kg m$^{-2}$ y$^{-1}$ near the port entrances. These values correspond to 384 and 113 g m$^{-2}$ day$^{-1}$ accounting for 18.4–104.7 mmol m$^{-2}$ day$^{-1}$ of nitrogen and 2.2–11.5 mmol m$^{-2}$ day$^{-1}$ of phosphorus.

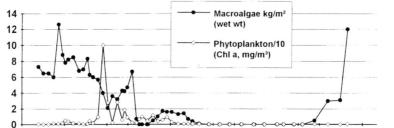

**Figure 8** Macroalgal and phytoplankton trends in a dystrophic area of the lagoon of Venice.

# N:P atomic ratios (1985-86)

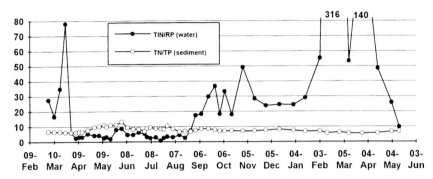

**Figure 9** Water and sediment N:P atomic ratios in a dystrophic area of the lagoon of Venice.

Laboratory experiments evaluating the contribution of sediment resuspension[80] and benthic macrofauna bioturbation[81] to the release of nutrients at the sediment-water interface indicated only minor nutrient increases in the overlying waters after normal shear stresses applications,[80] but benthic macrofauna addition to sediment increased the inorganic nitrogen release rate from the sediment by an order of magnitude (compared with the values obtained in the undisturbed experimental system, 1.7 to 23 mmol $m^{-2}$ $day^{-1}$); while a two- to threefold increase was observed for reactive phosphorus (from 0.26 to 0.7 mmol $m^{-2}$ $day^{-1}$).[81]

An estimate of the nutrient amounts exchanged annually with the sea was recently reported. Such an estimate was obtained by combining the nutrient concentration results of the recent investigation,[21] together with the dissolved/particulate ratio for organic nitrogen and phosphorus previously reported[82] and the hydrological data at the three port entrances.[21] Ammonium and total organic nitrogen, as well as orthophosphate and total phosphorus, were reported to be removed from the lagoon; nitrate to be imported. This influx of nitrate, convoyed particularly through the Chioggia port entrance, was ascribed to the coastal inputs from the rivers (Brenta and Adige) flowing into the Northern Adriatic Sea, as well as to the oxidation of ammonium, and mineralization of organic nitrogen removed from the lagoon. This conclusion appears questionable because of both the different times that chemical and hydrological data were acquired and the organic nutrient dissolved/particulate ratio. Moreover, the nitrate import would require concentrations of this nitrogen form in the nearshore Venice gulf higher than in the lagoon areas near the port entrances, which does not occur. Table 6 shows the average specific loads of nitrogen and phosphorus calculated by taking into account the amounts carried by inflowing rivers (A) and those discharged directly, including atmospheric fall-out (B); nitrogen and phosphorus exchanged with the sea (C) and released from the sediment (D) are also reported. The main contributor of both nitrogen and phosphorus to the lagoon water appears to be the sediment, a consequence of the effects of regeneration processes following the deposition of particulate material of both biogenic and terrestrial origin. The discrepancy between the sum of the overall inflowing nutrient specific loads (Table 6A,B) and those exchanged with the sea (Table 6C) could be ascribed to the underestimation of the lagoon-sea transfer of micro- and macrophyte biomass, as well as to the uncertainty of experimental measurements.

## VIII. CONCLUSIONS

Before drawing final conclusions, it should be emphasized that a full understanding of the eutrophication of the Venice lagoon is hampered by deficiencies in important areas of knowledge, notably the hydrodynamics (in terms of the water mean residence time in different areas and of the water fraction left in ebb and re-entering in flood), the organic nutrient dissolved/particulate ratio in the inflowing rivers and in the sea exchanged water, and the estimations of $N_2$ and $CO_2$ losses by denitrification and mineralization, respectively. Nonetheless, based on the available data, the following conclusions can be drawn.

**Table 6**  Specific Loads (A, B, C) and Sediment-Water Exchange (D) of Nutrients in the Whole Lagoon, According to Different Estimates (Positive Values Indicate Inputs; Negative Values Indicate Outputs)

| References | g m$P_2$ y$^{-1}$ | g m$N_2$ y$^{-1}$ |
|---|---|---|
| A. From the inflowing rivers | | |
| ENEA (1990)[21] | 1.7 | 7.4 |
| B. From direct inputs, including atmosphere | | |
| Andreottola et al. (1990)[20] | 1.2 | 19.2 |
| ENEA (1990)[21] | 2.4 | 19.6 |
| C. From the lagoon to the sea | | |
| ENEA (1990)[21] | −1.8 | −11.8 |
| D. From the sediment | | |
| Tecneco (1978)[58] | 1.2 | 18.5 |
| Battaglia et al. (1983)[44] | 7.0 | 59.3 |
| Orio and Donazzolo (1987)[63] | 3.3 | 42.7 |
| ENEA (1990)[21] | 1.1 | 26.3 |

1. The location of the most productive areas is not explainable on the basis of the nutrient concentration ranges both in water and sediment, but depends primarily on hydrodynamic and morphologic factors.
2. At least 60% of primary production in the whole lagoon and 90% in the Central lagoon, is accounted for by macroalgae.
3. The comparison between the nutrient loads entering and leaving the lagoon (though affected by questionable assumptions and omissions) and the amounts of nutrients associated to the net primary production point out the role of macroalgae as a major factor responsible for the trophic level of the lagoon.
4. Organic matter decomposition, physicochemical conditions and nutrient regeneration occurring in the spring-summer period on the surface sediment is the main factor affecting the composition of the lagoon waters and prompts sedimentary phosphorus and nitrogen to behave quite differently. The former nutrient is released from the sediment more quickly than the latter; thus nitrogen can be the limiting constituent during the massive macroalgal blooms.
5. Lagoon restoration measures should concentrate on decreasing the nutrient turnover in the lagoon areas displaying the highest biomass density. This can be accomplished by: (a) improving water circulation during the productive months; (b) reducing the nutrient loads entering the lagoon; (c) removing the biomass.

The most important factor affecting the quantity and distribution of macroalgal biomass is the water renewal. Both dredging and enlarging the existing canals, as well as the opening across the Lido island canals regulated by mobile dams, could be used to improve the water exchange of the lagoon areas to the south of the Venice historical center. The diversion of the canals entering the lagoon north to the translagoon bridge (particularly the Osellino) into the Sile River, through artificial streams (some of which already exist), would avoid the discharge of polluted sewage in a hydrodynamically critical area. In addition, might the mobile barriers formerly conceived to prevent the occurrence of "high water" phenomena be used to flood the hydrodynamically critical areas? A recent work,[83] based on the application of advanced mathematical models, reinforces this hypothesis.

The area in front of the airport, north to the translagoon bridge, would benefit by the reduction of point source nutrient loads from the inland-lagoon border through the completion of the Mestre sewerage system to the sewage treatment plants. The latter should be improved concurrently in order to reduce nitrogen and phosphorus concentrations in the effluents. A significant fraction of nutrients carried by the rivers could be trapped by aquatic plants or macroalgae cultivated near the mouths of the major inflowing rivers.[14] Moreover, the historical center of Venice could be served by a sewerage system conveying the wastewater offshore in the Northern Adriatic Sea. As an aside, the shortage of nutrient inputs to the shallow waters (20–30 m) of this sea, together with the stratification of the upper water layer impoverished by phytoplankton blooms, are considered major causes of the condensed mucous material appearing in the Northern Adriatic since 1988.[84]

The *in situ* interventions consist mainly of the mechanical removal of biomass by harvesting machines. Based on the main growth process (by fragments) of macroalgae, it is extremely important to remove residual biomass surviving the cold season in order to minimize the overgrowth when climate conditions become favorable. This allows a biomass growth delay of approximately 1 month. Any biomass collection should be carried out with the primary objective of preventing the occurrence of minimum biomass levels (usually, 3–5 kg m$^{-2}$) leading to distrophic conditions and hindering the water circulation. The benefits of this type of intervention are limited to restricted lagoon areas and to the period corresponding to the initial biomass growth.

6. Once appropriate actions are taken, additional time must be allowed to elapse before a decrease in the eutrophication level of the sediment is likely to be seen.

## REFERENCES

1. **Rosa Salva, P.,** Origin and formation of the lagoon of Venice, (in Italian), in *The Lagoon of Venice: Conservation of an Ecosystem,* Arsenale, Venezia, 1987, 13.
2. **Miozzi, E.,** *Venice in the Centuries: the Lagoon* (in Italian), Libeccio, Venezia, 1980.
3. **Rosa Salva, P.,** The processes of anthropic transformation (in Italian), in *The Lagoon of Venice: Conservation of an Ecosystem*, Arsenale, Venezia, 1987, 14.
4. **Carbognin, L. and Gatto, P.,** An overview of the subsidence of Venice (in Italian), I.A.H.S. Report N. 151, 321, 1985.
5. **Gatto, P. and Carbognin, L.,** The lagoon of Venice: natural environmental trend and man-induced modification, *Hydrol. Sci. Bull. des Sci. Hydrol.*, 26, 4, 1981.
6. **Rabagliati, R.,** Hydrodynamic characteristics and evolution trends of the Venice lagoon, (in Italian), in *The Lagoon of Venice: Conservation of an Ecosystem*, Arsenale, Venezia, 1987, 19.
7. **Sfriso, A., Marcomini, A., Pavoni, B., and Orio, A. A.,** Eutrophication and macroalgae: the lagoon of Venice as study case (in Italian), *Inquinamento*, 32(4), 62, 1990.
8. **Damiani, A.,** The eutrophication of the lagoon of Venice. Part I (in Italian), *Ing. Ambientale*, 19, 271, 1990.
9. **ENEA,** Sampling campaigns July 1987–July 1988. Investigation 1.3.1, Internal report, Venezia Nuova, 1988.
10. **Del Turco, A.,** Air quality (in Italian), in *Portomarghera, Venice, and Its Environment*, Ente Zona Industriale di Porto Marghera e Associazione degli Industriali della Provincia di Venezia, Eds., Canova, Treviso, 1990, 105.
11. **Fossato, V.,** Lagoon waters (in Italian), in *Portomarghera, Venice, and Its Environment*, Ente Zona Industriale di Porto Marghera e Associazione degli Industriali della Provincia di Venezia, Eds., Canova, Treviso, 1990, 81.
12. **Bernardi, S., Cecchi, R., Costa, F., Ghermandi, G., and Vazzoler,** Fresh water and pollutant transfer in the lagoon of Venice (in Italian), *Inquinamento*, 28(1/2), 46, 1986.
13. **Cavazzoni, S.,** Freshwaters in the lagoon of Venice, (in Italian), Report no. 64, Laboratorio per lo Studio della Dinamica delle Grandi Masse, Venzia, 1973.
14. **Appi, A., Basili, M., Bendoricchio, G., Berbenni, P., Bortolussi, R., Fontanive, A., Franco, M., Marchesi, R., Perruccio, L., and Prokopowicz, J.,** *Agriculture Run-off Pollution in the Venice Lagoon* (in Italian), Consorzio Venezia Nuova, Ed., Soc. Coop. Tipogr., Padova, 1989.
15. **De Fraja Frangipane, E.,** *Fulfillment of the Guide Plan for the Venice Lagoon Restoration* (in Italian), Consorzio Venezia Nuova, Ed., Venezia, 1988.
16. **Lombardo, S. and Barbieri, P.,** *Statistical Information. 1987 Year-Book* (in Italian), Comune di Venezia, Ufficio Statistiche, 1988.
17. **Lee, N., Matarrese, G., and Miani, P.,** Porto Marghera industrial zone and the study area, in *Portomarghera, Venice, and Its Environment*, Ente Zona Industriale di Porto Marghera e Associazione degli Industriali della Provincia di Venezia, Eds., Canova, Treviso, 1990, 19.
18. **Bendoricchio, G., Agostinello, L., and Alessandrini, S.,** Nutrients of non-point sources from the catchments of the rivers inflowing into the lagoon (in Italian), *Acqua Aria*, 2, 177, 1985.
19. **Cossu, R., and de Fraja Frangipane, E.,** *The Available Knowledge of the Venice Lagoon Pollution* (in Italian), Consorzio Venezia Nuova, Ed., Venezia, 1985.
20. **Andreottola, G., Cossu, R., and Ragazzi, M.,** Nutrients loads from the Venice lagoon catchment area: comparison between direct and indirect assessment methods (in Italian), *Ing. Ambientale*, 19(3–4), 176, 1990.
21. **ENEA,** *Final Report on the Actual Conditions of the Lagoon Ecosystem* (in Italian), Consorzio Venezia Nuova, Ed., Venezia, 1990.
22. **Vazzoler, S., Costa, F., and Bernardi, S.,** Lagoon of Venice and transfer of freshwaters and pollutants (in Italian), *Istituto Veneto di Scienze, Lettere ed Arti. Rapporti e Studi*, XI, 81, 1987.
23. **Ghetti, A.,** Hydraulic problems of the Venice lagoon (in Italian), *Atti Ist. Veneto Sci. Lett. Arti*, 139, 123, 1981.
24. **Ministry of Public Works,** The tide currents in the lagoon of Venice (in Italian), Istituto di Idraulica di Padova, Padova, 1979.

25. **Consorzio Venezia Nuova**, Study 1.3.3, Distribution of the contaminant concentrations and of current speeds at different sections of the port entrances, *Internal Report*, Venezia, 1988.

26. **Alberotanza, L. and Marino, C. M.,** Remote sensing by thermal infrared of the Venice lagoon and its littoral, in *Proc. Conf. Lagoon, Rivers and Beaches: Five Centuries of Lagoon Water Management* (in Italian), Venezia, 10–12 June, Magistrato alle Acque and Ministry of Public Works, Eds., Vol. II, 4, 1983.

27. **Battiston, L., Giommoni, A., Pilan, L., and Vincenzi, S.,** Salinity exchange induced by tide in the Venice lagoon (in Italian), *Proc. Conf. Five Centuries of Water Management in the Venice Territory*, II-11, 2, 1983.

28. **Barillari, A. and Rosso, A.,** First information on the distribution of surface sediment in the northern basin of the Venice lagoon (in Italian), *Mem. Biogeografia Adriatica*, 9, 13, 1975.

29. **Barillari, A.,** First information on the distribution of surface sediment in the central basin of the Venice lagoon (in Italian), *Atti Ist. Veneto Sci. Lett. Arti*, 136, 125, 1978.

30. **Barillari, A.,** Distribution of surface sediment in the southern Venice lagoon (in Italian), *Atti Ist. Veneto Sci. Lett. Arti*, 139, 87, 1981.

31. **Barillari, A.,** The lagoon bottoms (in Italian), in *The Lagoon of Venice: Conservation of an Ecosystem*, Arsenale, Venezia, 1987, 24.

32. **Hieke Merlin, O., Menegazzo Vitturi, L., and Semenzato, G.,** Contribution to the knowledge of the surface sediment of the Venice lagoon (in Italian), *Atti Ist. Veneto Sci. Lett. Arti*, 137, 35, 1979.

33. **Menegazzo Vitturi, L., and Molinaroli, E.,** Role of the sediment mineralogical and physical characteristics on the pollution processes in a typical area of the Venice lagoon (in Italian), *Rapp. Stud. Ist. Veneto Sci. Lett. Arti,* 9, 354, 1984.

34. **Menegazzo Vitturi, L., Molinaroli, E., Pistolato, M., and Rampazzo, G.,** Geochemistry of recent sediments in the lagoon of Venice, *Rend. Soc. Ital. Mineral. Petrol.*, 42, 59, 1987.

35. **Battiston, G. A., Croatto, U., Degetto, S., Sbrignadello, G., and Tositti, L.,** Radioisotopic techniques for dating recent sediments and their use in reconstructing the pollution dynamics (in Italian), *Inquinamento*, 6, 49, 1985.

36. **Batttiston, G. A., Degetto, S., Gerbasi, R., and Sbrignadello, G.,** Determination of sediment composition and chronology as a tool for environmental impact investigations, *Mar. Chem.*, 26, 91, 1989.

37. **Vatova, A.,** Animal associations in the lagoon of Venice (in Italian), *Istituto Italo-Germanico di Biologia Marina*, 3, 1, 1940.

38. **Schiffner, V. and Vatova, A.,** The lagoon algae: *Chlorophyceae, Phaeophyceae, Rhodophyceae, Myxophyceae* (in Italian), in *The Lagoon of Venice*, Minio, Venezia, 1938.

39. **Pignatti, S.,** Marine algae associations on the Venetian coast (in Italian), *Mem. Ist. Veneto Sci. Lett. Arti*, 32, 1, 1962.

40. **Sfriso, A.,** Flora and vertical distribution of macroalgae in the lagoon of Venice: a comparison with previous studies, *G. Bot. Ital.*, 121, 65, 1987.

41. **Sfriso, A., Marcomini, A., Pavoni, B., and Orio, A. A.,** Macroalgal biomass production and nutrient recycling in the lagoon of Venice, *Ing. Sanitaria*, 5, 255, 1988.

42. **Giordani Soika, A. and Perin, G.,** The pollution of the Venice lagoon: study of chemical changes and population variations in the lagoon sediments in the last twenty years (in Italian), *Boll. Mus. Civ. Stor. Nat. Venezia*, 26(1), 25, 1974.

43. **Sfriso, A., Marcomini, A., and Pavoni, B.,** Relationships between macroalgal biomass and nutrient concentrations in a hypertrophic area of the Venice lagoon, *Mar. Environ. Res.*, 22, 297, 1987.

44. **Battaglia, B., Datei, C., Dejak, C., Gambaretto, G., Guarise, G. B., Perin, G., Vianello, E., and Zingales, F.,** *Hydrothermodynamic and Biological Investigations to Determine the Environmental Consequences of the ENEL Thermoelectric Plant at Full Capability Working* (in Italian), Regione Veneto, Venezia, 1983.

45. **Degobbis, D., Gilmartin, M., and Orio, A. A.,** The relation of nutrient regeneration in the sediments of the Northern Adriatic to eutrophication, with special reference to the lagoon of Venice, *Sci. Total Environ.*, 56, 201, 1986.

46. **Sfriso, A., Pavoni, B., and Marcomini, A.,** Macroalgae and phytoplancton standing crops in the central Venice lagoon. Primary production and nutrient balance, *Sci. Total Environ.*, 80, 139, 1989.

47. **Ceretti, G., Ferrarese, U., and Scattolin, M.,** *Chironomids in the Venice Lagoon* (in Italian), Arsenale, Venezia, 1985.

48. **Gilmartin, M. and Relevante, N.,** The phytoplankton of the Adriatic sea: standing crop and primary production, *Thalassia Yugosl.*, 19, 173, 1983.

49. **Socal, G., Bianchi, F., Cioce, F., and Alberighi, L.,** Evolution of phytoplankton blooms in the central Venice lagoon (in Italian), Final Report CNR-ENEL, Venezia, 1987.

50. **Barillari, A., Bianchi, F., Boldrin, A., Cioce, F., Comaschi Scaramuzza, A., Rabitti, S., and Socal, G.,** Variations of hydrological parameters, particulate and plankton biomass during a tidal cycle in the Venice lagoon (in Italian), Proceedings of the 6th A.I.O.L. Conference, Trieste, 1986, 227.

51. **Socal, G. and Bianchi, F.,** Comaschi Scaramuzza, A., and Cioce, F., Spatial distribution of plankton communities along a salinity gradient in the Venice lagoon, *Arch. Ocean. Limnol.*, 21(1), 19, 1987.

52. **Socal, G., Ghetti, L., Boldrin, A., and Bianchi, F.,** Annual cycle and phytoplankton diversity in the Malamocco channel, lagoon of Venice (in Italian), *Atti Ist. Veneto Sci. Lett. Arti,* 143, 15, 1985.

53. **Socal, G., Pellizzato, M., and da Ros, L.,** Qualitative analysis of phytoplankton in waters for mussel aquaculture (lagoon of Venice, Southern Basin) (in Italian), *Soc. Veneziana Sci. Nat.*, 11, 143, 1986.

54. **Comune di Venezia,** *The Algae in the Venice Lagoon* (in Italian), Arsenale, Venezia, 1991.

55. **Tolomio, C.,** The phytoplankton in the Brenta pond (lagoon of Venice). Seasonal investigation: March 1980–March 1982 (in Italian) *Arco Oceanogr. Limnol.*, 21, 117, 1988.

56. **Tiso, A.,** Pollution of the lagoon waters produced by the industrial effluents of Porto Marghera (in Italian), *Atti Ist. Veneto Sci. Lett. Arti,* 3, 1966.

57. **Perin, G.,** The Chemical pollution of the Venice lagoon, (in Italian), in *Problemi dell'Inquinamento Lagunare*, Consorzio per la Depurazione delle Acque e Ente della Zona Industriale di Porto Marghera, Venezia, 1975, 47–89.

58. **TECNECO,** Evaluation of the maximum acceptable concentrations of pollutants discharged in the Venice lagoon. Definition of a control system for the lagoon water quality monitoring, Technical Report (in Italian), Comune di Venezia, 1978.

59. **Comune di Venezia,** Environmental study of the Venice lagoon to define a plan of remediation and of quality control of the lagoon waters (in Italian), Technical Report, Venezia, 1979.

60. **Zucchetta, G.,** The pollution of the Venice lagoon (in Italian), *Mem. Ist. Veneto Sci. Lett. Arti,* 21, 1, 1983.

61. **Degobbis, D., Homme-Maslowska, E., Orio., A.A., Donazzolo, R., and Pavoni, B.,** The role of alkaline phosphatase in sediments of the Venice lagoon on nutrient regeneration, *Estuarine Coastal Shelf Sci.*, 22, 425, 1986.

62. **Bianchi, F., Boldrin, A., Cioce, F., and Socal, G.,** Nutrient concentrations in the Venice lagoon. The northern basin (in Italian), Proceedings 7th A.I.O.L. Conference, Trieste, 1986, 155.

63. **Orio, A. A. and Donazzolo, R.,** Toxic and eutrophicating substances in the lagoon and the gulf of Venice (in Italian), *Istituto Veneto di Scienze, Lettere ed Arti. Rapporti e Studi. XI,* 149, 1987.

64. **Bianchi, F., Boldrin, A., Cioce, F., Rabitti, S., and Socal, G.,** Seasonal variations of nutrients and particulate material in the lagoon of Venice. The northern basin (in Italian), *Mem. Ist. Veneto Sci. Lett. Arti,* XI, 49, 1987.

65. **Alberotanza, L. and Zucchetta, G.,** *Characteristics of the Waters of the Venice Lagoon (Central basin). 1984–1985 survey* (in Italian), CCID, CNR-ISDGM, Regione Veneto, Eds., Venezia, 1986.

66. **Alberotanza, L. and Zucchetta, G.,** *Characteristics of the Waters of the Venice Lagoon (Central Basin). 1986 Survey* (in Italian), CCID, CNR-ISDGM, Regione Veneto, Eds., Venezia, 1987.

67. **Alberotanza, L. and Zucchetta, G.,** *Characteristics of the Waters of the Venice Lagoon (Central Basin). 1987 Survey* (in Italian), CCID, CNR-ISDGM, Regione Veneto, Eds., Venezia, 1988.

68. **Alberotanza, L. and Zucchetta, G.,** *Characteristics of the Waters of the Venice Lagoon (Central Basin). 1988 Survey* (in Italian), CCID, CNR-ISDGM, Regione Veneto, Eds., Venezia, 1989.

69. **Bianchi, F., Cioce, F., Comaschi Scaramuzza, A., and Socal, G.,** Dissolved nutrients distribution in the central basin of the Venice lagoon. Autumn 1979 (in Italian), *Boll. Mus. Civ. Venezia,* 39, 7, 1990.

70. **Pavoni, B., Donazzolo, R., Sfriso, A., and Orio, A. A.,** The influence of wastewaters from the city of Venice and the hinterland on the eutrophication of the lagoon, *Sci. Total Environ.,* 96, 235, 1990.

71. **Chiaudani, G. and Vighi, M.,** Multistep approach to identification of limiting nutrients in Northern Adriatic eutrophied coastal waters, *Water Res.,* 16, 185, 1982.

72. **Damiani, A., Degobbis, D., and Precali, R.,** Main mechanisms governing the spatial and temporal variations of the nutrient concentrations in the Venice lagoon, *Ing. Ambientale* (in Italian), 19, 554, 1990.

73. **Perin, G., Orio, A. A., Pavoni, B., Donazzolo, R., Pastre, B., Carniel, A., Gabelli, A., and Pasquetto, A.,** Nutrients and heavy metals in the sediment (in Italian), *Acqua Aria,* 6, 623, 1983.

74. **Pavoni, B., Donazzolo, R., Marcomini, A., Degobbis, D., and Orio, A. A.,** Historical development of the Venice lagoon contamination as recorded in radiodated sediment cores, *Mar. Poll. Bull.,* 18(1), 17, 1987.

75. **Sfriso, A., Pavoni, B., Marcomini, A., and Orio, A. A.,** Annual variations of nutrients in the lagoon of Venice, *Mar. Poll. Bull.,* 19(2), 54, 1988.

76. **Krom, M. D. and Berner, R. A.,** Adsorption of phosphate in anoxic marine sediment, *Limnol. Oceanogr.,* 25, 797, 1980.

77. **Sfriso, A. and Marcomini, A.,** Macrophyte typology, production, and nutrient recycle in the lagoon of Venice, in *Marine Benthic Vegetation in European Coastal Waters: a Documentation of the Effects of Eutrophication*, W. Schramm, R. L. Fletcher, and P. N. Nienhuis, Eds., EC-COST 48, Brussels, 1992, 000.

78. **Solazzi, A.,** Report on the census of macro-algae in the Venice lagoon, Internal Report, C.S.A.R.E., Rovigo, 1981.

79. **Sfriso, A., Pavoni, B., Marcomini, A., Raccanelli, S., and Orio, A. A.,** Particulate matter deposition and nutrient fluxes onto the sediments of the Venice lagoon, *Environ. Technol.,* 13, 473, 1992.

80. **Sfriso, A., Donazzolo, R., Calvo, C., and Orio, A. A.,** Field resuspension of sediments in the Venice lagoon, *Environ. Technol.,* 12, 371, 1990.

81. **Donazzolo, R., Degobbis, D., Sfriso, A., Pavoni, B., and Orio, A. A.,** Influence of Venice lagoon macrofauna on nutrient exchange at the sediment-water interface, *Sci. Total Environ.,* 86, 223, 1989.

82. **Rossi, A.,** Distribution of Chlorophyll a, Carbon, Nitrogen and Total Phosphorus in the Particulate Matter and Surface Waters of the Northern Basin of the Venice Lagoon, Ph.D. thesis, Faculty of Science, Padua, 1979.

83. **Rusconi, A., Tomasin, A., Rossi, G., and Umgiesser, G.,** Enriched flooding of the Venice lagoon by the mobile barriers (in Italian), *A.I.I. National Conference*, Taormina, 1988.

84. **Tomasino, M.,** A model to forecast the formation of condensed mucous matter in the Adriatic sea (in Italian), *G.N.F.A.O. Workshop*, Naples, 1991.

Chapter 7

# The Ems Estuary, The Netherlands

*Victor N. de Jonge*

## CONTENTS

I. Introduction ..................................................................................................................81
II. Physical Properties .......................................................................................................82
   A. Climate ...............................................................................................................82
   B. Topography and Morphology ...........................................................................83
   C. Tide .....................................................................................................................84
III. Catchment ....................................................................................................................85
   A. Population and Industry ....................................................................................85
   B. Freshwater Discharge ........................................................................................85
   C. Nutrients in Freshwater .....................................................................................86
IV. Water Movement and Transport ..................................................................................88
   A. Tidal Currents ....................................................................................................88
   B. Water Circulation ..............................................................................................88
   C. Residual Water Circulation ..............................................................................88
   D. Time Scales of the Water ..................................................................................88
   E. Salinity ...............................................................................................................89
V. Nutrient Concentrations ..............................................................................................89
   A. Concentrations in the Water .............................................................................89
   B. Processes Regulating Concentration Levels .....................................................90
   C. Nutrients in the Sediment .................................................................................93
VI. Sediment Properties .....................................................................................................93
   A. Composition of Bottom Sediments ...................................................................93
   B. Accumulation and Transport of Suspended Matter ..........................................95
   C. Resuspension of Sediments and Microbenthos ................................................96
   D. Gross and Net Fluxes of Sediment and Biota ..................................................98
   E. Implications of Resuspension Dynamics for Ecological Functioning ...............100
VII. Symptoms of Nutrient Enrichment .............................................................................102
VIII. Consequences of Nutrient Enrichment ........................................................................102
IX. Management Options for Restoration ..........................................................................103
References ...............................................................................................................................105

## I. INTRODUCTION

The estuary of the River Ems is situated at the border between the Netherlands and Germany (Figure 1). It was formed during the Pleistocene.[1-3] Originally, the estuary had a funnel shape, with the River Ems more or less at the present place, while the second river, the Westerwoldsche Aa, reached the estuary near the Punt van Reide (Figure 1). During a number of floods in the 14th and 15th centuries the natural form changed drastically and a large area became inundated.[4] During the period of the extension of the Dollard area accretion began, followed by land reclamation from the sea by diking.[5,6] Diking continued until the present century, resulting in the size of the Dollard as it is today.

The Ems estuary is an important navigation route for sea-going vessels and river ships. The estuary connects the northeastern part of the Netherlands and the northwestern part of Germany with the North Sea. There are three important harbors, one in Germany (Emden) and two in the Netherlands (Eemshaven and Delfzijl). The estuary is situated in an agricultural area.

Fishing occurs in the estuary but does not, on a structural basis, play an important role. Twenty-two ships are licensed for the shrimp fishery. During August and September, fishing of the edible cockle

0-8493-6839-1/95/$0.00+$.50
© 1995 by CRC Press, Inc.

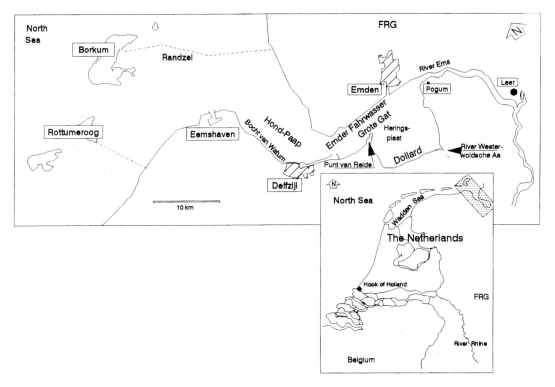

**Figure 1**  Map of the Ems estuary at the border between The Netherlands and the Federal Republic of Germany (northwestern part of Europe; cf. also Figure 2).

(*Cerastoderma edule*) is allowed. When spat stocks in the western Wadden Sea do not suffice, seedling mussels of *Mytilus edulis* are fished from the Ems estuary during May/June and during some days in October. Mussel culturing itself takes place in the western Dutch Wadden Sea. Eel are also fished.

## II. PHYSICAL PROPERTIES

### A. CLIMATE

The climate of northwestern Europe, in which the Ems estuary is situated, is temperate with mild winters and cool summers. The air temperature and consequently the water temperature is regulated by the seasonally fluctuating insolation. The seasonal curves for the air temperature in the different parts of the estuary differ slightly. In summer, the air temperature in the upper part of the estuary is a few degrees higher than near the barrier islands, while in winter the opposite occurs. This is caused by the moderating influence of the sea, which rapidly diminishes in the upstream direction. This marine influence results in a characteristic gradient in temperature along the axis of the estuary which during spring is opposite to that in fall. In early spring, a rapid temperature rise is achieved in the Dollard by the warming up of the intertidal flats during exposure. During the flood this heat is transferred to the water. The percentage of intertidal flats increases in an upstream direction concomitantly with a decrease in water volume. This leads to a faster rise in water temperature in the upper parts than in the lower parts of the estuary. In late fall the water in the upper parts cools faster, so the opposite gradient is found.

The River Ems is a rain-fed river. The annual precipitation varies between years with a coefficient of variation of roughly 18%.[7] The annual precipitation increases from 72.5 cm near the barrier islands to 74 cm at Emden near the river proper.[8] Further upstream, in the drainage basin, 82 cm has been measured.[7] On an annual basis the precipitation in the estuary is approximately balanced by evaporation.[9] Precipitation-induced salinity changes in the main water masses are thus negligible.

In the region of the estuary, winds are predominantly from a westerly direction. This can indirectly influence the small-scale morphology of the intertidal flats and the composition of its sediments. The mean annual wind speed near the barrier islands is ca. 7 m s$^{-1}$ and in the upstream direction in the south eastern part of the Dollard is less than 4.5 m s$^{-1}$.

There is a seasonal cycle in the mean monthly wind speed. Lowest monthly averages are found in summer and highest values in winter. The strongest winds are west southwesterly, while the weakest are northeast to southeasterly.

Depending on direction and speed, the wind raises or lowers the water level and generates waves. Large waves are formed in the deep water of the North Sea. These enter the estuary by the tidal inlet. They can be traced as far as Eemshaven. Most waves, however, are smaller and formed within the estuary. These small waves play a crucial role in the resuspension and the sorting of sediments from the tidal flats. The wind also generates drift currents.

## B. TOPOGRAPHY AND MORPHOLOGY

The estuary of the River Ems is part of the 600 km long European Wadden Sea (Figure 2). This shallow sea consists of a chain of tidal basins that are protected from the direct influence of the North Sea waves by a girdle of barrier islands. Tidal inlets allow water movement to and from the North Sea. The different tidal basins are partly separated from each other by high tidal flats (tidal watersheds) so that the water masses of the different basins are only connected with each other during high tide. Further, the Wadden Sea is intersected by a number of rivers, the River Ems being one of them.

The Ems estuary, without its outer delta, has a surface of approximately 500 km$^2$. Without the tidal reach of the river, this area is 463 km$^2$. Roughly one half of this surface is covered by intertidal flats.

Topographically and hydrographically, the estuary can be divided into three parts. The upper part covers 100 km$^2$ and is situated between Pogum and the mouth of the Dollard (Figure 1). The Emder Fahrwasser consists of a canalized and heavily dredged tidal channel of which the embankments (50% of the surface area) consist of intertidal flats. In the Dollard, 85% of the total surface consists of intertidal flats.[10] The total water volume of this upper part at mean sea level is approximately $120 \times 10^6$ m$^3$.

Downstream of the Dollard inlet (Figure 1) is the middle part which has the typical funnel shape of estuaries in this part of the world. This middle part covers 155 km$^2$ of which 35% are intertidal flats. This region extends to the harbor "Eemshaven" where the estuary starts to intersect the Wadden Sea. A large intertidal flat (Hond-Paap) divides the estuary longitudinally creating two channels. The mean depth of

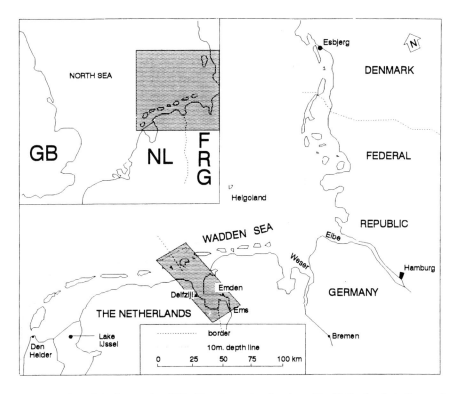

**Figure 2** Map of the international Wadden Sea extending from The Netherlands to Denmark.

this section increases gradually in the seaward direction. The water volume of this area at mean sea level is approximately $550 \times 10^6$ m³.

The most seaward part of the estuary is the intersection between the Wadden Sea and the estuary. The boundary with the North Sea is formed by two barrier islands and the tidal inlet between them. The high tidal flats between the barrier islands and the mainland (dashed line in Figure 1) form the hydraulic boundaries between the estuary and the adjacent tidal basins. The boundary with the mainland is formed by dikes as for the rest of the estuary. The area comprises 215 km² of which over 45% consists of intertidal flats. This area is also a two-channel system due to a series of shoals. The volume of this part of the estuary is approximately $770 \times 10^6$ m³ at mean sea level.

The morphology of intertidal flats, channels and gullies is complex and unstable. Slow changes occur through the natural processes of sediment supply, meandering of channels and gullies, but also through human influences such as land reclamation, building of harbors, dredging and sand mining. Samu[11] concluded that major morphological changes in the lower reaches occur in a cycle with a period of ca. 25 years. Similar morphological changes due to dredging and sand mining were clearly noticeable within 1 year.[10]

## C. TIDE

The tide, with its consequent water movement, is very important for the transport and mixing of the water and hence, for the biology of the intertidal zone. It is also indispensable for navigation purposes. Vertical movement in the estuary is mainly caused by tidal waves of which the range can increase or decrease through wind stress. The tides are dominated by the semi-diurnal lunar tide with a mean period of nearly 12 hours, 25 min. The cycle in tidal ranges between spring tide and neap tide (Figure 3) is caused by the semi-diurnal solar tide. Moreover, there is a significant diurnal inequality in the tidal range (Figure 3). In Delfzijl (Figure 1), the diurnal averaged tidal periods approximately fluctuate between 12.2 and 12.6 hours (Figure 3). The highest tidal ranges at spring tide coincide with the shortest tidal period, while at neap tide the opposite occurs (Figure 3). Consequently, current velocities at spring are much larger than at neap tides.

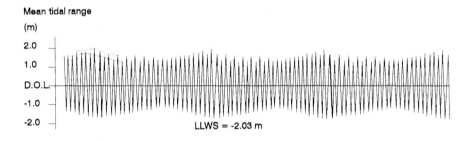

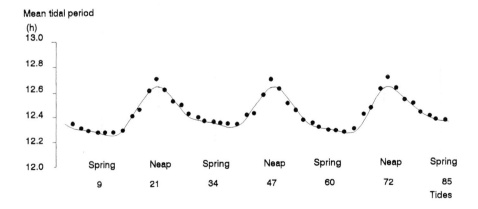

**Figure 3** Variation in the mean tidal range and in the mean tidal period due to the lunar cycle. Values calculated from Dutch tide tables, Staatsuitgeverij, The Hague, The Netherlands.

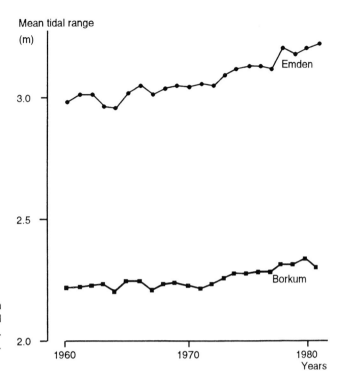

**Figure 4** Long-term change in mean tidal range for stations Emden and Borkum. [After de Jonge, V. N., *Can. J. Fish. Aquat. Sci.*, (Suppl. 1), 289, 1983. With permission.]

The tidal range along the axis of the estuary increases upstream due to resonance. In 1971, the mean annual tidal range at Borkum was 2.20 m. In Delfzijl, the range was 2.79 m and at Emden 3.03 m. In the southeastern part of the Dollard the mean tidal range is usually somewhat higher than at Emden. Upstream of Emden on the River Ems the mean tidal range slowly decreases again.

The mean annual tidal range is not constant (Figure 4), having increased since 1971. This increase appears to consist of two components. A local component is due to the heavy dredging activities reported by the author,[10] while the origin of the second component is unknown but seems to operate over the entire Atlantic Ocean.[12]

The tidal curve is asymmetrical. For example at Delfzijl, the mean period of the flood is 55 minutes shorter than the ebb period, hence flood currents are stronger than ebb currents. This has a large impact on the transport of sediments and biota.

## III. CATCHMENT

### A. POPULATION AND INDUSTRY

The River Ems crosses an area with a population density between 100–200 km$^{-2}$. The drainage basin is, however, enclosed by an area with a population density of over 200 people km$^{-2}$. Approximately 35% of the total population is working in industry and ca. 10% in agriculture. The main part of the area is used for agricultural purposes.

### B. FRESHWATER DISCHARGE

Freshwater enters the Ems estuary by different sources of which the most important is the German River Ems. This is a rain-fed river of which the drainage basin comprises an area of 12,650 km$^2$ with a length of approximately 200 km (Figure 2).

The freshwater discharge of the river varies strongly both seasonally and annually (Figure 5). The highest monthly discharge at Pogum for the period 1970–1979 was 390 m$^3$ s$^{-1}$ and the lowest 25 m$^3$ s$^{-1}$.

A second river supplying freshwater to the estuary is the Westerwoldsche Aa (Figure 1). The discharge point of this small canalized river is situated at Nieuwe Statenzijl in the southeastern part of the Dollard (Figure 1). Controlled quantities of water can be sluiced out only during low tide. The Westerwoldsche Aa does not have a well-defined drainage basin because the river is connected with other water courses in the northern part of the Netherlands. The connections are used for flushing the extensive canal system

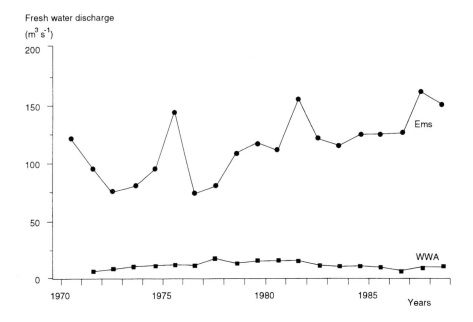

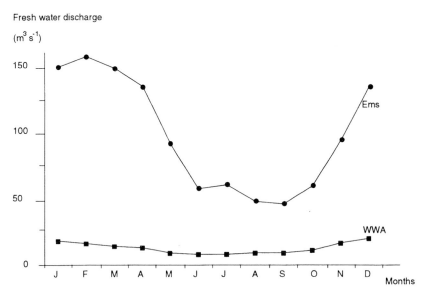

**Figure 5** Mean annual discharge for the River Ems and Westerwoldsche Aa (WWA) over the period 1970–1979 and mean monthly discharges calculated for the same period. Values from local authorities in The Netherlands and Germany.

in the northeastern part of the country. The freshwater discharge of this canalized river is roughly 10% of that of the River Ems. The mean seasonal discharges (Figure 5) as well as the year-to-year variations are given, because of the impact on the estuary exerted by the dissolved components and particulate organic matter in the discharged water volumes.

## C. NUTRIENTS IN FRESHWATER
Different nutrients undergo different patterns of seasonal variation in the water column (e.g., Figure 6). Nitrate values are high all the time, although the concentrations in summer are lower than in the other seasons. Ammonia values are very high in early spring. Reactive silicate in the river is very high in spring, decreasing during the summer and then increasing again. Soluble reactive phosphate values tend to increase from January to September followed by a decrease.

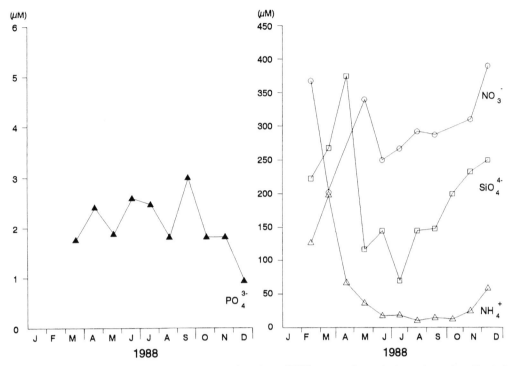

**Figure 6** Seasonal variation in soluble reactive phosphate (SRP), ammonium, nitrate and reactive silicate for the river Ems near the town Leer. Open circles: $NO_3^-$ open squares: $SiO_4^{4-}$, open triangles: $NH_4^+$, closed triangles: $PO_4^{3-}$.

Nutrients enter the Ems estuary from different sources. In total, the River Ems and the Westerwoldsche Aa, as well as some small canals, bring large amounts of nutrients (both dissolved and particulate forms) into the estuary (Table 1). The values in the table indicate that the River Ems is the main contributor of nutrients to this estuary. Precipitation contributes only a small amount of nutrients to the estuary. Large amounts of particulate nitrogen and phosphorus are also transported from the Wadden Sea area and the North Sea coastal waters into the estuary. The particulate nitrogen supply from these areas can not be quantified at present. The annual particulate phosphorus supply from these coastal areas to the estuary was estimated using data on particulate phosphorus in suspended matter, the accretion rate of the estuary

**Table 1** Annual Supply and Concentration Ranges of Nutrients from the River Ems, the Westerwoldsche Aa and Other Sources

| Load (metric tonnes a⁻¹) | | Borrowed from | Parameter | Minimum (μmol l⁻¹) | Maximum (μmol l⁻¹) |
|---|---|---|---|---|---|
| **Ems** | | | | | |
| Total N | 18,500 | a,b,c | Dissolved inorganic N | 70 | 460 |
| Total P | 2,190 | a,b | Orthophosphate | 3 | 20 |
| Silicate | 13,700 | a,b | Silicate | 40 | 240 |
| **Westerwoldsche Aa** | | | | | |
| Total N | 6,550 | d,f | Dissolved N | 430 | 5,000 |
| Total P | 890 | d,f | Orthophosphate | 3 | 100 |
| Silicate | 1,625 | a,d,e | Silicate | 80 | 315 |
| **Other sources** | | | | | |
| Total N | 2,780 | d,f,g,h | | | |
| Total P | 430 | d,f,g,h | | | |
| Silicate | >2,400 | a,d,g | | | |

*Note:* Values calculated using: (a) BOEDE measurements; (b) river discharges (Wasser und Schiffahrtsamt Emden); (c) Reference 64; (d) river discharges (Rijkswaterstaat); (e) Reference 65; (f) measurements (Rijkswaterstaat); (g) Reference 66; (h) precipitation.[9]

and the mineral composition of the sediments. This results in an annual influx ranging from 9?? to 3950 tonnes phosphorus per annum. The mean annual influx amounts to ca. 2250 tonnes P, corresponding with 4.8 g P m$^{-2}$ year$^{-1}$. Application of the procedure used by de Jonge and Postma[13] results in a higher value of 3500 tonnes P.

## IV. WATER MOVEMENT AND TRANSPORT

### A. TIDAL CURRENTS

The rise and fall of the water, the vertical tide, generates the oscillating tidal currents (horizontal tide) in the Ems estuary. This horizontal water movement is the dominant feature of the tide in the Ems estuary.

The tidal currents result in a displacement of water which can be indicated by different terms, one of them being the tidal prism. The tidal prism near Borkum is approximately $1000 \times 10^6$ m$^3$. In the upstream direction the tidal prism decreases strongly due to the decreasing basin surface area and depth. In the Dollard inlet the tidal prism is reduced to only $115 \times 10^6$ m$^3$. The value for the lower end of the river Ems, the Emder Fahrwasser, is approximately $75 \times 10^6$ m$^3$. Another feature of the tide, the tidal traveling distance, also varies over the estuary, although less than the tidal prism. Near the tidal inlet the tidal traveling distance is approximately 17 km and in the Dollard it amounts to 12 km. Both the tidal prism and the tidal traveling distance vary with changes in the vertical tide.

### B. WATER CIRCULATION

The oscillating water movement indicated above leads to typical circulation and to mixing. In the Ems estuary three important circulation mechanisms are present: a tide-induced water circulation, a river-induced water circulation and wind-driven water circulations or drift currents. The magnitude of drift currents is governed by wind direction and the morphology of the basin. In the Ems estuary drift currents play a role in the lower reaches. This is the part of the estuary that is situated in the Wadden Sea. Here large water masses can be transported by wind from the estuary to the adjacent tidal basins, whereby the direction depends on the wind direction. Drift currents also occur in the Dollard. Together the different circulation processes result in changing residual movements of water at certain localities. For a physical description of different types of water circulation the reader is referred to Dronkers and Zimmerman.[14]

### C. RESIDUAL WATER CIRCULATION

Residual circulations are present at different scales. They occur between pairly organized tidal channels as well as within channels (Figure 6). They are caused by cross-sectional pressure gradients of the flowing water and the Coriolis force.[15] In the paired system a flood surplus may occur in one of the channels. Consequently, this channel is named the flood channel and the other one the ebb channel.

In the Ems estuary the volumes of the residual circulations amount to 1–18% of the tidal volumes.[16] The magnitude of residual currents and a qualitative distribution pattern of the residual currents in the Ems estuary is presented in Figure 15.

### D. TIME SCALES OF THE WATER

The end effect of the described water circulations is mixing and consequently transport. This water transport and its dissolved and suspended substances are of paramount importance because they give a continuous supply and exchange of substances between different estuarine parts. However, the different processes contributing to this transport are very complex, so for calculations simplification is required.

Dorrestein and Otto[17] described in the early 1960s a simple one-dimensional transport model for the Ems estuary which was improved in the early 1980s by Helder and Ruardij.[18]

A selection of results from the latter authors (residence-time, turnover time and mean age of water) is given in Table 2.

Here the residence time (r) is the average time interval a parcel of water or its dissolved constituents needs to wend its path from the source until its arrival at the point or part of the estuary for which the residence time is calculated. When this time interval is calculated for the total pathway from inlet to outlet, this interval is called transit time (t). The transit time of freshwater equals the flushing time (f), which is the ratio between the total freshwater volume in the estuary and the discharge rate of freshwater.

The turnover time (e) is the time interval needed to decrease the volume of water or mass of dissolved constituent present at t = 0 in a part of the estuary, to a fraction e$^{-1}$.

**Table 2**  Comparison of Time Scales (Days) at High Discharge (351 m³ s⁻¹) and Low Discharge (34 × 1 m³ s⁻¹) of Freshwater for the Ems Estuary as a Whole and in the Dollard Separately at High Discharge (31.0 m³ s⁻¹) and Low Discharge 5 × 1 m³ s⁻¹) of WWA Water

| Time scale | Time (days) | |
| --- | --- | --- |
| | High discharge | Low discharge |
| Ems estuary | | |
| Sea water | | |
|    Mean age | 14.0 | 36.3 |
| Ems water | | |
|    Mean age | 19.3 | 64.7 |
|    Flushing time | 12.1 | 72.1 |
| WWA water | | |
|    Mean age | 12.9 | 32.4 |
| Turnover time of basin | 17.7 | 35.7 |
| Dollard | | |
| Water entering at mouth | | |
|    Mean age | 5.9 | 17.9 |
| WWA water | | |
|    Flushing time | 8.7 | 31.2 |
| Turnover time of basin | 10.0 | 21.0 |

Data from Helder, W. and Ruardij, P., *Neth. J. Sea Res.*, 15, 293, 1982.

The mean age (a) is the time that water parcels located in a certain part are present in the estuary since arrival from their source (viz., sea, Ems, Westerwoldsche Aa).

## E. SALINITY

Due to strong mixing processes, vertical differences in salinity are usually very small.[18] Only longitudinal and cross-sectional differences in salinity are therefore discussed here.

Shape and slope of the longitudinal salinity gradient in the estuary are continuously changing through the varying river discharge. At low river discharges (mainly during summer) the salinity gradient is relatively steep and short. During the winter at high river discharge the slope decrease and the gradient are longer.

At a fixed station the salinity changes due to tidal water transport. Near the River Ems these changes are relatively large and near the tidal inlet the changes in salinity are relatively small (Figure 7).

## V. NUTRIENT CONCENTRATIONS

### A. CONCENTRATIONS IN THE WATER

Once in the estuary, dissolved components supplied by the River Ems and Westerwoldsche Aa are dispersed and diluted in basically the same way as described for the freshwater. This results in a concentration gradient (Figure 8) with high concentrations in the river water and low near the tidal inlet. A comparison of the nutrient curves shows remarkable differences in shape. Reactive silicate concentrations gradually decrease from the river (town Leer) in a seaward direction. In general, soluble reactive phosphate follows the same distribution pattern, although high phosphate loads from the River Westerwoldsche Aa alter the phosphate curve considerably from that of reactive silicate. The effect of the discharges from the Westerwoldsche Aa is even more pronounced for ammonia. This curve shows that the distribution of ammonia in the estuary is dominated by the loads from the River Westerwoldsche Aa. The nitrate concentrations in the estuary are dominated by the discharges from the River Ems. All together the curves illustrate poor oxygen (low redox) conditions in the water discharged from the canalized River Westerwoldsche Aa some 15 years ago.

Biologically, the most important areas are those where growth conditions are favorable for algae and where the dissolved nutrients reach low concentration levels so that nutrient limitation might occur, that is, in the downstream parts of the estuary (Figure 8).

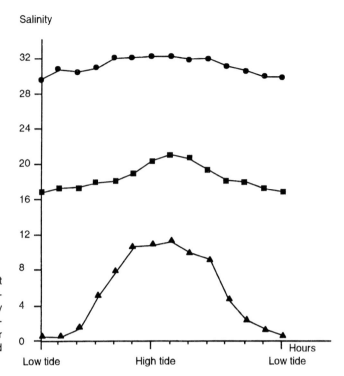

**Figure 7** Tidal variation in salinity at stations Eemshaven (lower reaches, represented by closed circle), halfway through the Dollard (upper reaches, represented by closed square) and the river Ems near Pogum (represented by closed triangles).

## B. PROCESSES REGULATING CONCENTRATION LEVELS

During transport of nutrients from the river to the sea, biological processes are responsible for the disappearance of dissolved nutrients from the water. Nutrients are mainly taken up by algae and phanerogams while mineralization processes are responsible for the production of dissolved nutrients.

Physicochemical processes also play a role in the dynamics of nutrients, e.g., phosphorus. Particulate nutrients can pass into solution or dissolved nutrients can change to a particulate phase due to physicochemical processes. Experiments have shown that dissolved reactive phosphorus is in equilibrium with a releasable phosphate fraction in the suspended matter.[19-21] However, under field conditions this equilibrium was less clear, and this was ascribed to the operation in the field of several processes together. Under field conditions phosphate-salinity plots showed nonconservative behavior of this nutrient with high summer values in the middle reaches of the estuary.[19] Both biological and physicochemical processes could serve as an explanation. Available data indicated a significant role of calcite in the adsorption or coprecipitation of phosphorus in the Ems estuary. New research revealed that the hypothesized role of calcite in this estuary possibly was overestimated.[22] In the suspended matter near the inlet of the Dollard only ca. 5% of the phosphorus turned out to be calcite associated, ca. 70% was iron-bound, 20% was incorporated in organic matter and consequently the remaining 5% was thought to be adsorbed to or incorporated in clay minerals. Thus, recent investigations suggest the main processes to be connected with iron compounds and organic matter. Concentration measurements indicate that both iron and organic matter (partly as humic substances) are supplied in large amounts from the River Ems. At low redox potential the iron-bound phosphate is mobilized.[23] This situation may occur during high oxygen consumption in especially the bottom sediments.[24] Although not measured, the shallow Dollard with its sediments rich in fine particles, possibly plays a significant role in this process. Also, suspended organic matter as humic substances can play a role in phosphate adsorption.[25-27] Experimental evidence is available for the role of variations in redox potential and organic matter on the flux of phosphate from and to suspended sediment fractions in the Ems estuary.[22] The net result of all the operating processes is a seasonal nutrient cycle (Figure 9). Examples are given for the tidal inlet and a station halfway through the main channel Grote Gat in the Dollard. In the Dollard limitation of algal growth is unlikely to occur as the concentrations stay high during all seasons. However, in the outer area (Figure 9), seasonally low levels are reached so that nutrient limitation may occur for diatoms in spring and for other phytoplankton in summer.

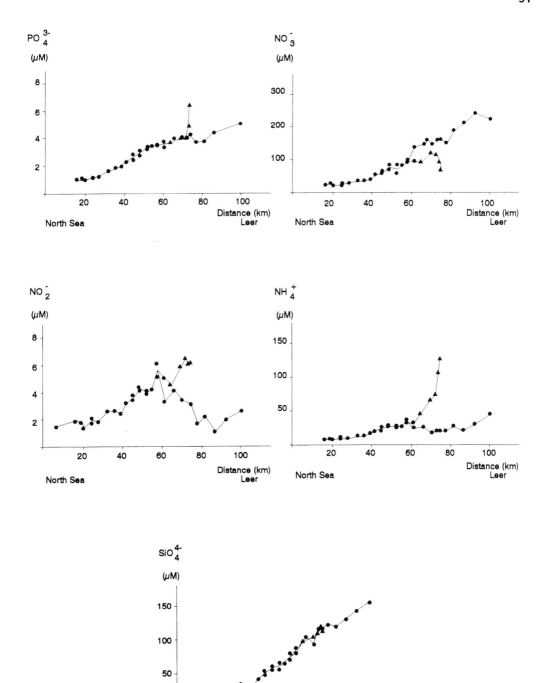

**Figure 8** Mean annual gradient over 1975–1976 of the concentrations of soluble reactive phosphate, nitrate, nitrite, ammonia and reactive silicate. Solid dots represent the main axis of the Ems Estuary and solid triangles the Dollard. The barrier islands are situated at the position of 20 km.

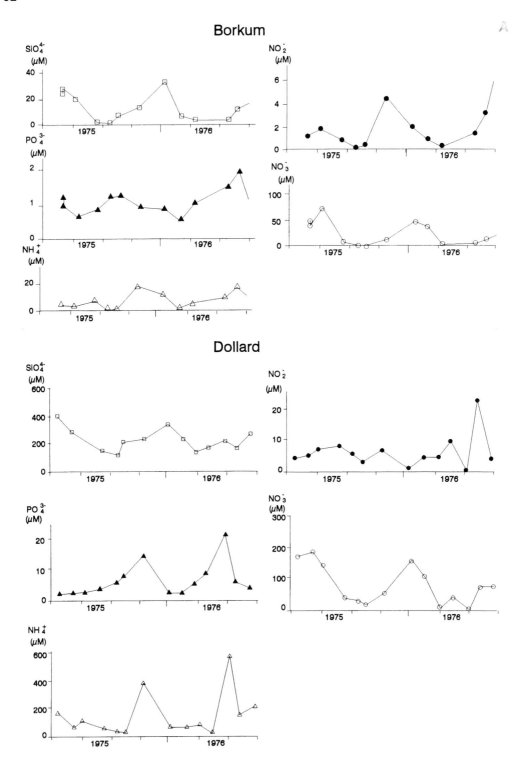

**Figure 9** (A) The seasonal cycle of soluble reactive phosphate, nitrate, nitrite, ammonia and reactive silicate in 1975–1976 for the stations Borkum and Grote Gat (Dollard). (B) The seasonal cycle of soluble reactive phosphate, nitrate, ammonia and reactive silicate for station Borkum in 1988. For A and B, open circles: $NO_3^-$, open squares, $SiO_4^{4-}$, open triangles: $NH_4^+$, closed triangle: $PO_4^{3-}$.

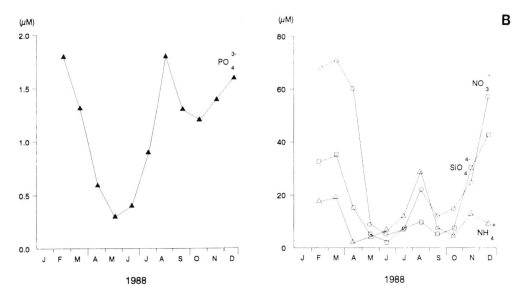

Figure 9 (B)

## C. NUTRIENTS IN THE SEDIMENT

As a considerable part of the total primary production occurs by benthic diatoms (see below), the nutrient levels in the pore water of the surface layers of the intertidal flats are important.

During low tide the concentration of reactive phosphate and reactive silicate in the thin water layer on the sediment and in the pore water of the surface layer of the intertidal flats is usually higher than 1 $\mu M$ (Figure 10). Only in the outer region do reactive silicate values drop below 1 $\mu M$ during the phytobenthos bloom in spring. This may indicate growth limitation due to low nutrient values in the sandy sediments there. In the other areas growth limitation due to low nutrient values probably rarely occurs. Typical mean annual concentration gradients of soluble reactive phosphate and reactive silicate are given in Figure 11. The nitrogen cycle in the bottom sediments of the estuary was studied by Rutgers van der Loeff et al.[28] and Helder.[29] The nitrogen values published by these authors and my own data indicate that this nutrient possibly does not regulate the benthic diatom growth. There are even indications that ammonia concentrations above 500 $\mu M$ inhibit diatom growth.[30]

## VI. SEDIMENT PROPERTIES

### A. COMPOSITION OF BOTTOM SEDIMENTS

In the estuary of the River Ems a great diversity in sediment types exists because of the differences in mineral composition and organic matter content (e.g., peat). The basic process responsible for these differences is physical sorting of the sediments which results in gradients of grain size and consequently organic matter content. Sediments and benthic organisms together determine the physicochemical conditions, although these conditions in turn also influence the composition and activity of the benthic flora and fauna.

The grain size distribution and the clay content of the sediments in the Dollard area was investigated in the past by Maschhaupt[31] and more recently by Wiggers.[5] Wiggers also investigated a number of localities in the lower reaches of the Ems estuary. In 1990, an extensive survey[32] was carried out to map the sediment composition of the main part of the estuary following a grid of 1 km². The results of this mapping are presented in Figure 12. Wiggers[5] concludes that in the long run the granular composition of the sediments in the Dollard is conservative. For clay content, Schröder and Kop (personal communication) observed that in fact the same holds during shorter periods although within-year fluctuations occur due to changing wind conditions.[33]

The values given below are borrowed from the authors referred to above. The percentages given below refer to w/w.

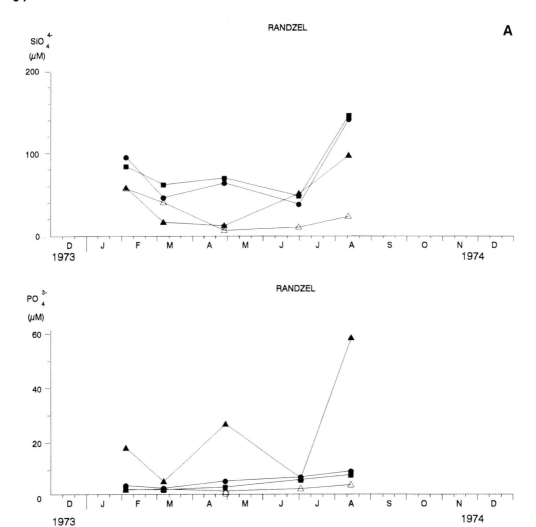

**Figure 10** Seasonal cycle of soluble reactive phosphate and reactive silicate in pore water for stations Randzel (lower reaches near Borkum) and Heringsplaat (tidal flat in central part of the Dollard). Open squares: channel water; open triangles: water on tidal flat; closed triangles: sampled sediment layer 0–1 cm; closed circle: sampled sediment layer 0–5 cm; closed square: sampled sediment layer 5–10 cm.

The sediments in the lower reaches are mainly composed of sand (87%). The median grain size is 169 µm and the silt content is on average 13%. The clay content in these exposed lower reaches is low and on average 1.4%. However, near the land reclamation works along the main coastline the clay content is significantly higher.

In the middle reaches of the estuary the silt content (33%) and the clay content (4.5%) increase in comparison with the lower reaches. Consequently, the sand content (67%) and the median grain size (102 µm) is lower than in the lower reaches. The sediments of the Paap tidal flat in the middle reaches are 79% sand and 2.6% clay. The median grain size is 113 µm. The narrow elongated stretch along the Bocht van Watum contains fine sediments with a median grain size of only 68 µm.

In the Dollard, uniform gradients are present from a clay content of less than 5% (w/w) in the central part, steadily increasing in all coastal directions.[5,31] Near the salt marshes along the dikes the clay content may reach values up to 35%. The sand content in the Dollard is on average ca. 70%.[31]

Most clay is found in the sediments of sheltered tidal flats, that is in the neighborhood of islands and dikes, whereas the sediments along the main tidal channels are predominantly sandy. Nevertheless, the main constituent of the tidal flats is clearly sand. The channel beds are usually very sandy with the exception of the Emder Fahrwasser, the lower reach of the River Ems and the Dollard.

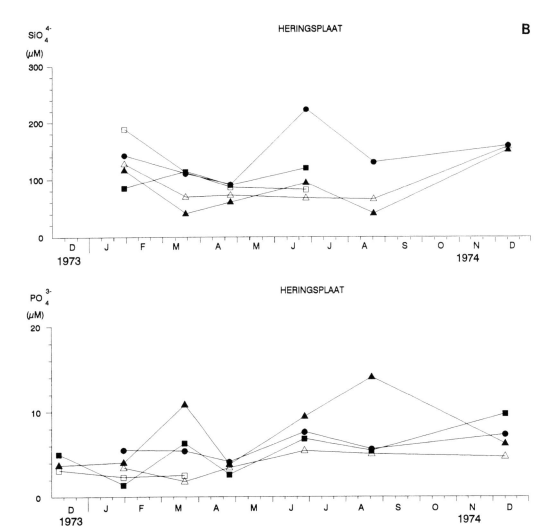

Figure 10 (B)

The quantity and the quality of the organic carbon present in the sediment is of decisive significance for the biological activities there. The oxygen content in the sediments determines the nature of the activities. The soils of the catchment area and consequently the sediments in the Ems estuary contain peat.[2] This material is present as particles, pieces and layers. Other organic matter is also present, the majority of which is present in aggregates composed of sand, clay and organic material.[34-36] The organic matter content of the sediments closely follows its clay content.[5,31] This indicates that the main part of the organic carbon is associated with the clay mineral fraction of the sediment. On average, 6.5–7.0% of the clay mineral fraction consists of organic carbon.

## B. ACCUMULATION AND TRANSPORT OF SUSPENDED MATTER

The concentrations of suspended matter in the Ems estuary show a gradient which is typical for most coastal plain estuaries. Low concentrations occur near the tidal inlet and very high values are present at the salt intrusion limit in the River Ems. The zone with high values is called the "turbidity maximum".[37] Upstream of the turbidity maximum the concentration of suspended matter decreases again.

The longitudinal gradient in suspended matter is formed and maintained by two processes,[37] resulting in the typical turbidity maximum.[38] During the neap tide and spring tide cycle a substantial part of the accumulated suspended matter is deposited as shoals and resuspended again.[39] Usually two shoals are found, one near the HW location of the salt intrusion limit and one at the LW location.[39]

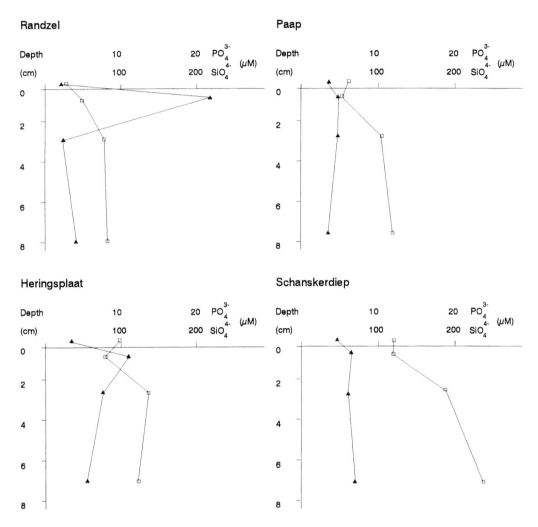

**Figure 11** Mean annual vertical distribution of concentrations of soluble reactive phosphate and reactive silicate at stations Randzel (lower reaches), Paap (middle reaches), Heringsplaat (central Dollard upper reaches) and Schanskerdiep (close to the freshwater discharge point of the River Westerwoldsche Aa). Closed triangle: $PO_4^{3-}$; open square: $SiO_4^{4-}$.

The suspended matter concentrations in the turbidity maximum are also inferred by river discharge.[40] The peak values of suspended matter in the turbidity region of the Ems estuary are inversely correlated with the river discharge. The present author found that in the turbidity region between Emden and Leer the mean annual suspended matter concentrations between 1970 and 1979 varied little from 200 g m$^{-3}$.

The role of tide-induced, residual sediment transport was first recognized by Postma.[41] Later, the process was analyzed experimentally as well as theoretically.[42-44] A description of the development of the theory has recently been published.[45] The theory consists of a complex of factors, each of them favoring the net upstream transport and accumulation or deposition of suspended matter.

## C. RESUSPENSION OF SEDIMENTS AND MICROBENTHOS

Resuspension is caused by tidal currents and wind-induced waves. The transport of resuspended material is caused by tidal currents. As discussed above, the tidal change of suspended matter in the channels is mainly caused by resuspension of sediment material from the channel bed, which mainly consists of sand. Although the current velocity of the water above the tidal flats is much weaker than in the channels it also causes resuspension (Figure 13).

The effect of differences in wind speed on the resuspension of "mud" (fraction <55 μm) in the lower reaches and in the Dollard is significant. Recently, a relation was found between "effective wind speed"

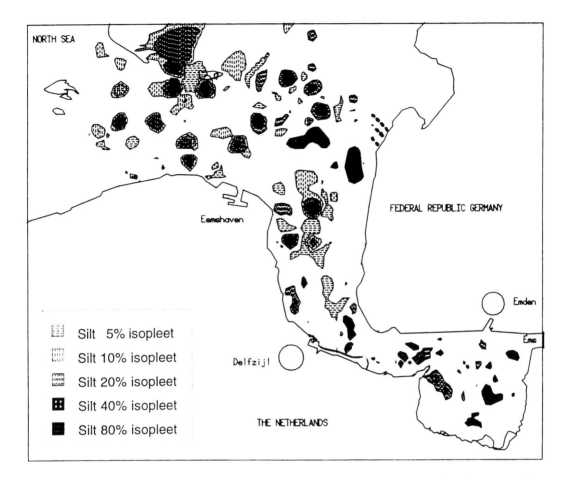

**Figure 12** Map of the sediment composition in the Ems estuary. The diagram gives the silt content of the sediments.

(being the mean wind speed over three high water periods preceding sampling) and the mud concentration in the lower reaches and in the Dollard (Figure 14). The linear regression functions were used to calculate the wind-induced transport of mud between tidal flats and channels within the estuary. The curves in Figure 15 illustrate for the lower reaches that the wind contributes strongly to resuspension while current velocities contribute only moderately. When the effective wind speed significantly decreases, mud concentrations return to background within only 2 days.

Resuspension has a high impact. Resuspended mud is responsible for short-term changes in turbidity and thus is a major factor governing algal growth. An example of the calculated wind-induced fluctuations in suspended matter per tide over a full year is given in Figure 16.

Resuspension of mud may also be responsible for a high turnover rate of nutrients between tidal flats and channels in the estuary because an important proportion of total phosphorus is available as particulate phosphorus in the suspended matter.[13,19] Part of this fraction is readily biologically available.[19] The high exchange rate of mud between tidal flats and channels may thus guarantee a steady phosphate supply to the water column and possibly a certain partitioning of phosphate between the channel water and the sediments.

Together with the mud fraction, microorganisms (benthic diatoms and other microphytobenthos) are resuspended. The resuspended fraction of total microphytobenthos can be described as a function of the effective wind speed (Figure 17). Resuspension of microphytobenthos is very important because a significant part of the phytoplankton in the Ems estuary proved to be resuspended microphytobenthos (Table 3). Based on chlorophyll $a$, in the lower reaches 30% of the phytoplankton is in fact microphytobenthos. In the Dollard this value is much higher and over 90%.[46] Resuspended microphytobenthos provides a significant part of the total annual primary production.

98

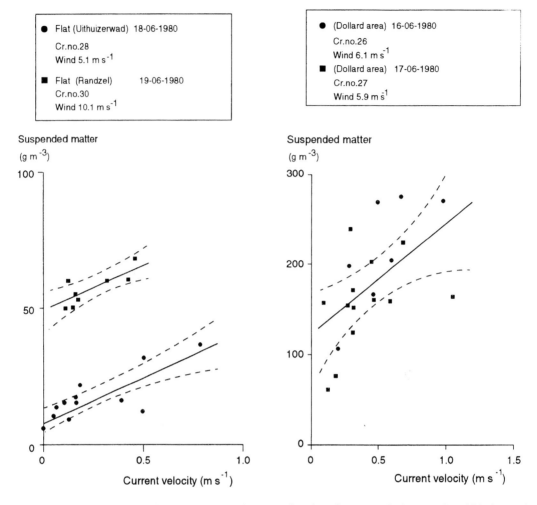

**Figure 13** Plots of suspended matter concentrations as a function of current velocity at stations Uithuizerwad (lower reaches), Randzel (lower reaches) and Dollard.

## D. GROSS AND NET FLUXES OF SEDIMENT AND BIOTA

Most suspended matter in the water and by far the most sediments of the estuary are of marine origin,[47-49] and so the import and sedimentation of marine particulate material predominates in the Ems estuary.[45]

The net transport of material through the estuary is extremely low in comparison with the amount that is tidally deposited and resuspended due to changes in effective wind speed.[46] With each tide a considerable transport and exchange of different sediment fractions occurs between tidal flats and channels. The annual gross fluxes of mud between tidal flats and estuary channels in two areas, and the lateral fluxes of microphytobenthos carbon for the same areas, are plotted in Figure 18.

In the tidal channels gross fluxes of mud and organic carbon also occur. These amounts can be easily estimated from the mean annual concentrations of mud and organic carbon in the water and the mean tidal water transport at a certain cross section. Further, there is the net import of material from the rivers Ems and Westerwoldsche Aa and the local carbon production by the algae.

The net import of mud and organic carbon from the North Sea coastal zone was estimated using the mean silt content of the sediment in the different parts of the estuary and the annual accretion of the Dollard of 0.008 m a$^{-1}$.[50] An annual import of $1.2 \times 10^6$ tonnes of mud and $0.06 \times 10^6$ tonnes of organic carbon was calculated.

Annual gross fluxes between the tidal flats and the channels and longitudinal transport through the channels are substantial when these figures are compared with the net import of particulates by the river and the calculated net import of particulate material from the coastal area. These values indicate the total

Suspended matter in the main channels

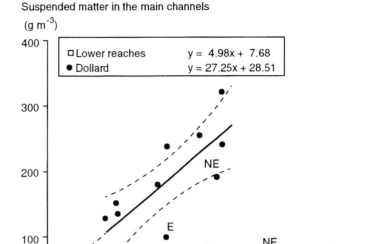

**Figure 14** Plot of mean suspended matter concentrations in two parts of the Ems estuary (lower reaches and Dollard) as a function of "effective wind speed".

net fluxes of material from the North Sea coastal zone to the Ems estuary amount to only a few percent of the gross fluxes. Thus, it must be assumed that net annual fluxes are not measurable when conventional sampling strategies are applied. In using current meters, the water transport in the estuary can be determined with an accuracy of 5–10%.[16] This means that a net flux of suspended sediment which

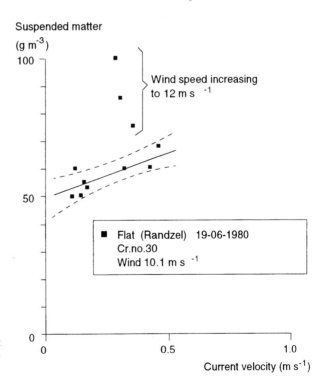

**Figure 15** Example of the combined effect of moderate wind and current velocity on the concentrations of suspended matter.

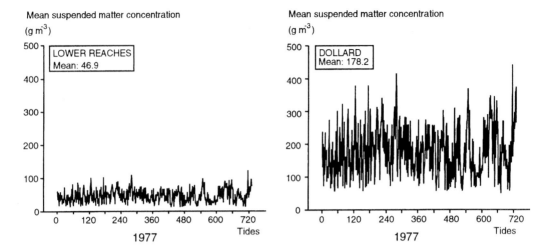

Mean suspended matter concentration
(g m⁻³)

LOWER REACHES
Mean: 46.9

1977

Mean suspended matter concentration
(g m⁻³)

DOLLARD
Mean: 178.2

1977

**Figure 16** Wind-induced fluctuations in suspended matter for lower reaches and Dollard as calculated from the equations in Figure 14.

amounts to only 4% (tidal inlet) or 1.7% (Dollard inlet) of the annual gross flux can therefore not be determined with the above mentioned methods. An alternative strategy for determining such mud transports was recently published.[51]

With reference to the high correlation between suspended matter concentrations and the organic carbon concentrations, the same holds for organic carbon.

## E. IMPLICATIONS OF RESUSPENSION DYNAMICS FOR ECOLOGICAL FUNCTIONING

In Table 3 values are given for the primary production of phytoplankton and microphytobenthos for the water column and the tidal flats.[46] Expressed in tonnes per unit area the suspended microphytobenthos

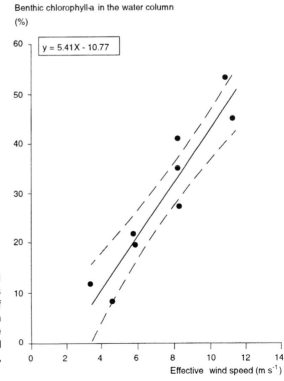

Benthic chlorophyll-a in the water column
(%)

$y = 5.41X - 10.77$

Effective wind speed (m s⁻¹)

**Figure 17** Relation between effective wind speed and the fraction of resuspended microphytobenthos expressed as chlorophyll *a*. Total amount of microphytobenthos is defined here as the amount in the uppermost 0.5 cm of the sediment and the resuspended amount. (From de Jonge, V. N. and van Beusekom, J. E. E., *Neth. J. Sea Res.*, 30, 91, 1994. With permission.)

**Table 3** Total Annual Primary Production in Tonnes a⁻¹ Per Area for Lower Reaches (Area Seaward of the Harbor Eemshaven), Middle Reaches (Area Between Eemshaven and the Dollard Inlet) and the Dollard

| Primary production | Lower reaches | | Middle reaches | | Dollard |
|---|---|---|---|---|---|
| True phytoplankton | 48,000 | (60,000) | 8,600 | (9,400) | 200 |
| Resuspended microphytobenthos | 21,000 | | 3,300 | | 2,600 |
| Microphytobenthos on tidal flats | 15,000 | | 2,600 | | 7,000 |
| Total | 84,000 | (96,000) | 14,500 | (15,300) | 9,800 |

*Note:* Values between brackets include excretion.

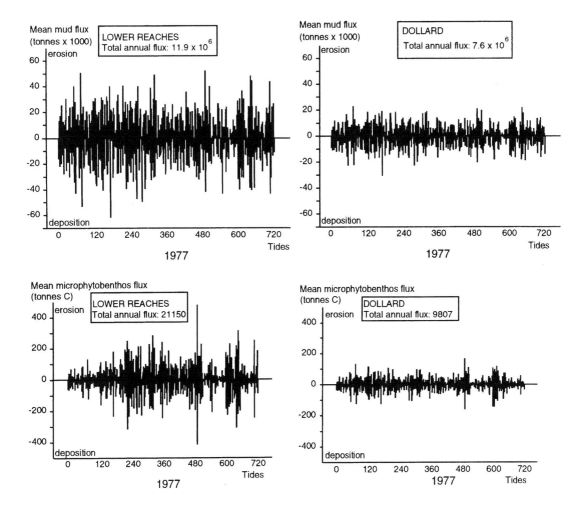

**Figure 18** (A) Tidal fluxes of mud (fraction <55m) between tidal flats and channels in the lower reaches and the Dollard. Calculations done for the year 1977. (B) Tidal fluxes of microphytobenthos carbon between tidal flats and channels in the lower reaches and the Dollard for the year 1977.

contributes significantly to the annual primary production. Of the total production in the estuary it is concluded that ca. 25% of the primary production is provided by microphytobenthos on the tidal flats, ca. another 25% is supplied by suspended microphytobenthos and ca. 50% of the total primary production is provided by true phytoplankton. This means the tidal flats are of greater significance to subtidal areas than previously thought.

## VII. SYMPTOMS OF NUTRIENT ENRICHMENT

The Ems estuary is a system importing mud and organic matter. This material is mineralized within the system. For that reason this area can be classified as a naturally eutrophic system. However, when nutrients limit algal growth, enrichment with the limiting compound(s) will increase the primary production.

For the Ems estuary primary production data for phytoplankton are available for the years 1972 and the period 1976–1980.[52,53] Data on microphytobenthos are restricted to only a couple of years. Information on macroalgae is not available, except that macroalgae do not play an important role in this system.

The available primary production data for phytoplankton roughly show an increase since 1972. The trend, tested by Spearman's Rank Correlation Test, is not statistically significant for any of the three reaches. Increased primary production was thought to be caused by nutrient discharge from both the River Ems and the small River Westerwoldsche Aa. However, detailed data of the nutrient loads, such as those available for a comparable study of other areas, were not available for the River Ems.[54-56] For that reason, and assuming that nutrient concentrations do not vary strongly between years, the annual primary production was related to the mean annual freshwater discharges. The correlation is statistically significant for the lower reaches and the middle reaches, but not for the Dollard. Support for this approach was recently also provided by Schaub and Gieskes,[57] who showed chlorophyll $a$ concentrations in the North Sea coastal zone to be correlated very well with local freshwater discharges (implicitly representing nutrient discharges). The result for the Ems estuary is given in Figure 19, and agrees with findings for the western Dutch Wadden Sea where the annual primary production and chlorophyll $a$ appeared to be significantly by correlated with the mean annual phosphate loads from Lake IJssel.[55,56] The results given in Figure 19 indicate that in the middle reaches and the lower reaches the annual primary production is, at least partly, governed by nutrient loads, a conclusion which does not hold for the very turbid Dollard. The remainder of the organic matter is transported in an upstream direction, where it is mineralized.[58]

A particular symptom due to carbon enrichment, and were the low oxygen values in the Dollard during autumn when the potato-starch industry was active. A surface of several square kilometers was then covered by water with very low oxygen content due to the discharge from the Westerwoldsche Aa of freshwater which was depleted in oxygen. This results in local mortality of fish and shrimps. Since a couple of years this phenomenon is history, due to new wastewater plants in this industry. Other clear symptoms of increased nutrient enrichment are not recorded in this estuary.

## VIII. CONSEQUENCES OF NUTRIENT ENRICHMENT

Because of the moderate flow of freshwater into the Ems estuary (see above) the consequences of nutrient enrichment are limited. Effects of nutrient enrichment were only detected on the primary production of phytoplankton. This means that in terms of fisheries no clear effects can be expected. In fact the same holds for recreation because the recreation in this estuary is not significant. Recreation is only important along the coast of the North Sea just outside the estuary.

By Dutch law, the Dollard is a conservation area. However, there is much debate between The Netherlands and the Federal Republic of Germany on the location of the border which crosses the Dollard (Figure 1). Because differences in policy occur between FRG and the Netherlands, the Dollard is not protected from all foreseeable claims.

Recent studies have been conducted into the effects of phosphate loads on the biomasses of several groups of organisms in different parts of the estuary.[59,60] This was done in applying the ecosystem model of the Ems estuary.[24] The scenarios were a halving and a doubling of the phosphate loads (Figure 20). Generally, the biomass of algae (diatoms and flagellates) and herbivores in an upstream direction reacted most strongly because in the applied model the North Sea acts as an infinite source of nutrients (note that in compartment 1 no benthic filter feeders are present due to low oxygen conditions). This is inconsistent with the results of Figure 19 where annual primary production in the lower reaches mean annual freshwater discharge, but not in the turbid Dollard.

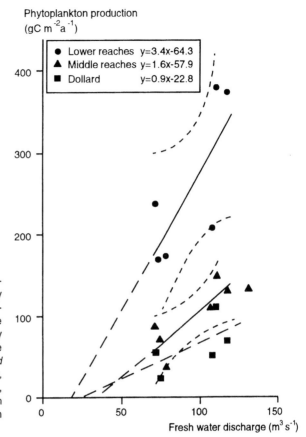

Phytoplankton production
$(gC\ m^{-2}a^{-1})$

| | | |
|---|---|---|
| ● | Lower reaches | y=3.4x-64.3 |
| ▲ | Middle reaches | y=1.6x-57.9 |
| ■ | Dollard | y=0.9x-22.8 |

Fresh water discharge $(m^3\ s^{-1})$

**Figure 19** Annual primary production of phytoplankton in three areas of the Ems estuary plotted against the sum of the freshwater discharge of the River Ems and Westerwoldsche Aa. The changes in freshwater discharge roughly represent changes in nutrient supply. (From de Jonge, V. N. and Essink, K., *Estuaries and Coasts: Spatial and Temporal Intercomparisons*, ESCA 19 Symp., Elliot, M. and Ducrotoy, J.-P., Eds., Olsen & Olsen International Symposium Series, Fredensborg, Denmark, 1991, 307. With permission.)

## IX. MANAGEMENT OPTIONS FOR RESTORATION

In general, improvement of the water quality in the Wadden Sea depends heavily upon reduction in the pollution loads in all rivers draining into the North Sea or directly into the Wadden Sea. For the short term, the best management option with regard to nutrient concentrations is a reduction in the loads of nutrients to the estuary, including the coastal areas from which particulates are transported to the area. This plan is implemented in the North Sea countries by the International Conferences on the Protection of the North Sea held in 1984, 1987, and 1990 and the meetings planned for 1993 and 1995. It was agreed to reduce the loads of nutrients by 50% before 1995.

In the longer term, a fundamental reorientation is required of the way policy makers and planners approach the development of natural systems. This may be done by the implementation of the Guiding Principle. This principle consists of a number of supporting principles.[70] The most important are

- The "wise use principle" in which sustainable development of the ecosystem is the main goal
- Treatment of the estuary as one management unit
- Incorporation into the management of the maintenance of natural biological and physical linkages among the system and surrounding land, the catchment area included
- Subjecting the development of activities to a comprehensive environmental impact assessment (EIA) procedure
- Application of the precautionary principle

Effects of the reduction of nutrient discharges can be investigated in several ways. One possible technique is the application of computer simulation models, as in the ecosystem model of the Ems estuary from which some results were given above.[24]

A serious problem in all available ecosystem models is the transport of mud and sand. Some models are available for sediment transport in estuaries, but most are exclusively based on river-induced

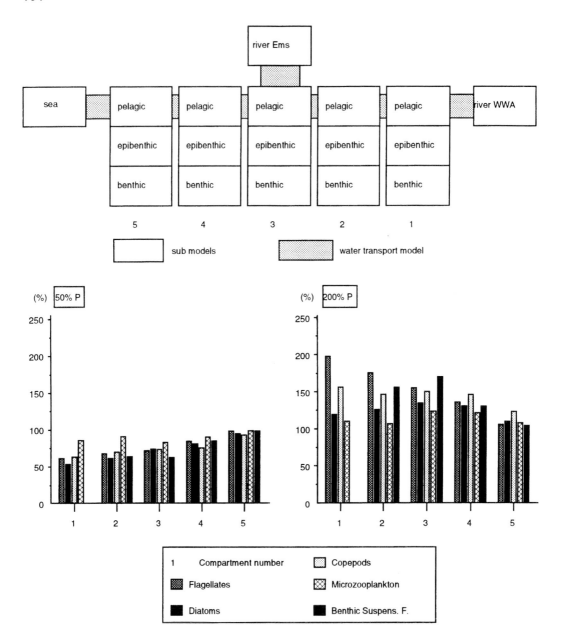

**Figure 20** Relative changes in mean annual biomass in a simulation model of the five groups of organisms that reacted the most when the phosphate loads of the Ems and Westerwoldsche Aa were halved (50%) or doubled (200%). The diagrammatic representation of the set-up of the ecosystem model of the estuary is given in the top part of the figure.

circulation and not on the tide-induced residual transport mechanism. This is understandable because water transport has been much more studied and is physically better understood than sediment transport. Recently a new attempt was made to model the tide-induced residual sediment transport.[63] The best approach would be a model based on a combination of river-induced and tide-induced transport mechanisms. Combination of a correct transport of sediments with the available modules to model benthos, epibenthos and pelagians offers the possibility for answering questions about effects of increased as well as decreased nutrient loads more accurately than has been achieved until now. This is partly due to the fact that the transport of nutrients as particulates cannot be satisfactorily simulated.

# REFERENCES

1. **Jelgersma, S.,** Die Palynologische und C14-Untersuchung einiger Torfprofile aus dem N.S.-Profil Meedhuizen-Farmsum. In *Das Ems-Estuarium (Nordsee),* Voorthuijsen, J. H., Ed., *Verh. Kon. Ned. Geol.-Mijnbouwk. Genoot. Geol. Ser.,* 19, 25, 1960.

2. **de Smet, L. A. H.,** Die holoz ne Entwicklung der niederländ- dischen Randgebietes des Dollarts und der Ems. In *Das Ems-Estuarium (Nordsee),* Voorthuijsen, J. H., Ed., *Verh. Kon. Ned. Geol.-Mijnbouwk. Genoot. Geol. Ser.,* 19, 15, 1960.

3. **van Voorthuijsen, J. H.,** Tektonische Vorgeschichte. In *Das Ems-Estuarium (Nordsee),* Voorthuysen, J. H., Ed., *Verh. Kon. Ned. Geol.-Mijnbouwk. Genoot. Geol. Ser.,* 19, 11, 1960.

4. **Stratingh, G. A., Venema, G. A., J. Oomkes, J. Zoon, and R. J. Schierbeek,** *De Dollard,* Groningen, 1–333, 1855.

5. **Wiggers, A. J.,** Die Korngrössenverteilung der Holoz nen Sedimenteim Dollart-Ems Estuarium. In *Das Ems-Estuarium (Nordsee),* Voorthuijsen, J. H., Ed., *Verh. Kon. Ned. Geol.-Mijnbouwk. Genoot. Geol. Ser.,* 19, 111, 1960.

6. **de Smet, L. A. H. and Wiggers, A. J.,** Einige Bemerkungen Hüber die Herkuns und die Sedimentationsgeschwindigkeit der Dollartablagerungen. In *Das Ems-Estuarium (Nordsee),* Voorthuijsen, J. H., Ed., *Verh. Kon. Ned. Geol.-Mijnbouwk. Genoot. Geol. Ser.,* 9, 129, 1960.

7. **Hinrich, H.,** Schwebstoffgehalt, Gebietsniederschlag, Abfluss und Schwebstoff fracht der Ems bei Rheine und Versen. In den Jahren 1965 bis 1971, *D. Gewasserk. Mitt.,* 18, 85, 1974.

8. **Dorrestein, R.,** Einige klimatologische und hydrologische Daten für das Ems-Estuarium. In *Das Ems-Estuarium (Nordsee),* Voorthuijsen, J.H., Ed., *Verh. Kon. Ned. Geol.-Mijnbouwk. Genoot. Geol. Ser.,* 19, 39, 1960.

9. **Klimaatatlas van Nederland,** KNMI, Staatsuitgeverij, Den Haag, 1972.

10. **de Jonge, V. N.,** Relations between annual dredging activities, suspended matter concentrations, and the development of the tidal regime in the Ems estuary. *Can. J. Fish. Aquat. Sci.,* 40 (Suppl. 1), 289, 1983.

11. **Samu, G.,** Die morphologische Entwicklung der Aussen Ems vom Duke-gat bis zur See, Report Bundesantstalt für Wasserbau, Hamburg, 1979.

12. **Führböter, A.,** Changes of the tidal water levels at the German North Sea coast. *Helgol. Meeresunters.,* 43, 325, 1989.

13. **de Jonge, V. N. and Postma, H.,** Phosphorus compounds in the Dutch Wadden Sea, Neth. *J. Sea Res.,* 8, 139, 1974.

14. **Dronkers, J. and Zimmerman, J. T. F.,** Some principles of mixing in tidal lagoons. *Oceanologica Acta.,* No. SP, 107, 1982.

15. **Zimmerman, J. T. F.,** Mixing and flushing of tidal embayments in the western Dutch Wadden Sea, Part II. Analysis of mixing processes. *Neth. J. Sea Res.,* 10, 397, 1976.

16. **de Jonge, V. N.,** Tidal flow and residual flow in the Ems Estuary. *Estuarine Coastal Shelf Sci.* 34, 1, 1992.

17. **Dorrestein, R., and Otto, L.,** On the mixing and flushing of the water in the Ems-estuary. In *Das Ems Estuarium,* Voorthuijsen, J. H., Ed., *Verh. Kon. Ned. Geol.-Mijnbouwk. Genoot. Geol. Ser.,* 19, 43, 1960.

18. **Helder, W. and Ruardij, P.,** A one-dimensional mixing and flushing model of the Ems-Dollard estuary: calculation of time scales at different river discharges. *Neth. J. Sea Res.,* 15, 293, 1982.

19. **de Jonge, V. N. and Villerius, L. A.,** Possible role of estuarine processes on phosphate-mineral interactions. *Limnol. Oceanogr.,* 34, 330, 1989.

20. **Aston, S. R.,** Nutrients, dissolved gases and general biochemistry in estuaries, in Chemistry and Biochemistry of Estuaries, Olausson, E. and Cato, I., Eds., John Wiley & Sons, New York, 1980, 233.

21. **Pomeroy, L. R., Smith, E. E., and Grant, C. M.,** The exchange of phosphate between estuarine water and sediments. *Limnol. Oceanogr.,* 10, 167, 1965.

22. **de Jonge, V. N. and Engelkes, M. M.,** The role of mineral compounds and chemical conditions in the binding of phosphate in the Ems estuary, *Neth. J. Aquat. Ecol.,* 27, 227, 1993.

23. **Balzer, W., Erlenkeuser, H., Hartmann, M., Muller, P. J., and Pollehne, F.,** In *Lecture Notes on Coastal and Estuarine Studies,* Rumohr, J., Walger, E., and Zeitzschel, B., Eds., Springer-Verlag, Berlin, 1987, 112.

24. **Baretta, J. W. and Ruardij, P.,** *Tidal Flat Estuaries: Simulation and Analysis of the Ems Estuary.* Ecological Studies No. 71, Springer-Verlag, Heidelberg, 1988, 1–353.

25. **Sholkovitz, E. R.,** Flocculation of dissolved organic and inorganic matter during the mixing of river water and seawater. Geochim. *Cosmochim. Acta,* 40, 831, 1976.

26. **Smith, J. D. and Longmore, A. R.,** Behaviour of phosphate in estuarine water. *Nature,* 287, 532, 1980.

27. **Carpenter, P. D. and Smith, J. D.,** Effect of pH, iron and humic acid on the estuarine behavior of phosphate. *Environ. Technol. Lett.* 6, 65, 1984.

28. **Rutgers van der Loeff, M. M., van Es, F. B., Helder, W., and de Vries, R. T. P.,** Sediment water exchanges of nutrients and oxygen on tidal flats in the Ems-Dollard estuary. *Neth. J. Sea Res.,* 15, 113, 1981.

29. **Helder, W.,** Aspects of the Nitrogen Cycle in Wadden Sea and Ems-Dollard Estuary with Emphasis on Nitrification Ph.D. thesis, State University of Groningen, Groningen, 1983.

30. **Admiraal, W.,** Tolerance of estuarine benthic diatoms to high concentrations of ammonia, nitrite ion, nitrate ion and orthophosphate. *Mar. Biol.,* 43, 307, 1977.

31. **Maschhaupt, J. G.,** Soil survey in the Dollard area, Verslagen van landbouwkundige onderzoekingen, No. 54. Gravenhage, 1948.
32. **van Heuvel, Tj.,** Sedimenttransport in het Eems-Dollard estuarium, volgens de methode McLaren, Tidal Waters Division, Report GWWS-91.002, Rijkswaterstaat, 1991.
33. **Kamps, L. F.,** *Mud Distribution and Land Reclamation in the Eastern Wadden Shallows,* Publ. 9, International Institute for Land Reclamation and Improvement, Veenman, Wageningen, 1963.
34. **Meadows, P. W. and Anderson, J. G.,** Micro-organisms attached to marine sand grains. *J. Mar. Biol. Assoc. U.K.,* 48, 161, 1968.
35. **Frankel, L. and Mead, D. J.,** Mucilaginous matrix of some estuarine sands in Connecticut. *J. Sediment. Petrol.,* 43, 1090, 1973.
36. **Englington, G. and Barnes, P. J.,** Organic matter in aquatic sediments. In *Environmental Biogeochemistry and Geomicrobiology,* Vol. 1, Krumbein, W. E., Ed., Ann Arbor Science, Ann Arbor, MI, 1978, 25.
37. **Postma, H.,** Sediment transport and sedimentation in the estuarine environment. In *Estuaries,* Lauff, G. H., Ed., American Association for the Advancement of Science, 83, 158, 1967.
38. **Postma, H. and Kalle, K.,** Die Entstehung von Trübungszonen im Unterlauf der Flüsse, speziell im Hinblick auf die Verhältnisse in der Unterelbe, *Dsche. Hydrogr. Z.,* 8, 137, 1955.
39. **Wellershaus, S.,** Turbidity maximum and mud shoaling in the Weser estuary. *Arch. Hydrobiol.,* 92, 161, 1981.
40. **Postma, H.,** Exchange of materials between the North Sea and the Wadden Sea. *Mar. Geol.,* 40, 199, 1981.
41. **Postma, H.,** Hydrography of the Dutch Wadden Sea. *Arch. Néerl. Zool.,* 10, 1, 1954.
42. **Postma, H.,** Transport and accumulation of suspended matter in the Dutch Wadden Sea. *Neth. J. Sea Res.,* 1, 148, 1961.
43. **Straaten, L. M. J. U. and van Kuenen, Ph. H.,** Tidal action as a cause for clay accumulation. *J. Sediment. Petrol.,* 28, 406, 1958.
44. **Groen, P.,** On the residual transport of suspended matter, by an alternating tidal current. *Neth. J. Sea Res.,* 3/4, 564, 1967.
45. **Postma, H.,** Hydrography of the Wadden Sea: Movements and properties of water and particulate matter. Wadden Sea Working Group, Report 2, Leiden, The Netherlands, 1982.
46. **de Jonge, V. N.,** Wind Driven Tidal and Annual Gross Transports of Mud and Microphytobenthos in the Ems Estuary and Its Importance for the Ecosystem, Ph. D. thesis, State University of Groningen, Groningen, 1992.
47. **Favejee, J. Ch. L.,** On the origin of the mud deposits in the Ems estuary. In *Das Ems-Estuarium (Nordsee),* Voorthuijsen, J. H., Ed., *Verh. Kon. Ned. Geol.-Mijnbouwk. Genoot. Geol. Ser.,* 19, 151, 1960.
48. **Crommelin, R. D.,** A contribution to the sedimentary petrology of the Dollard as compared with adjoining areas. In *Das Ems-Estuarium (Nordsee),* Voorthuijsen, J. H. van, Ed., *Verh. Kon. Ned. Geol.-Mijnbouwk. Genoot. Geol. Ser.,* 19, 135, 1960.
49. **Salomons, W.,** Chemical and isotopic composition of carbonates in recent sediments and soils from Western Europe, *J. Sediment. Petrol.,* 45, 440, 1975.
50. **Reenders, R. and van der Meulen, D. H.,** De ontwikkeling van de Dollard. Over de periode 1952-1969/'70, Rijkswaterstaat, direktie Groningen, afdeling Studiedienst, nota 72.1, 1972.
51. **Pejrup, M.,** Suspended sediment transport across a tidal flat. *Mar. Geol.,* 82, 187, 1988.
52. **Cadée, G. C. and Hegeman, J.,** Primary production of phytoplankton in the Dutch Wadden Sea. *Neth. J. Sea Res.,* 8, 240, 1974.
53. **Colijn, F.,** Primary Production in the Ems-Dollard Estuary. Ph.D. thesis, State University of Groningen, Groningen, 1983.
54. **Cloern, J. E.,** Annual variations in river flow and primary production in the South Francisco Bay Estuary (USA). In *Estuaries and Coasts: Spatial and Temporal Intercomparisons.* ECSA 19 Symp., Elliott, M. and Ducrotoy, J.-P., Eds., Olsen & Olsen: International Symposium Series, Fredensborg, Denmark, 1991, 91.
55. **de Jonge, V. N.,** Response of the Dutch Wadden Sea ecosystem to phosphorus discharges from the River Rhine. *Hydrobiologia,* 195, 49, 1990.
56. **de Jonge V. N. and Essink, K.,** Long-term changes in nutrient loads and primary and secondary production in the Dutch Wadden Sea. In *Estuaries and Coasts: Spatial and Temporal Intercomparisons,* ECSA 19 Symp., Elliott, M. and Ducrotoy, J.-P., Eds., Olsen & Olsen International Symposium Series, Fredensborg, Denmark, 1991, 307.
57. **Schaub, B. E. M. and Gieskes, W. W. C.,** Trends in eutrophication of Dutch coastal waters: the relation between Rhine river discharge and chlorophyll-*a* concentrations, in Estuaries and coasts: Spatial and temporal intercomparisons. ECSA 19 Symp., Elliott, M., and Ducrotoy, J.-P., Eds., Olsen & Olsen, International Symposium Series, Fredensborg, Denmark, 1991, 85.
58. **van Es, F.B. and Laane, R. W. P. M.,** The utility of organic matter in the Ems-Dollard estuary. *Neth. J. Sea Res.,* 16, 300, 1982.
59. **de Jonge, V. N. and DeGroodt E. G.,** The mutual relationship between the monitoring and modelling of estuarine ecosystems. *Helgol. Meeresunters.,* 43, 537, 1989.
60. **DeGroodt, E. G. and de Jonge, V. N.,** Effects of changes in turbidity and phosphate influx on the ecosystem of the Ems estuary as obtained by a computer simulation study. *Hydrobiologia,* 195, 39, 1990.

61. **Festa, J. F. and Hansen, D. V.,** Turbidity maxima in partially mixed estuaries: a two-dimensional numerical model. *Estuarine Coastal Mar. Sci.,* 7, 347, 1978.

62. **Odd, N. V. M.,** Mathematical modelling of siltation in tidal channels. In *Hydraulic Modelling in Maritime Engineering,* Proc. Conf. Institution of Civil Engineers, London, 1982, 39.

63. **Dronkers, J.,** Tidal asymmetry and estuarine morphology. *Neth. J. Sea Res.,* 20, 117, 1986.

64. **Preston, A.,** Input of pollutants to the Oslo Commission area. ICES, Cooperative Research Report, Copenhagen, Denmark, 1978, 77.

65. **Helder, W., de Vries, R. T. P., and Rutgers van der Loeff, M. M.,** Behaviour of nitrogen nutrients and dissolved silica in the Ems-Dollard estuary. *Can. J. Fish. Aquat. Sci.,* 40(Suppl. 1), 188, 1983.

66. **Dankers, N., Binsbergen, M., Zegers, K., Laane, R., and Rutgers van der Loeff, M.,** Transportation of water, particulate and dissolved organic and inorganic matter between a salt marsh and the Ems-Dollard estuary. *Neth. Estuarine Coast. Shelf Sci.,* 19, 143, 1984.

67. **de Jonge, V. N. and van Beusekom, J. E. E.,** Wind and tide induced resuspension and transport of sediment and microphytobenthos in the Ems estuary. In Physical Processes and Dynamics of Microphyto-benthos in the Ems Estuary (The Netherlands). Ph.D. thesis, State University of Groningen, Groningen, 1992.

68. **Waddenzeecommissie,** Rapport van de Waddenzeecommissie, The Hague, 1974.

69. **Wolters-Noordhoff,** De Grote Geïllustreerde Bosatlas, Wolters-Noordhoff Atlasproducties, Groningen, The Netherlands, 1983.

70. **World Wide Fund for Nature,** The Common Future of the Wadden Sea, Husum, The Netherlands, 1991.

# Chapter 8

# The Ecology and Management of Zandvlei (Cape Province, South Africa), an Enriched Shallow African Estuary

*J. A. Thornton, H. Beekman, G. Boddington, R. Dick, W. R. Harding, M. Lief, I. R. Morrison, and A. J. R. Quick*

## CONTENTS

I. Introduction ........................................................................................................... 109
II. Physical Properties ............................................................................................... 111
   A. Climate .......................................................................................................... 111
   B. Morphometry ................................................................................................ 111
III. Catchment Description ......................................................................................... 111
   A. Geography .................................................................................................... 111
   B. Demography and Land Usage ...................................................................... 113
   C. Hydrology ..................................................................................................... 113
   D. Nutrient Status and Loadings ...................................................................... 116
IV. Water Movement and Exchange ......................................................................... 120
V. Nutrients in the Estuary and Estuarine Sediments ............................................. 120
   A. The Water Column ....................................................................................... 120
   B. The Sediments .............................................................................................. 120
VI. Symptoms and Consequences of Nutrient Enrichment ...................................... 120
   A. Identification of Symptoms .......................................................................... 120
   B. Macrophytes and Their Consequences ........................................................ 121
   C. The Algal Community and Its Consequences ............................................. 122
   D. The Development of Eutrophication at Zandvlei .......................................... 122
VII. Discussion ............................................................................................................ 122
   A. Development of a Management Philosophy ................................................. 122
   B. The Management Approach .......................................................................... 123
      1. Catchment-Based Planning Options ........................................................ 123
      2. Lake-Based Planning Options .................................................................. 124
      3. Use Modification Options ........................................................................ 125
   C. The Management Plan .................................................................................. 126
Acknowledgments ........................................................................................................ 126
References .................................................................................................................... 126

## I. INTRODUCTION

Capetonians have long evinced an appreciation of the natural environment seldom equaled in South Africa. Initially, their concern was predicated upon the provision of safe and reliable sources of drinking water to commerce, a growing resident population, and to agriculture. Over the years, this depth of concern has broadened to include recreational waters, and waters associated with power generation, storms, and domestic and industrial processes (wastewaters).

Zandvlei (34° 05′S, 18° 28′E) is fairly typical of the recreational waters of the southwestern Cape Province, being located in southern Cape Town adjacent to Muizenberg on the shores of False Bay (Figure 1). Originally an inlet,[1] the vlei (similar to a pond or lakelet; a perennial body of shallow, standing water in Cape parlance) has developed as the result of human manipulation of the original estuarine system. Named in 1650 (reputedly) by Jan van Riebeeck of the Dutch East India Company (probably in recognition of its situation on the edge of a recent marine coastal plain), the vlei began its long association

0-8493-6839-1/95/$0.00+$.50

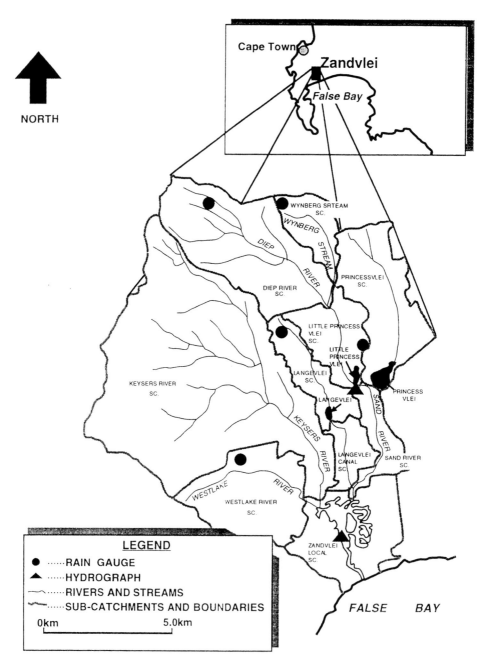

NORTH

**Figure 1**  Locality plan showing the location of Zandvlei and its principal influent rivers, their subcatchments and the location of the rainfall and flow/level gauging stations.

with low intensity agriculture in 1673 when a cattle post was established on its shores. During the next 200 years, the nature of the agricultural operations gradually changed from cattle rearing to market gardening, but human interference with the natural system remained marginal. Then, in 1866, after a long period of drought and in response to the resultant, markedly lowered water level, came the proposal to reclaim the waterbody totally for agricultural purposes. The mouth of the vlei was duly closed at this time,[1] but, due to an inadequate provision for the transmission of the winter floodwaters, the reclaimed land was almost immediately re-flooded.

However, even larger changes were in store for the waterbody in the years that followed. Foremost among these was the arrival of the railway at Muizenberg from Cape Town in 1882. This single event inaugurated more than a century of recreational use of the vlei's waters and engendered a sequence of user-related events that has changed the vlei out of all recognition. The vlei has been dredged to dampen seasonal water level fluctuations (originally planned for 1939, but ultimately carried out in 1947 due to the intervention of World War II), deepened to permit recreational boating (in 1961), and developed to accomodate the ever increasing lakeside residential population that has replaced the former agro-industry (the Marina da Gama canal system and housing estate was built between 1969 and 1973; a second phase, planned for northern Zandvlei, was cancelled due to the economic downturn associated with the 1973 "oil crisis").

Today, Zandvlei remains a popular recreational waterbody. A recent survey enumerated between 2000 to 3000 persons per day visiting the vlei during a typical peak holiday period.[2] About half of these people came from outside of the Cape Town metropolitan area, and most visited the waterbody for the better part of a day — boardsailing, braaiing (i.e., barbequing), picnicking and walking through the surrounding parklands during the course of their visit. In addition to these very popular activities, the vlei plays host to persons engaging in other water-based recreational pursuits, including fishing, which highlights the role of the vlei as a conservation and nursery area for juvenile fishes that form the basis for the commercial catches taken from the neighboring False Bay. Ironically, while the railway has virtually ceased to be their principal mode of transportation, most of these visitors cited ease of access as the primary reason for visiting this waterbody.

## II. PHYSICAL PROPERTIES

### A. CLIMATE
Zandvlei is situated in the South African winter rainfall area,[3] and lies in the narrow coastal strip characterized by a Mediterranean climate[4] and its associated fynbos biome.[5,6] Annual rainfall[3] occurs between May and September, and averages between 400 and 600 mm, although annual precipitation in the vicinity of the vlei during the 5-year period from 1983 to 1988 was in the order of 1100 mm. This latter deviation is probably due to local orographic effects which favor higher rainfalls in areas adjacent to the mountains.[7] Annual evaporation[3] in the region is given as $\leq$1,400 mm; it has been measured at a neighboring vlei (Zeekoevlei, approximately 5 km east of Zandvlei) as 1,423 mm, averaged over a 20-year period from 1969 to 1988. This relationship between rainfall and evaporation is typical of the region.[8] Annual surface air temperatures average 15.0 to 17.5°C,[3] and vary diurnally by an average of 8.5°C. Temperatures range from 15 to 30°C in summer, and from 8° to 15°C in winter[9]: there is no freeze-thaw cycle typical of many northern hemisphere estuaries.

### B. MORPHOMETRY
Zandvlei is a small (surface area = 1 km²), shallow (mean depth = 1 m) waterbody composed of three principal water masses (Figure 2), namely, the main basin (0.6 km²), the Marina da Gama canal system on the eastern shore, separated from the main basin by the mass of Park Island (0.3 km²), and the Westlake Wetland at the northwest corner of the vlei and isolated by the railway viaduct (0.1 km²). Table 1 gives additional morphometric data for the vlei, and Figure 3 presents the bathymetry of the waterbody. Water levels in the vlei are strictly controlled by means of a "rubble" weir in the outlet channel, to maintain a relatively constant water level suitable for boating and boardsailing, and to prevent flooding of the Marina da Gama estate. This weir is constructed of loosely packed paving stones, which are replaced on approximately a 5-yearly cycle. In addition, the vlei is naturally isolated from the sea for much of the year by a sandbar which forms across the vlei mouth in summer[1,10] (although, today, this sandbar is also artificially manipulated to assist in water level and flood control). The outlet channel, itself, is a 20 m wide, concrete-lined structure, approximately 1 to 2 m deep (depending on the phase of the tidal cycle), linking the main body of the vlei to False Bay (Figure 3).

## III. CATCHMENT DESCRIPTION

### A. GEOGRAPHY
The Zandvlei catchment extends over 92 km², encompassing as subcatchments the drainage basins of the Diep River-Sand River system, the Keysers River, and a number of local tributaries including the Westlake-Raapkraal Stream and several unnamed, seasonal streams (Figure 1). The catchment is characterized by a low, relatively flat sandy plain to the east, rising only some 20 m above mean sea

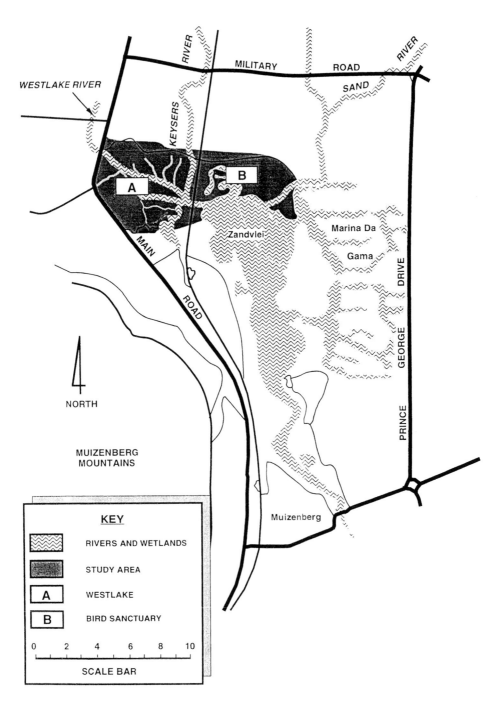

**Figure 2** Site plan of the Zandvlei estuary showing the location of the Westlake Wetland, main vlei basin and Marina da Gama canals.

level, and a steep, mountainous topography to the west, rising some 800 m over a distance of 1 or 2 km (Figure 3). These two features, respectively, form part of the Cape Flats quartzitic sands of recent marine origin, and the Peninsula Mountain Chain sandstones, shales and granites of the Table Mountain and Malmsbury Series and are typical of the southwestern Cape Province of South Africa.

**Table 1** Selected Morphometric Data
for the Zandvlei Estuary, Southwestern
Cape Province, South Africa

| Characteristic | Magnitude | |
|---|---|---|
| Volume | 1.455 | $10^6$ m$^2$ |
| Surface area (SA) | 0.966 | km$^2$ |
| Catchment area (CA) | 92.0 | km$^2$ |
| CA:SA | 95:1 | |
| Maximum depth | 2.1 | m |
| Mean depth | 1.14 | m |

## B. DEMOGRAPHY AND LAND USAGE

The catchment extends over areas under the jurisdiction of the City Council of Cape Town (46% of the catchment), the Regional Services Council of the Western Cape (a "county" type of regional government responsible for 33% of the catchment), and the State (21% of the catchment) (Figure 4). Some 35,000 to 45,000 people have been estimated to work and/or reside in these areas.[13] Table 2 gives the breakdown of current (1988) and projected (based on zoning) land uses in the catchment area: land uses are categorized on the basis of the South African National Land Use Classification System (SANLUC).[11] The major land uses are residential, forestry and open space, the last including areas of vleis and waterbodies in addition to public open space, parkland and undeveloped real estate. Agriculture, principally viticulture (the Constantia region being renowned for its red and fortified wines[12]), still accounts for a large percentage of the land usage in the catchment, as does the provision of infrastructure (in the form of highways, roads and rail lines). Commerce and industry form a minor class of land usage, but the location of the Retreat Light Industrial Park near the headwaters of the vlei weighs heavily in terms of the potential impact of industrial activity on the system.

Planned future developments are primarily residential in nature.[13] Lower income (to the east and north of the vlei) and middle to upper income (to the west) housing developments will subsume a further 5% of the total catchment in the next decade. This development will occur largely at the expense of the forestry and open space land uses. In addition, there are likely to be some "unplanned" changes in land usage as the result of individual rezonings or other small-scale land use changes. Current projections of population in the area of the Zandvlei catchment suggest that the area will continue to grow by some 7% over this same period,[13] primarily in the upper and lower income brackets. Some 40,000 persons are expected to move into the planned Muizenberg East development adjacent to Marina da Gama during the coming decade, increasing both the population density and catchment area (stormwater in this middle income residential area is to be drained, in part, into Zandvlei).[14]

## C. HYDROLOGY

Two gauging stations exist in the Zandvlei catchment; namely, the flow gauging station at the outlet of the Little Princess Vlei on the Diep-Sand River (equipped with a continuous recorder), and the level recorder located in Zandvlei. In addition, four rainfall gauging stations are located in the catchment at Cecelia, Wynberg, Kendal and Southfield (Figure 1). A water balance for the vlei has been constructed using 5 years of records (1983–1987) from these stations. The results of this analysis are given in Table 3.

Mean annual runoff for the Zandvlei catchment averaged approximately $22 \times 10^6$ m$^3$, distributed between the Diep-Sand Rivers (43% of the inflow), Keysers River (45% of the inflow), and Westlake-Raapkraal Stream and local catchment (12% of the inflow). This volume amounted to about 24% of the mean annual rainfall during the period from 1983 to 1987. Surface runoff (principally in the austral winter) accounted for about 15% of the total rainfall volume, with the balance being made up as groundwater flows, primarily in the perennial rivers, during the dry season (the austral summer) (Figure 5). Use of these data provided the opportunity for development of a U.S. Soil Conservation Service (SCS)-based, rainfall-runoff model.[15-17] Tidal seawater inflows amounted to approximately $3.1 \times 10^6$ m$^3$ annually.

## D. NUTRIENT STATUS AND LOADINGS

Mean annual nutrient concentrations in the influent rivers are typical of other southern African catchments influenced by urban development.[19-22,22a] Mean annual total phosphorus concentrations between 1988 and

114

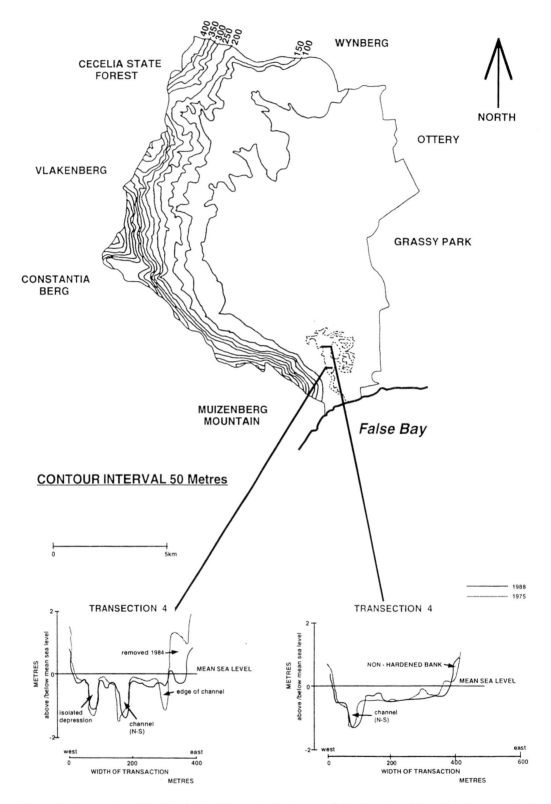

**Figure 3** Topography of the Zandvlei catchment, including a partial bathymetry of Zandvlei showing typical bottom contours for this waterbody.

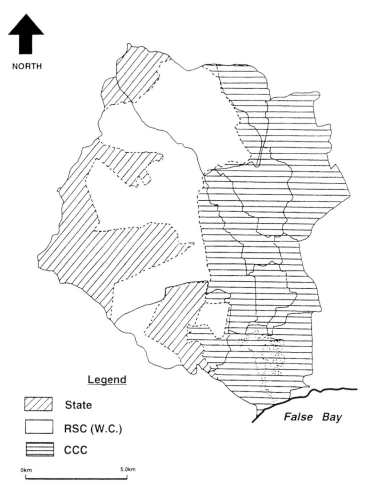

**NORTH**

**Legend**

State

RSC (W.C.)

CCC

0km        5.0km

**Figure 4**  Land use plan showing jurisdictional boundaries. Lands under the direct control of South African government departments are shown as State land (diagonal cross-hatching), those under the jurisdiction of the Western Cape Regional Services Council as RSC (WC) land (no shading), and those under the jurisdiction of the City of Cape Town as CCC land (horizontal cross-hatching).

**Table 2**  Present (1988) and Projected Future Land Use, in km² , in the Zandvlei Catchment

| Land use class | Present area | Future area | % Change |
|---|---|---|---|
| Residential | 20.9 | 25.3 | + 4.4 |
| Industrial | 0.6 | 0.95 | + 0.4 |
| Commercial | 3.7 | 3.25 | − 0.5 |
| Agriculture | 13.2 | 14.0 | + 0.8 |
| Forestry | 20.1 | 20.1 | 0 |
| Open space[a] | 24.2 | 19.8 | − 4.4 |
| Roads and reserves | 9.6 | 9.8 | + 0.2 |

*Note:*  Projections based on the 1988 Zoning Schemes of the City of Cape Town and Western Cape Regional Services Council.

[a]  Open space includes the area of the waterways and estuary, in addition to parklands and undeveloped lands.

**Table 3**  Annual Water Budget (in 10⁶ m³) and Water Residence Time (in years) for the Zandvlei Estuary

| Inflow/outflow | Catchment area (km²) | Volume |
|---|---|---|
| Inflows | | |
| Diep-Sand River | 46.8 | 7.24 |
| Keysers River | 39.8 | 10.78 |
| Westlake Stream | 9.5 | 8.00 |
| False Bay | 0.49 | |
| Direct precipitation | 0.9 | 0.89 |
| Outflows | | |
| Zandvlei Estuary | 96.1 | 23.63 |
| Evaporation | 0.9 | 1.65 |
| Net gain/loss (over annual cycle) | | 0.00 |
| Residence time | | 0.06 years |

1990 averaged about 100 mg m⁻³ in the Diep-Sand and Keysers rivers on an annual basis (within ranges of 59 to 577 mg m⁻³ and 55 to 632 mg m⁻³, respectively), while slightly higher values were recorded in the Westlake-Raapkraal Stream (about 150 mg m⁻³ in a range of 30 to 577 mg m⁻³). Oceanic total phosphorus concentrations were in the order of 75 mg m⁻³. Mean annual soluble phosphorus (SRP) concentrations averaged about 50 mg m⁻³ in the two major rivers and about 80 mg m⁻³ in the local catchment, which was primarily fed by stormwater drains. Mean annual nitrogen values in the influent rivers varied considerably between systems, with total nitrogen concentrations of 1,500 mg m⁻³ (range = 1,040 to 2,639 mg m⁻³) being measured in the Diep-Sand River, 400 mg m⁻³ (range = 43 to 1,193 mg m⁻³) in the Keysers River and 300 mg m⁻³ (range = 60 to 837 mg m⁻³) in the Westlake-Raapkraal system. Oceanic total inorganic nitrogen (TIN) inputs averaged 100 mg m⁻³. All of the inflows carried significant total dissolved solids (TDS) loads, reflecting both the proximity of the waterbody to the sea (with the associated deposition of salts of aeolian origin) and the presence of remnant salt deposits in the recent marine sediments of the Cape Flats sands. Concentrations of between 400 and 900 mg m⁻³ were recorded in the three influent river systems, while a concentration of 34.8 × 10⁶ mg m⁻³ (= 34.8 ppt) was observed in the nearshore waters of False Bay. The Diep-Sand River system carried the highest concentrations of fecal coliforms, averaging 2,300 per 100 ml (median values reported), compared with 1,000 per 100 ml in the Keysers and 390 per 100 ml in the Westlake-Raapkraal, and, in addition, deposited approximately 700 m³ of inorganic suspended sediments in the vlei annually.

The foregoing flow and concentration values were subsequently used as an input to the U.S. Army Corps of Engineers' nutrient budget models, known collectively as "BATHTUB".[18] The resultant nutrient and dissolved solids mass-balances are shown in Table 4. Although this suite of computer models was designed for use in large reservoirs,[18] the calculated nutrient and dissolved solids concentrations and associated chlorophyll *a* concentration were well within the observed standard deviations of the measured values (Figure 6). (The Secchi disc transparency value fell outside the observed range, but this is probably an artifact of the shallow nature of the system and its greater potential for resuspension of bottom sediments, etc.; see below.)

All of these values are within the ranges subsequently reported by Harding in his analyses of a more extensive data base.[57,58] In these studies, Harding showed total phosphorus concentrations that ranged from 115 to 1,080 mg m⁻³ in the Diep-Sand River, from 106 to 632 mg m⁻³ in the Keysers, and from 160 to 1,401 mg m⁻³ in the Westlake-Raapkraal during the 13 year period between 1978 and 1991.[57] Total nitrogen values ranged from 1,800 to 6,200 mg m⁻³, from 1,500 to 5,500 mg m⁻³, and from 1,900 to 9,300 mg m⁻³ in these rivers, respectively, during this same period.[57] Median fecal coliform counts ranged from a low of 80 per 100 ml in the Westlake-Raapkraal Stream to a high of 9,500 per ml in the Diep-Sand River during the 9 year period between 1983 and 1991.[58]

## IV. WATER MOVEMENT AND EXCHANGE

Although Zandvlei is a highly manipulated system, with numerous human-imposed limitations on "natural" patterns of circulation in both the horizontal and vertical planes, evidence from lesser disturbed systems along the southwestern African coastline suggest that the oceanic-estuarine exchanges have been

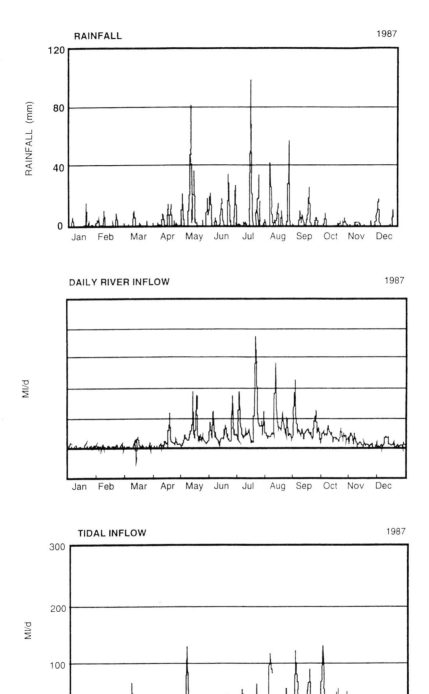

**Figure 5** Annual rainfall, runoff, and seawater inflow distributions during a typical year (1987).

**Table 4**  Annual Chemical Mass-Balances (in kg) Measured at Zandvlei During 1988

| Inflow/outflow | Total P | Total N | TDS |
|---|---|---|---|
| Inflows | | | |
| Diep-Sand River | 711.5 (15) | 5,625.3 (120) | 6,393.8 |
| Keysers River | 1,127.6 (28) | 4,941.2 (124) | 4,184.0 |
| Westlake Stream | 1,172.0 (123) | 3,881.3 (408) | 3,401.2 |
| False Bay | 35.9 | 48.1 | 170,557.7 |
| Direct precipitation | 143.6 | 612.8 | 5.4 |
| Outflows | | | |
| Zandvlei | 2,925.4 | 9,038.5 | 222,074.7 |
| Net gain/loss | 265.2 | 6,070.2 | −37,532.6 |

*Note:*  Nutrient export coefficients in mg m$^{-2}$ are given in parentheses.

least affected by the level controls in comparison to other anthropogenic disturbances.[10] Estuaries in this region typically have episodic periods of oceanic-estuarine exchange, the estuaries being isolated for much of the year from the ocean by naturally occurring sandbars. In the case of Zandvlei, this natural regime has been modified by both the human manipulation of the sandbar and the construction of an artificial weir at the outlet resulting in an artificially elevated water level with consequent interference in Salinity Cycling (see above). The net effect of these impedances is that ocean inflows during winter are usually associated with storms in combination with spring tides, while summer inflows are predominantly tidal. This seasonal pattern is the inverse of the typical pattern of river inflows, maximum inflows occurring during the winter rains. Thus, the salinity of the estuary varies most on an annual, rather than the classical daily basis, with peak salinities being observed in midsummer (year-end) and minima being recorded in midwinter (midyear) (Figure 7). Diurnal (tidal) effects are visible in Figure 7 as scatter in the seasonal profile. Average hydraulic residence times in the vlei approximate 0.065 years (Table 3).

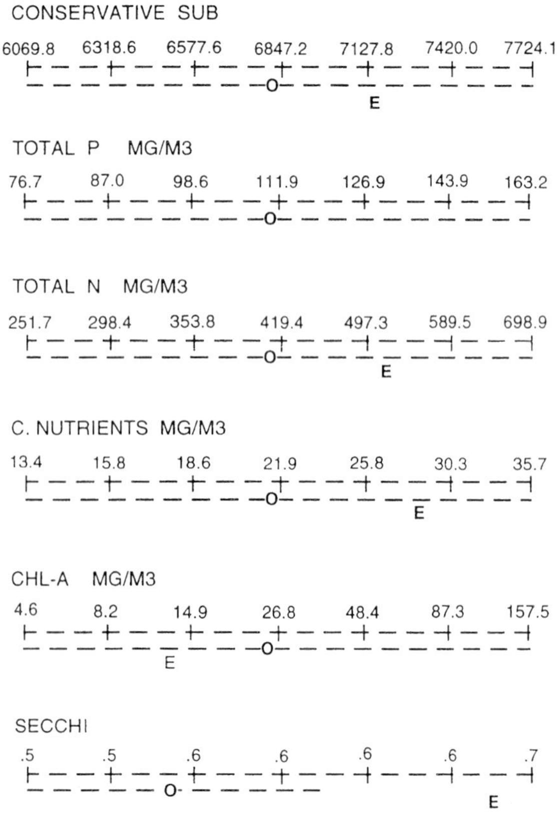

**Figure 6**  Estimated (E) and observed (O) total dissolved solids (shown as the conservative substance), total phosphorus, total nitrogen, composite nutrient (a measure of the effective nutrient supply independent of the limiting nutrient; equivalent to total P at high N:P ratios) and chlorophyll *a* concentrations, and Secchi disc transparencies, in the main basin of Zandvlei during 1987–88. Horizontal bars show the confidence limits around the observed values (as two standard errors of the annual mean), while the scale gives the relevant range of concentrations or depths for each parameter relative to the observed values. The predicted values were generated using the "BATHTUB" suite of models.[18,19]

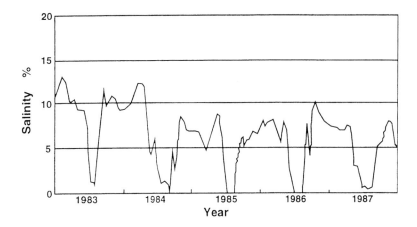

**Figure 7**  Annual distribution of salinity (ppt) in Zandvlei during the period 1983–1987.

Rapid mixing under a favorable wind regime of SE winds occurs between incoming oceanic water and the vlei water, resulting in a rather homogeneous distribution of salinities across the main body of the vlei (averaging 6.8 ppt). Nonetheless, the typical estuarine salinity gradient from near oceanic salinities at the mouth to near freshwater salinities (conductivities) in the upper reaches has been observed in an attenuated form in the extremities of the vlei. Before 1989, this distribution of salinities was partly aided by the existence of a saltwater barrier at the railway causeway that restricted, but did not eliminate, the entry of waters of higher salinity into the Westlake Wetland. This barrier was probably built in the 1930s to enhance the use of waters from the Wetland for crop irrigation: Azorin[23] presents strong evidence of this limitation in his study of the vegetation in the wetland areas at the northern end of Zandvlei. His analysis of aerial photographs, taken over the vlei at 20-year intervals since 1935, suggests that the flora of the wetlands changed from one that was characteristic of estuarine conditions throughout to two distinct floras — one becoming more freshwater in nature and distinct from that of the wetland on the northern edge of the main body of the vlei. The barrier was removed in 1989 in an attempt to control the rampant growth of *Myriophyllum aquaticum* in the Wetland's waterways, this plant having been shown to be susceptible to fungal/bacterial infection and senescence at salinities in excess of 5 to 10 ppt.[38] Nevertheless, salinities of up to 20 ppt have been recorded in the northern reaches of the vlei, grading from about 35 ppt at the mouth. The high salinities in the vicinity of the estuary mouth are maintained, in part, by the existence of the sandbar and rubble weir, which trap seawater between them and act as a rachet to retain sea-borne sands (approximately 600 m³ year⁻¹) which accumulate in the narrow, lower reaches of the waterbody (and act as further barrier to mixing).

A curious side effect of the human manipulation of the system is the occurence of strong haloclines in the Marina da Gama canals.[24] The artificial canals form the deepest point in the vlei, being 2 m in depth, and are linked to the main body of the vlei through two narrow inlets. The southernmost inlet was dredged to about 2 m in depth during the early 1970s, while the northernmost is <1 m in depth. Density gradient effects permit marine waters entering the vlei to accumulate in the canals with little chance of escape — the canal design and the architecture of the houses with wind-deflecting roofs being such that wind effects at the water surface are all but eliminated by the parallel rows of townhouses in the Estate, with limited possibility of escape through other inlets/outlets. Similar haloclines rarely develop in the main basin of the vlei due to the intensity of the wind-induced mixing, even though portions of the vlei are up to 2 m in depth; average wind speeds in this area (measured at Strandfontein) exceed 5 m s⁻¹ for much of the year, with the dominant direction (35% of the year) being from the south, directly across the vlei surface with a relatively unimpeded fetch.

A further consequence of this highly manipulated state appears to have been the collapse of the *Potamogeton pectinatus* population in 1991. The effective isolation of the Vlei from the sea between 1989 and 1991, and the subsequent decline in salinity levels, apparently caused a shift in dominance among the principal primary producers in the system from the macrophyte to planktonic algae.[57] Restoration of the regular (annual) opening of the Vlei to the sea during 1992–93 is expected to encourage the reestablishment of the *Potamogeton* populations.[57,60]

## V. NUTRIENTS IN THE ESTUARY AND ESTUARINE SEDIMENTS

### A. THE WATER COLUMN

Zandvlei is usually a well-mixed estuary, lacking thermal and chemical stratification except in the artificial canals of Marina da Gama as noted above. The annual distribution of nutrient concentrations, similar to the distribution of salinity, is relatively uniform throughout the main basin of the waterbody. Annual total phosphorus concentrations in the vlei averaged 110 mg m$^{-3}$ in a range from 50 to 500 mg m$^{-3}$, with SRP concentrations of 50 mg m$^{-3}$ (range = 3 to 117 mg m$^{-3}$) being recorded in the vlei over the annual cycle. Total nitrogen concentrations averaged 415 mg m$^{-3}$, with DIN concentrations of 200 mg m$^{-3}$ in a range of 13 to 945 mg m$^{-3}$. The N:P ratios of about 4:1 are reminiscent of both nearshore marine ratios[25] and eutrophic freshwater values[20,21] observed in this region. Chlorophyll *a* concentrations averaged 25 mg m$^{-3}$ annually, while fecal coliform counts averaged 100 per 100 ml, in the main basin of the vlei. Fecal coliform counts were somewhat lower in the Marina da Gama canals, where values of 50 per 100 ml have been recorded on average throughout the year. [The significantly lower fecal coliform values in the vlei, compared with those in the influent rivers, are consistent with the observed antibacterial effects of seawater and sunlight.[26] Significant intra-annual (seasonal) variability in fecal coliform counts has also been observed in response to seasonal runoff events.[58]]

### B. THE SEDIMENTS

The sediments of Zandvlei are dominated by the sand component and typically have a low organic content, a fact not unexpected, given the situation of the vlei basin in a landscape dominated by recent marine sands. The sediments of both the wetlands to the north of the vlei and the canals of Marina da Gama to the east have higher concentrations of organic matter, which reflect their recent histories. The wetland sediments, naturally, reflect the accumulation of decayed and decaying plant material which has contributed to their sandy loam texture (organic carbon content >10%).[23] The Marina da Gama canals, on the other hand, were constructed in part on soils that were once associated with a land fill operation (and, in part, on soils that were once wetland that had been reclaimed for agricultural use), and have typically supported higher growths of aquatic macrophytes (dominated by *Potamogeton pectinatus*) than other parts of the estuary. These sediments, too, have a higher organic content (>10% organic carbon) and have been described as slightly silty sand. In addition, these peripheral sediments reflect the lack of mixing and sorting processes that operate in the well-mixed main vlei basin, where the organic component of the sediments is typically lower.[27] Such distributions in sediment texture and organic content are common in this type of coastal vlei.[28]

Sediment nutrient concentrations in coastal lakes generally parallel sediment organic contents.[28] N:P ratios of >10:1 are usually found in association with sediments having organic carbon contents in excess of 12%, and measurable differences exist between the sediments of the main vlei, bordering wetlands and man-made canals.[60] Nitrogen concentrations in these sediments averaged 1,150 mg kg$^{-1}$ (total Kjeldahl nitrogen) in the main vlei and 2,790 mg kg$^{-1}$ in the canals; ammonium nitrogen was undetectable in the main basin but present in the canals (averaging 56.2 mg kg$^{-1}$). Combined nitrate and nitrite nitrogen concentrations in these sediments were 6.3 mg kg$^{-1}$ and 27.4 mg kg$^{-1}$, respectively. The total phosphorus concentrations in these sediments averaged 167 mg kg$^{-1}$ and 358 mg kg$^{-1}$, which are similar to concentrations measured in other southern African waterbodies.[20,29] The higher total phosphorus concentrations noted in the Zandvlei canals are indicative of enriched conditions.[29,30]

## VI. SYMPTOMS AND CONSEQUENCES OF NUTRIENT ENRICHMENT

### A. IDENTIFICATION OF SYMPTOMS

In a survey of over 2,000 vlei users,[2] 60% of the respondents classed Zandvlei as a clean water vlei, despite the fact that the water quality indicators typically used to determine trophic state[21,31] suggested otherwise (i.e., phosphorus concentrations and chlorophyll *a* values exceeded the threshold values established in the OECD[32] studies of Vollenweider and his colleagues). Of the nearly 40% who responded to the contrary, most objected to the presence of "water weeds"; others objected to litter and debris on the water surface, obstacles under the water surface, and mud and coloration of the water. While none of the 2,000 respondents appeared to be deterred in their utilization of the estuary for recreational purposes (over 25% of respondents engaged in some form of active contact recreation on the vlei),[2] the unexpected perception of the vlei as a marginally impaired waterbody led to a reevaluation of the use of classical descriptors of enrichment in

**Table 5** Water Quality Indicators Used to Define Eutrophy in Semi-Arid Zone Waterbodies[31,33,34]

| Indicator | Threshold value | |
|---|---|---|
| Total P | 50 | mg m$^{-3}$ |
| Total inorganic N | 200 | mg m$^{-3}$ |
| Chlorophyll *a* | 15 | mg m$^{-3}$ |
| Secchi transparency | 1 | m |
| Coliform bacteria | 100 | (100 ml)$^{-1}$ |
| Oxygen | 4 | mg l$^{-1}$ |
| Hydrogen sulfide | 0.4 | mg l$^{-1}$ |
| Primary productivity | 2 | g C m$^{-2}$ d$^{-1}$ |

semiarid climates.[33-35] The result was that "new" standards were proposed and used in this study (Table 5), in terms of which Zandvlei is, indeed, borderline impaired/eutrophic.

## B. MACROPHYTES AND THEIR CONSEQUENCES

The most frequently cited water quality problem observed in Zandvlei was the presence of excessive macrophytic biomass.[2] The main basin of Zandvlei is dominated by the macrophyte *Potamogeton pectinatus*, and its associated epiphytic/epizooic flora and fauna, consisting mainly of the epiphytes *Cladophora* sp. and *Enteromorpha intestinalis,* and the epizooite *Ficopomatus enigmaticus*.[1,36,37] Growth of this macrophyte in the vlei has been controlled by mechanical harvesting since 1976, with an average of 224 g dry biomass m$^{-2}$ being removed annually. Annual yields varied little during this period (1983 to 1988); however, the plant does exhibit a distinct seasonality, with peak biomass production in late summer (January to April and population crashes have occurred — in 1978 and later in 1992). Because of the differences in substrate composition noted above, estimates of total yield (e.g., harvested biomass plus standing crop) vary from between 120 and 450 g m$^{-2}$ year$^{-1}$ in the main basin of the vlei to between 280 and 690 g m$^{-2}$ year$^{-1}$ in the canals of Marina da Gama.[38] These levels of production are similar to those measured elsewhere in the Cape Province[28] and in inland waters in southern Africa,[39] but such productivity is modest when compared with that measured in Swartvlei (2,506 g m$^{-2}$ year$^{-1}$),[28,40,41] an estuary located in the summer rainfall region (in which the opening of the sandbar coincides with the period of maximum land runoff) of the eastern Cape Province. Dick[38] notes a number of climatic factors (including a harsher temperature regime, lesser degree of light penetration, greater degree of turbulence and less than ideal substrate in Zandvlei) that could account for this observation, especially given the higher yields in the sheltered canals of Marina da Gama. Interestingly, the macrophyte appears to obtain its primary supply of nutrients, in both systems, from the sediments,[28,38,40-42] a factor which becomes significant when the nutrient input (at least to Zandvlei) consists of almost 50% particulate phosphorus which contributes to this sediment phosphorus pool.

Despite the nuisance value for recreational users of the vlei of *Potamogeton* and its associated epiphytic community, recent evidence suggests that the plants and their attached flora and fauna also play an essential, beneficial role in the vlei.[59] With an annual mean Secchi disc transparency of 0.5 m, Zandvlei is a relatively clear water vlei when compared with its surrounding brown water lakelets.[22,22a,43-45] While such clarity may be unexpected in a shallow waterbody subject to extreme wind conditions (see above), it is important to the public's perception of the vlei as a clean water lakelet.[2,35] Davies et al.[36] postulate that this clarity is due to the extremely efficient removal of fine particulates (2 to 16 μm) by the epizooite, *Ficopomatus enigmaticus*. This serpulid polychaete occurs in the waterbody in densities of between 5.23 and 84.9 g dry body mass m$^{-2}$ (the former reflecting the density observed in the main body of the vlei, and the latter the density in the Marina da Gama canals). With their filtering rate of 8.6 ml mg$^{-1}$ dry body mass h$^{-1}$, these animals can turn over the entire volume of the estuary in slightly over 26 h, reducing the particle mass suspended in the water column by approximately 50% d$^{-1}$ (exclusive of feces production and resuspension). In addition the plants, with their associated flora and fauna, provide an essential refuge and food resource for juvenile fishes (principally, the mullet, *Mugil cephalus*, and steenbras, *Lithognathus lithognathus*) that use the waterbody as a "nursery".[46,59] These fishes, which form the basis of a lucrative fishing industry in False Bay, also contribute to both recreational and subsistence fishing in the estuary, and to the diets of the numerous species of piscivorous birds in the Wetland and adjacent bird sanctuary.[46,47]

In areas further removed from the influence of the sea, the estuary is dominated by the submersed macrophytes *Myriophyllum aquaticum* and *Ceratophyllum demersum* and the emergent macrophytes *Typha capensis*, *Phragmites australis*, *Scirpus* spp. and *Juncus kraussii*. Recently, the floating macrophytes *Eichhornia crassipes* and *Azolla filliculoides* have been observed in the upper reaches of the waterbody, primarily in the Westlake Wetland. Of these macrophytes, most (excepting the *Scirpus* and *Juncus* species) have developed in nuisance proportions in various areas of the Zandvlei drainage basin, necessitating an on-going program of river rehabilitation to minimize the risk of flooding. *Myriophyllum* and *Ceratophyllum*, in particular, have created nuisance conditions in the Westlake Wetland where they have infested virtually all of the "open" waterways, closing them to recreational use.

## C. THE ALGAL COMMUNITY AND ITS CONSEQUENCES

Phytoplankton are present in the system, but not usually in nuisance quantities. The epiphyte *Cladophora* forms undesirable mats along the vlei edge at times during the summer months.[36,37] Zandvlei, unlike neighboring Zeekoevlei, does not regularly undergo periodic switches in floral dominance between algae and macrophytes[22,22a,57,59] (although, in recent years, the algal flora of Zeekoevlei has been stabilized — and is now dominated almost entirely by the cyanophyte, *Microcystis aeruginosa* — due to the use of an arsenic-based herbicide used to control rooted plant growth in that lake during 1951). Chlorophyll *a* values are typically low, averaging 25 mg m$^{-3}$, compared with values of over 500 mg m$^{-3}$ recorded in Zeekoevlei. The latter waterbody, formerly considered to be a sister estuary to Zandvlei, is now entirely isolated from the sea and is predominantly fresh in character. While numerous hypotheses have been advanced concerning the differences between these two neighboring waterbodies, this, together with its higher nutrient loads and wind-induced mixing regime, may be the determining factors.[22,22a,45,57,59] Recent evidence[57,59] also suggests a major role for a salinity in triggering the switch in dominant primary producer in those systems.

## D. THE DEVELOPMENT OF EUTROPHICATION AT ZANDVLEI

While the available evidence, based on written and oral histories, suggests that Zandvlei has only relatively recently developed the symptoms of eutrophication,[1] the lack of previous limnological/oceanographic data prevents a conclusive statement to this effect. Nevertheless, the development of the rainfall-runoff relationship and validation of the "BATHTUB" model provided an opportunity to hindcast the "original" water quality in the estuary. Using nutrient export coefficients calculated from a nearby mountain catchment (surrounding Steenbras Dam which has been subjected to only minimal human disturbance), the total phosphorus and nitrogen concentrations of the "pristine" estuary were calculated. The resultant data suggested that the water quality in the estuary has changed only marginally (predicted concentrations being 50 mg P m$^{-3}$, 750 mg N m$^{-3}$, 8 mg Chl m$^{-3}$, and a Secchi disc transparency of 0.7 m). Such calculations could not take into account the potential effects of a vastly larger marshland area in reducing the influent nutrient concentrations to the headwaters of the estuary (i.e., the predicted nutrient loads were probably higher than those that would have been actually observed at the time), nor could the hindcast river flows account for the greater degree of water retention that could be expected in an undeveloped catchment lacking the impervious surfaces of the present catchment (leading to an underestimate of the potential for the nutrient loads to generate an algal response). Hence, the constraints imposed by the modeling process probably offset each other and resulted in the predicted concentrations being not unreasonable, even given the changed morphometry and geography of the estuary (the original configuration of which would probably have enhanced the oceanic influence, the chemical composition of which has remained for the last 500 years[48]). Thus, it is highly probable that this particular inlet/estuary was, and always has been, a relatively productive system. Nevertheless, it is equally probable that, given the propensity for the general public to judge pollution status based on aesthetic criteria,[34,35] the observer-assessed water quality of the system has declined over the years.

## VII. DISCUSSION

### A. DEVELOPMENT OF A MANAGEMENT PHILOSOPHY

Table 6 summarizes the nature and magnitude of the various symptoms of impaired water quality observed in Zandvlei and identified by both the user survey and City of Cape Town staff. Those related to nutrient enrichment have been discussed above. Others include litter and flooding, but most, if not all, are the direct consequence of human manipulation of the system, and should be addressed by any management program adopted for this vlei. Given the importance of this estuary as a recreational venue

**Table 6** Summary of Water Quality Problems Identified in Zandvlei by Vlei Users and City of Cape Town Staff

| Problem | Description | Magnitude |
|---|---|---|
| Flooding | Occurs principally along the Diep River | ±5 homes annually |
| Weeds | *Myriophyllum, Ceratophyllum,* and *Eichhornia* in the catchment; *Potamogeton* in the vlei | Identified by 15% of respondents; >1,400 tonnes wet mass removed annually |
| Silt | Creation of deltas at the inflows; deposition at the outflow | Identified by the Imperial Yacht Club; ± 700 m³ annually at the Sand River inlet; ± 600 m³ annually at the Zandvlei outlet canal |
| Litter | Flotable plastics, glass, and other obstacles | Identified by 14% of respondents; + 2,500 bags annually plus >360 bags in *ad hoc* clean-ups |
| Bacteria | Fecal coliforms and skin irritants | Identified by 1% of respondents; 390–2,300 per 100 ml in the inflows; 50–100 per 100 ml in the vlei |
| Nutrients | Eutrophication | 300–1,500 mg N m⁻³ and 40–80 mg P m⁻³ in the inflows; 200 mg N m⁻³ and 50 mg P m⁻³ in the vlei |
| Odors | Hydrogen sulfide release in the Marina da Gama canals | Identified by 3% of respondents; unquantified |
| Colors | Unknown; possibly oil films | Identified by 6% of respondents; unquantified |
| Dead fish | Unknown; probably natural, seasonal mortality | Identified by 1% of respondents; unknown |

Data from City of Cape Town, *Zandvlei User Assessment Survey,* Cape Town, 1988.

for the Greater Cape Town region (and its related importance as a revenue generator for the City), the City Engineer and City Planner considered it to be important that each be recognized and ameliorated if possible (see below). The City of Cape Town staffers, therefore, adopted an holistic approach to problem solving at Zandvlei,[49] and the Inland Waters Management Team was created to evaluate the perceived problems and recommend an effective course of action to the City Council. This interdisciplinary (and interjurisdictional) approach continues to be employed to good effect within the City's boundaries, on its urban waterways and coastal marine waters.[50,61,62]

## B. THE MANAGEMENT APPROACH

Given the philosophy of integrated resource/catchment management adopted by the Inland Waters Management Team, both catchment-based and lake-based management techniques were evaluated. The objectives of the proposed management intervention had been determined by means of the user survey referred to earlier.[2] In summary, the water quality objectives could be stated as the achievement of an average annual chlorophyll *a* concentration of <15 mg m⁻³, an average Secchi disc transparency of between 1 and 2 m, a median fecal coliform count of <100 per 100 ml, average total nitrogen and phosphorus concentrations of <200 mg m⁻³ and <50 mg m⁻³, respectively, and a water residence time of between 0.2 and 1.0 years. In addition, certain other, primarily aesthetic, criteria were established, such as a reduction in the volume of flotables and other debris in and around the estuarine lake and some improvements in the number and type of facilities and services provided to the public.[2] Because conditions existing in Zandvlei were close to satisfying the water quality objectives, and because little further degradation was to be expected due to the highly developed state of the catchment (present development was at 95% of the total zoned development), it was decided that the immediate goals should be to address the aquatic plant and odor problems, and litter problem, identified by the vlei users.[2] In the longer term, however, the issues of continued urban growth in the watershed and nonpoint source pollution remain.[59,62]

## 1. Catchment-Based Planning Options

Seven catchment-based water quality management options were identified during 1989 as having potentially beneficial consequences for Zandvlei (these included the "do nothing" or null option). Of these, four were an extension of the City's "Greening the City" program[51] (which had been implemented some 4 to 5 years previously in 1984), and involved:

1. A citizen-based program of revegetating and upgrading various portions of the City's open space network — in this case, the river banks and buffer strips around the influent waterways.

2. A revitalized program of street-sweeping and litter collection, including the provision of additional litter bins and refuse dumping sites.
3. A renewed commitment to public education to inform and guide citizen-based action programs, including river cleaning (in connection with River Day), recycling campaigns and litter reduction campaigns (currently sponsored under the auspices of the Fairest Cape Association).
4. A stricter zoning code to better protect and rehabilitate the City's waterways.

Of these, approach 3 was already being implemented and had led to adoption of certain stretches of river by local service groups and/or schools acting in concert with the Fairest Cape Association and other organizations[52] (thus satisfying in part the demands of approach 1). Of the other two approaches, 2 could not be fully implemented due to financial constraints, and 4 could not easily be implemented given the highly developed state of the catchment. Implementing approach 4 would involve the compensation of large numbers of land owners for losses suffered through the imposition of additional restrictions on the use of their land, a consequence as cost-prohibitive as approach 2.

Two other approaches, based on engineering "solutions", involved:

5. The use of wetlands (natural or artificial) to act as both flood detention basins and in-stream nutrient filters.
6. The use of elaborate, in- and off-stream diversion and treatment structures, acting in concert with stricter nonpoint source pollution control measures.

Both of these approaches were capital-intensive, and given existing demands on City finances, could not be justified in a single (small) part of the City. In effect, the City had already constructed a series of artificial detention basins in the Diep-Sand River catchment for flood control purposes (approach 5).[52] Aside from upgrading and more regularly servicing of the litter traps associated with these structures,[59] no further engineering projects were contemplated for this system, at least not until the existing structures had been in place for some time and their effectiveness evaluated.

The final option, (7), the "do nothing" option, was not considered due to the clear call by vlei users for improvements in the water quality. In addition, the elimination of the catchment from the planning area would place an unfair burden on the lakeside residents, who would be most directly affected by the influx of particulates and dissolved matter from the catchment. Indeed, with the adoption of approaches (3) and (5), the City negated the "do nothing" option.

## 2. Lake-Based Planning Options

Nine in-lake options were considered during 1989, including the null option. Again, the null option was not considered for the reason given above. It was also determined that this option would lead to a worsening of the aquatic plant infestation in the vlei, which would eliminate its usefulness as a recreational venue. The remaining eight options consisted of engineering and other "solutions", including:

1. Dredging the lake to deepen the basin bottom to below the depth of penetration of light at the 1% level.
2. Controlling the oceanic and riverine inflows by means of sluices to manipulate the salinity of the system and to artificially select for coastal marine species of aquatic plants.
3. Artificially aerating the canal bottoms to eliminate anoxia and/or destabilize the halocline created during periods of seawater inflow.
4. Artificially enhancing the circulation of the vlei waters by redirecting the riverine and stormwater inflows to eliminate stagnation especially in the Marina da Gama canals.
5. Manipulating the vlei water level using an impervious weir at the outlet, in order to draw down and/or flood undesirable vegetation.
6. Maintaining the program of weed harvesting, and possibly extending or revising the current harvesting schedule or methodologies.
7. Using the existing wetland areas along the northern edge of the vlei as natural wetland filters/nutrient buffers for the vlei.
8. Imposing additional zoning and development controls on lakeside development.

Most of the structural options, like those in the catchment, were capital-intensive and therefore largely ruled out of consideration. Nevertheless, slumping of the outlet canal wall[1] did create a potential safety hazard, and funds had been budgeted for its replacement as part of the Diep-Sand River flood control scheme. Despite this, the installation of control structures was seen as extremely difficult in view of the pervious nature of the substrate that made it impossible to achieve a reasonable seal; approaches (2) and

(5) were, therefore, eliminated from further consideration. Aeration (approach [3]) was eliminated due to the potential noise and safety hazards that could arise in heavily populated and used waterways. Finally, approach (8) was subject to the same objections as had affected catchment-based approach (4); i.e., the high cost to the City of implementing such controls after the land had been occupied and developed, although in this case and in any event, only limited areas adjacent to the waterbody fell outside of the City's direct control (and use as parkland).

Of the remaining four approaches, each had been adopted to some extent in the management plans of the City. Approach (6), the weed harvesting program, was an on-going procedure, although it was decided that the cutting process could be made more effective by targeting the most heavily used recreational areas for more intensive cutting and using some rudimentary navigational procedures to allow the harvester driver to orient himself in the vlei basin. Approaches (4) and (7) were combined in the Westlake Wetland in an effort to control the spread of the water hyacinth (*Eichhornia crassipes*), and achieve a measure of control over the spread of the *Myriophyllum* community by increasing the salinity of the wetland to pre-1930s levels (i.e., to above 5 ppt, as in the rest of the vlei). Subsequent research has borne out the fact that it is the salinity in the vlei that provides the *Potamogeton* community with its competitive advantage over both *Myriophyllum* and planktonic algae.[57] Further, *Potamogeton* and its epizooic *Ficopomatus* community has been shown to play a major stabilizing role in the vlei, reducing the potential for resuspension of sediments, oxygenating the water column, removing nutrients and bacteria, and providing refugia and food sources for juvenile fishes.[59]

In the Marina da Gama Estate and the neighboring Muizenberg East development, plans are being prepared for the disposition of stormwater runoff from the Muizenberg East development, via detention/infiltration basins, into the blind canals of the Marina Estate in order to better circulate the estuarine waters in these areas (calculations of the density of stormwater suggested that it could penetrate the halocline and permit the desired destratification to occur or, at least, inject oxygen rich water into the hypolimnion of the canals). Finally, the deepening option (approach [1]) has been seriously considered as a two-phase project. The initial phase of the project will involve selective dredging of sandbars and other obstructions to enhance both circulation and recreational use of the waterbody (e.g., dredging of the vlei mouth and riverine inlet areas, and the northernmost inlet to the canal system).

In the longer term, consideration will be given to the deepening of the entire vlei bottom to restrict the areas colonized by the *Potamogeton* beds to those nearshore areas away from the recreational sites. This will have the dual effects of maintaining the fishery and reducing the need for mechanical harvesting. In addition, the dredge spoil, virtually pristine sand, can be used to meet the projected need for road fill once the planned False Bay arterial is constructed just to the north of the vlei.[53] Estimates of the depth to which the vlei will have to be deepened have been made using the Canfield relationships;[54] the vlei bottom will have to be dropped by approximately 1 m to a maximum depth of 3 m below full supply level. This deepening is calculated to have little effect on the water quality of the vlei, with virtually no differences being predicted should the additional volume be filled with either seawater or freshwater from the influent rivers. Further, hydrodynamic calculations[55] suggest that this deeper vlei will still be subject to full vertical mixing for all but about 7 days per year, given the prevailing wind speeds and directions. Nevertheless, extreme care will have to be exercised when deepening the vlei in order to maintain a viable *Potamogeton* community[57,59] and avoid further impairing water quality by triggering a shift to algal dominance which would seriously impact both the commercial fishery of False Bay and other recreational uses.

## 3. Use Modification Options

There remained one further set of options, often overlooked, that needed to be considered in compiling a management plan for the vlei; namely, the option of modifying the use of the vlei waters to one more suited to their quality. Three options were identified during 1989, and included:

1. Modifying the usage of the vlei to nonrecreational activities suited to a decreased water quality (for example, wastewater treatment, stormwater treatment or aquaculture of coarse fish species).
2. Modifying the usage of the vlei to nonrecreational activities suited to a higher quality water [such as water supply augmentation, this option demanding the implementation of catchment-based option (6) in order to be successful].
3. Modifying the recreational use of the vlei to either noncontact water-based recreation or passive (shore-based) recreation only, or maintaining the present, full contact water-based recreational activities.

However, given the high degree of public interest in, and use of, Zandvlei, none of the above approaches were given serious consideration.

## C. THE MANAGEMENT PLAN

The Inland Waters Management Team, having carefully weighed the various options (summarized above) open to the City of Cape Town in terms of both the City's needs and City's finances, then developed a plan which was ultimately recommended to, and adopted by, the Cape Town City Council during 1990. The elements of this plan were as follows:

1. Maintenance of the current vlei management programs of weed harvesting, litter collection, canal clearance, and vlei level management for the foreseeable future, as no practicable alternatives would appear to exist within the limitations imposed on the City by finance, topography and technology [the total annual cost of the program, in 1988 Rands, already approaches R2 million (= U.S. $800,000), or about 1% of the City's annual budget].

2. Augmentation of the litter collection and weed harvesting programs to remove floating debris from the vlei surface and to clear undesirable macrophytes from recreational areas more frequently, by redistributing current litter collection and weed harvesting efforts within existing budgetary limitations.

3. Participation in River Day and other citizen-based public awareness and participatory programs through financial subvention and contributions of in-kind support services such as special litter collections and staff time, within the budgetary constraints of existing programs such as "Greening the City".

4. Continuation of the water quality monitoring programs, started to gather data for this evaluation, in order to continue to monitor the progress (or otherwise) of the vlei, including bacteriological monitoring conducted on behalf of the City's Medical Officer of Health.

5. The conduct of supplementary investigations into the potential cost-benefits of the dredging/deepening proposals, including further assessment of various means to improve seawater circulation that could be implemented within existing budgets when the vlei mouth is reconstructed in 1992.

While this program was the "lowest-cost" option of all programs considered, it provided the initial direction necessary to meet the City's objective of maintaining an acceptable level of water quality at Zandvlei. Subsequent refinements of this management plan are being developed to guide this program into the foreseeable future. In addition, by not drawing on additional funds, the adopted plan should also permit the City the "luxury" (in a declining world and local economy) of continuing to service the needs and aspirations of "all our people, regardless of race, colour or creed .... [for] the heart of Cape Town is its people".[56]

## ACKNOWLEDGMENTS

The authors would like to express their appreciation to Drs. Alan Heydorn and Patrick Morant of the CSIR's Division of Earth, Marine and Atmospheric Sciences, who referred Professor McComb's request to us, and to Profs. Bryan Davies, Richard Fuggle and Barry Gasson, and their colleagues, of the University of Cape Town, who contributed their time and energies to the conduct of many aspects of the studies reviewed in this chapter. Andrew Darroch, now a graduate of the University of Cape Town, assisted with much of the computer programming and raw data analysis while serving as an intern with the City Planner's Department, Cape Town City Council. We would also like to recognize the many contributions made by staff from other City departments, principally the City Engineer's Parks and Forests Branch and the City Administrator's Department, whose work contributed to the many studies reviewed in this chapter. This paper is published with the permission of the City Engineer and City Planner, City of Cape Town. The opinions expressed herein are those of the authors, and do not necessarily reflect those of the Cape Town City Council, the Southeastern Wisconsin Regional Planning Commission, or the numerous agencies mentioned in this review.

## REFERENCES

1. **Morant, P. D. and Grindley, J. R.,** Rep. No. 14: Sand (CSW4), in *Estuaries of the Cape. Part II: Synopses of Available Information on Individual Systems,* Heydorn, A. E. F. and Grindley, J. R., Eds., Council for Scientific and Industrial Research, Pretoria, 1982, Research Rep. 413.
2. **City of Cape Town,** *Zandvlei User Assessment Survey,* City of Cape Town, Cape Town, 1988.

3. **Department of Water Affairs,** *Management of the Water Resources of the Republic of South Africa,* Government Printer, Pretoria, 1986.
4. **Day, J. A.,** Mineral Nutrients in Mediterranean Ecosystems, in *South African National Scientific Programmes Rep. No. 71,* Cooperative Scientific Programmes, Series Ed., Council for Scientific and Industrial Research, Pretoria, 1983.
5. **Macdonald, I. A. W., Jarman, M. L., and Beeston, P.,** Management of Invasive Alien Plants in the Fynbos Biome, in *South African National Scientific Programmes Rep. No. 111,* Foundation for Research Development, Series Ed., Council for Scientific and Industrial Research, Pretoria, 1985.
6. **Van Wilgen, B. W.,** Fynbos Terrestrial Ecosystems, in *Long-Term Data Series Relating to Southern Africa's Renewable Natural Resources,* Macdonald, I. A. W. and Crawford, R. J. M., Eds., in *South African National Scientific Programmes Rep. No. 157,* Foundation for Research Development, Series Ed., Council for Scientific and Industrial Research, Pretoria, 1988, chapter 6.
7. **Erasmus, D. A.,** Toward an understanding of the effects of orography on the distribution of precipitation over the Cape Peninsula, *S. Afr. J. Sci.,* 77, 295, 1981.
8. **Alexander, W. J. R.,** Hydrology of low latitude southern hemisphere landmasses, *Hydrobiologia,* 125, 75, 1985.
9. **Fullard, H.,** *Philips' Large Print Atlas for Southern Africa,* George Philip and Son, London, 1978.
10. **Whitfield, A. K.,** A characterization of Southern African estuarine systems, *S. Afr. J. Aquat. Sci.,* 18, 89, 1992.
11. **Department of Physical Planning,** *South African National Land Use Code,* Government Printer, Pretoria, 1977.
12. **Van der Post, Sir L.,** African Cooking, in *Foods of the World,* Williams, R. L., Series Ed., Time-Life Books, New York, 1970.
13. **The Royal Society of South Africa,** False Bay, *Trans. R. Soc. S. Afr.,* 47 (4/5), 501, 1991.
14. **City of Cape Town,** *Muizenberg East Draft Structure Plan,* City Planner's Department, Cape Town, 1989.
15. **Soil Conservation Service,** Urban Hydrology for Small Watersheds, *U.S. Dep. Agric. Tech. Release No. 55,* Washington, DC, 1986.
16. **Schmidt, E. J. and Schulze, R. E.,** SCS-based Design Runoff, in *WRC Project Rep. No. 155/TT31/87,* Water Research Commission, Series Ed., Creda Press, Cape Town, 1987.
17. **Darroch, A.,** SCS Based SHOWER Software, Town Planning Branch, City of Cape Town, unpublished report, 1988.
18. **Walker, W. W., Jr.,** Empirical Methods for Predicting Eutrophication in Impoundments. Rep. 4, Phase III: Applications Manual, *U.S. Dep. Army (Waterways Exp. Stn.) Tech. Rep. No. E-81-9,* Vicksburg, MS, 1986.
19. **Thornton, J. A. and Walmsley, R. D.,** Applicability of phosphorus budget models to southern African man-made lakes, *Hydrobiologia,* 89, 237, 1982.
20. **Thornton, J. A.,** Nutrients in African lake ecosystems: Do we know all?, *J. Limnol. Soc. S. Afr.,* 12, 6, 1986.
21. **Thornton, J. A.,** Aspects of eutrophication management in tropical/sub-tropical regions, *J. Limnol. Soc. South. Afr.,* 13, 25, 1987.
22. **Harding, W. R.,** Zeekoevlei — water chemistry and phytoplankton periodicity, *Water SA,* 18, 237, 1992.
22a. **Harding, W. R.,** A contribution to the knowledge of South African coastal vleis: the limnology and phytoplankton periodicity of Princess Vlei, Cape Peninsula, *Water SA,* 18, 121, 1992.
23. **Azorin, E. J.,** Zandvlei Wetland Mapping: Plant Communities, Unpublished report to the City of Cape Town, Botany Department, University of Cape Town, Cape Town, 1988.
24. **Davies, B. R. and Stewart, B. A.,** A note on the salinity and oxygen stratification in the Marina da Gama, Zandvlei, *J. Limnol. Soc. South. Afr.,* 10, 76, 1986.
25. **Chapman, P.,** Nutrient cycling in marine ecosystems, *J. Limnol. Soc. South. Afr.,* 12, 22, 1986.
26. **Grabow, W. O. K., Coubrough, P., Nupen, E. M., and Bateman, B. W.,** Evaluation of coliphages as indicators of the virological quality of sewage-polluted water, *Water SA,* 10, 7, 1984.
27. **Postma, H.,** Sediment Transport and Sedimentation, in *The Chemistry and Biogeochemistry of Estuaries,* Olausson, E. and Cato, I., Eds., John Wiley & Sons, New York, 1980.
28. **Howard Williams, C.,** Aquatic macrophyte communities of the Wilderness Lakes: Community structure and associated environmental conditions, *J. Limnol. Soc. South. Afr.,* 6, 85, 1980.
29. **Howard Williams, C.,** The distribution of nutrients in Swartvlei, a southern Cape coastal lake, *Water SA,* 3, 213, 1977.
30. **Grobler, D. C. and Davies, E.,** Sediments as a source of phosphates: A study of 38 impoundments, *Water SA,* 7, 54, 1981.
31. **Ryding, S.-O. and Rast, W.,** *The Control of Eutrophication of Lakes and Reservoirs,* UNESCO MAB Series Vol. 1, Parthenon Publishing, London, 1989.
32. **Organization for Economic Cooperation and Development,** *Eutrophication of Waters: Monitoring, Assessment, Control,* OECD, Paris, 1982.
33. **Thornton, J. A. and Rast, W.,** Preliminary observations on nutrient enrichment of semi-arid man-made lakes in the northern and southern hemispheres, *Lake Reservoir Manag.,* 5, 59, 1989.
34. **Thornton, J. A. and McMillan, P. H.,** Reconciling public opinion and water quality criteria in South Africa, *Water SA,* 15, 221, 1989.
35. **Thornton, J. A., McMillan, P. H., and Romanovsky, P.,** Perceptions of water pollution in South Africa: Case studies from two water bodies (Hartbeespoort Dam and Zandvlei), *S. Afr. J. Psychol.,* 19, 197, 1989.

36. **Davies, B. R., Stuart, V., and de Villiers, M.,** The filtration activity of a serpulid polychaete population (*Ficopomatus enigmaticus* (Fauvel)[sic] and its effects on water quality in a coastal marina, *Estuarine, Coastal Shelf Sci.*, 29, 613, 1989.

37. **Byren, B. A. and Davies, B. R.,** The influence of invertebrates on the breakdown of *Potamogeton pectinatus* L. in a coastal marina (Zandvlei, South Africa), *Hydrobiologia*, 137, 143, 1986.

38. **Dick, R.,** Salinity and *Myriophyllum* growth: augmentation of salinity as a possible control measure in the Keysers-Westlake Wetland near 2 Report to the Inland Waters Management Team, City Engineer's Department, Scientific Services Branch, Cape Town, 1988.

39. **Vermaak, J. F., Swanepoel, J. H., and Schoonbee, H. J.,** The phosphorus cycle in Germiston Lake. IV. The relationship between the absorption, accumulation and release of phosphorus and the metabolic rate and phosphorus contents of the tissues of *Potamogeton pectinatus L., Water SA*, 9, 155, 1983.

40. **Howard Williams, C. and Allanson, B. R.,** Swartvlei Project Reports. Part II. The limnology of Swartvlei with special reference to production and dynamics in the littoral zone, in *Institute for Freshwater Studies Rep. No. 78/3*, Rhodes University, Grahamstown.

41. **Allanson, B. R. and Howard Williams, C.,** A contribution to the physicochemical limnology of Swartvlei, *Arch. Hydrobiol.*, 99, 133, 1984.

42. **Carignan, R.,** An empirical model to estimate the relative importance of roots in phosphorus uptake by aquatic macrophytes, *Can. J. Fish. Aquat. Sci.*, 39, 243, 1981.

43. **Gardiner, A. J. C.,** A Study on the Water Chemistry and Plankton in the Blackwater Lakelets of the Southwestern Cape, Ph.D. dissertation, University of Cape Town, Cape Town, 1988.

44. **Semmelink, M. M.,** An Introduction to the Study of Phosphorus Dynamics in Rondevlei, M.Sc. thesis, University of Cape Town, Cape Town, 1991.

45. **Harding, W. D.,** Phytoplankton Periodicity in Coastal Vleis of the Cape Peninsula, M.Sc. thesis, University of Cape Town, Cape Town, 1991.

46. **Quick, A. J. R.,** A preliminary survey of the fishes of Zandvlei, Unpublished report to the City of Cape Town, Freshwater Research Unit, University of Cape Town, Cape Town, 1988.

47. **Day, J. A.,** Southern African Inland Water Ecosystems, in *Long-Term Data Series Relating to Southern Africa's Renewable Natural Resources,* Macdonald, I. A. W. and Crawford, R. J. M., Eds., in *South African National Scientific Programmes Rep. No. 157,* Foundation for Research Development, Series Ed., Council for Scientific and Industrial Research, Pretoria, 1988, chapter 4.

48. **Orren, M. J., Eagle, G. A., Fricke, A. H., Gledhill, W. J., Greenwood, P. J., and Hennig, H. F.-K. O.,** The chemistry and meiofauna of some unpolluted sandy beaches in South Africa, *Water SA*, 7, 203, 1981.

49. **Anon.,** Integrated catchment management program becoming a reality at Zandvlei, *S.A. Waterbulletin*, 14, 22, 1988.

50. **City of Cape Town,** [Draft] Water Quality Planning Policy for Cape Town, City of Cape Town, Cape Town, 1990.

51. **City of Cape Town,** *Greening the City: Open Space and Recreation Plan for Cape Town,* City Engineer's Department Rep. No. 214/1982, Cape Town, 1982.

52. **Anon.,** Rivers: Arteries of the Earth, *Cape Town Municipal Bulletin Supplement,* April, 1990.

53. **City of Cape Town,** *False Bay Arterial Environmental Impact Statement,* City Planner's Department, Cape Town, 1987.

54. **Environmental Protection Agency,** The Lake and Reservoir Restoration Guidance Manual, *U.S. Environ. Prot. Ag. Rep. No. EPA/4405-88-002,* Washington, DC, 1988.

55. **Robarts, R. D., Ashton, P. J., Thornton, J. A., Taussig, H. J., and Sephton, L. M.,** Overturn in a hypertrophic, warm, monomictic impoundment (Hartbeespoort Dam, South Africa), *Hydrobiologia*, 97, 209, 1982.

56. **City of Cape Town,** Credo, 1988.

57. **Harding, W. R.,** Water quality trends and the influence of salinity in a highly regulated estuary near Cape Town, South Africa, *S. Afr. J. Sci.*, 90, 240, 1994.

58. **Harding, W. R.,** Faecal coliform densities and water quality criteria in three coastal recreational lakes in the SW Cape, South Africa, *Water SA,* 19, 235, 1993.

59. **Quick, A. J. R. and Harding, W. R.,** Management of a shallow estuarine lake for recreation and as a fish nursery: Zandvlei, Cape Town, South Africa, *Water SA,* 20, 264, 1994.

60. **Dick, R.,** Zandvlei sediments: Nutrient levels and physical characteristics, Report to the Inland Waters Management Team, City Engineer's Department, Scientific Services Branch, Cape Town, 1988.

61. **Harding, W. R. and Quick, A. J. R.,** Management options for shallow hypertrophic lakes, with particular reference to Zeekoevlei, Cape Town, *S. Afr. J. Aquat. Sci.,* 18, 3, 1992.

62. **Quick, A. J. R.,** An holistic approach to the management of water quality in False Bay, Cape Town, South Africa, *S. Afr. J. Aquat. Sci.,* 19, 50, 1993.

# Chapter 9

# Eutrophication of the Dutch Wadden Sea (Western Europe), an Estuarine Area Controlled by the River Rhine

*Victor N. de Jonge and Wim van Raaphorst*

## CONTENTS

I. Introduction ........................................................................................................... 129
II. Physical Properties ................................................................................................ 131
   A. Climate ............................................................................................................ 131
   B. Topography and Morphology ......................................................................... 132
   C. Tide .................................................................................................................. 132
III. Catchment ............................................................................................................. 133
   A. Population and Industry .................................................................................. 133
   B. Freshwater Discharges ................................................................................... 133
   C. Nutrients in the Fresh Water .......................................................................... 134
      1. Concentrations ........................................................................................... 134
      2. Loads .......................................................................................................... 135
   D. Nutrients in the North Sea ............................................................................. 135
IV. Water Movement and Transport ........................................................................... 135
   A. Tidal Currents ................................................................................................. 135
   B. Water Circulation ........................................................................................... 137
   C. Residual Water Circulation ............................................................................ 138
   D. Time Scales of the Water ............................................................................... 139
V. Nutrient Concentrations ........................................................................................ 139
   A. Concentrations in the Water ........................................................................... 140
   B. Processes Regulating Concentrations ............................................................ 141
   C. Nutrients in the Sediments ............................................................................. 142
   D. Exchange with the Open Sea ......................................................................... 143
   E. Response to Changing External Nutrient Inputs ........................................... 143
VI. Sediment Properties .............................................................................................. 143
   A. Composition of Bottom Sediments ............................................................... 143
   B. Accumulation and Transport of Suspended Matter ....................................... 143
   C. Resuspension of Sediments and Microphytobenthos .................................... 144
VII. Symptoms of Nutrient Enrichment ...................................................................... 144
   A. Primary Producers .......................................................................................... 144
      1. Primary Production ..................................................................................... 144
      2. Biomass ...................................................................................................... 144
   B. Secondary Producers ...................................................................................... 145
   C. Fish .................................................................................................................. 145
VIII. Consequences of Nutrient Enrichment ................................................................. 146
IX. Management Options for Restoration .................................................................... 147
References ....................................................................................................................... 147

## I. INTRODUCTION

The Wadden Sea is a shallow inshore area in the northwestern part of Europe with a surface area of roughly 6000 km² extending from The Netherlands to Denmark. The Dutch part of the Wadden Sea covers approximately 3000 km² (Figure 1). The basic geological form, consisting of six tongue basins,

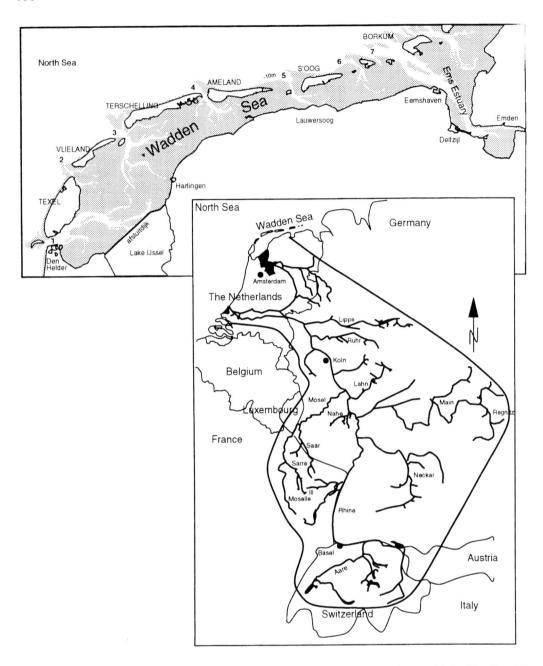

**Figure 1** Map of the study area with the River Rhine (including the lower branches and Lake IJssel) and its drainage basin, the coastal area of the North Sea and the Wadden Sea extending from The Netherlands, via the Federal Republic of Germany to the north of Denmark. The different tidal basins in the Dutch Wadden Sea are numbered (see Table 1 for explanation).

was formed during the Saalian Glaciation. Of these tongues, three are situated in The Netherlands and three in the Federal Republic of Germany (FRG).[1,2]

Since 1600, many areas in the Wadden Sea have been embanked or otherwise separated from the sea. During the present century the "Afsluitdijk" was constructed, closing off the Zuiderzee in 1932 (Figure 1). This decreased the former Wadden Sea by 3200 km², increased the tidal range by 50 cm (Figure 2),[3] and the tidal currents by 10–25%.[4] The slope of the salinity gradient increased dramatically due to the

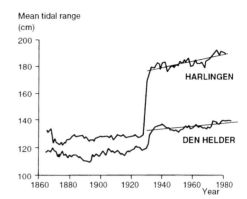

Mean tidal range
(cm)

HARLINGEN

DEN HELDER

**Figure 2** Changes in mean tidal range (cm) near Den Helder and Harlingen (western Dutch Wadden Sea). (Modified from de Jonge, V. N., Essink, K., and Boddeke, R., *Hydrobiologia*, 265, 45, 1993. With permission.)

closure dam.[5] Between 1932 and the early 1950s a rapid sediment accretion took place in the western Wadden Sea.[6]

In this paper, attention is restricted to the Dutch Wadden Sea, which can be subdivided into three parts, the western Dutch Wadden Sea, consisting of the tidal basins "Marsdiep", "Eijerlandsche Gat" and "Vlie", the eastern Dutch Wadden Sea consisting of "Borndiep", "Friesche Zeegat", "Lauwers" and "Schild", and the Ems estuary, situated on the border with the FRG (Figure 1). The Ems estuary will not be treated here, as it is the subject of Chapter 7.

In the Dutch part of the Wadden Sea there are three important seaports: Den Helder, situated at the border between the Wadden Sea and the North Sea; Harlingen, which is mainly used for coasters, fishing boats and recreational shipping; and Lauwersoog, which is mainly used by sea-going fishing boats.

The Wadden Sea is one of the most important wetlands of the world. The area is situated in the middle of the eastern Atlantic flyway (Figure 3). The area contains between 6 and 12 million birds representing over 50 species. Throughout the year these birds utilize the area for feeding, moulting, breeding and roosting.[7] The Dutch Wadden Sea is also an important nursery and feeding area for important species of fish (herring, *Clupea harengus*; sole, *Solea solea;* and plaice, *Pleuronectes platessa*), crustaceans and shellfish. The fishery for shrimps (*Crangon crangon*) has increased dramatically during the last 50 years. The Wadden Sea share of this species landed from the Dutch coastal area was only 20% in the mid 1970s, but increased to 50% in the mid-1980s. At present, approximately 90 vessels are licensed to fish for shrimps in the Dutch Wadden Sea.[3] The Wadden Sea is the main area for mussel (*Mytilus edulis*) breeding in The Netherlands. Annual production ranges from 30,000–70,000 t,[3] and is relatively stable. Cockles (*Cerastoderma edule*) are fished by hydraulic suction dredgers from natural beds. The catches depend on the present amount and licenses and range from zero to over 8000 tonnes annually.[3]

Secondary production in the Wadden Sea is high for two reasons. First, the Wadden Sea is an estuarine area where particulates (organic matter and sediments accumulate) and are mineralized.[8] This results in a natural nutrient enrichment of the area.[9] Second, the area is heavily influenced by nutrients from anthropogenic origin.[9-13] In this chapter the term eutrophication refers to the anthropogenic nutrient enrichment of the area.

## II. PHYSICAL PROPERTIES

The Wadden Sea is a highly dynamic area. Important natural factors influencing the dynamics are tide, wind and temperature. Important human activities include the mussel, cockle and shrimp fisheries, extraction of bottom deposits, dumping of harbor spoil and the engineering works (closures and land reclamation.

### A. CLIMATE

For information on the climate of the Dutch Wadden Sea, refer to Chapter 7 (the Ems estuary).[14] The figures given for the Ems estuary hold for the Dutch Wadden Sea.

Winds are predominantly from a westerly direction. The wind speed decreases landward and to the east. The mean annual wind speed near the barrier islands is 6.5–7 m s[-1] while near the mainland it is: 5.5–6 m s[-1].[15]

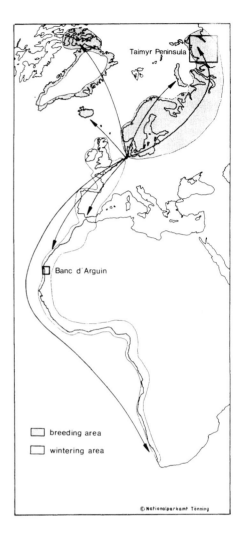

Figure 3  Map of the eastern Atlantic flyway showing breeding and wintering areas of birds that visit the Wadden Sea in the course of the year. (Modified from *Rep. 6th Trilat. Gov. Conf. Protection of the Wadden Sea,* Esbjerg, 1991, National Forest and Nature Agency, Tönning, Germany. With permission).

## B. TOPOGRAPHY AND MORPHOLOGY

The Dutch Wadden Sea is composed of seven tidal basins which are separated from the North Sea by a girdle of barrier islands. The tidal inlets between the different islands allow water movement to and from the North Sea. The tidal basins are separated from each other by high tidal flats (tidal watersheds). The Wadden Sea is separated from the mainland by dikes.

The total surface area of the different tidal basins, the surface area of tidal flats, the tidal prism, the mean water content during high tide, and water turnover times are listed in Table 1.

There is a shift in the total surface area of the tidal basins and in the total surface area of tidal flats going from west to east. The western basins have large surface areas and large water volumes while the percentage of tidal flats may be as low as 15% (Marsdiep basin). The low proportion of tidal flats in the western Wadden Sea is compensated by a relatively large subtidal surface area (approximately 30% on average). The surface area and volume of the eastern basins decline while the proportion of tidal flats reach values of 70 and 80%.

## C. TIDE

The tide plays a crucial role in the transport and mixing of the water. Moreover, it is highly significant for the transport of suspended material and for the biotic processes in the intertidal zone.

The vertical movement of the water in the Wadden Sea is dominated by the semidiurnal lunar tide with a mean period of approximately 12 hours, 25 minutes. The cycle in tidal ranges between spring tide and neap tide is caused by the semidiurnal solar tide. The tidal range can be increased or decreased by wind. For further details see de Jonge[14] and Postma.[16]

**Table 1**  Some Characteristics of the Different Tidal Basins of the Dutch Wadden Sea

| Number of the tidal basin | Name of the tidal basin | Total surface area ($10^6$ m$^2$) | Water content at HW[a] ($10^6$ m$^3$) | Tidal volume ($10^6$ m$^3$) | Turnover time water $t_e$ (tides) | Surface area of tidal flats ($10^6$ m$^2$) |
|---|---|---|---|---|---|---|
| 1 | Marsdiep | 712 | 3,357 | 1,140 | 17 | 107 (15%) |
| 2 | Eijerlandsche Gat | 153 | 313 | 207 | 3 | 99 (65%) |
| 3 | Vlie | 668 | 2,248 | 1,060 | 13 | 267 (40%) |
| 4 | Borndiep | 309 | 812 | 478 | 10 | 139 (45%) |
| 5 | Friesche zeegat | 195 | 350 | 300 | 9 | 136 (70%) |
| 6 | Lauwers | 200 | 300 | 240 | 8 | 130 (65%) |
| 7 | Schild | 29 | 41 | 31 | 8 | 23 (80%) |
|  | Total Dutch Wadden Sea | 2,266 | 7,400 | 3,388 |  | 901 (40%) |

*Note:* Turnover times ($t_e$) of tidal basins of the Dutch Wadden Sea are defined as the time interval necessary to reduce the mass present in a basin to a fraction $e^{-1}$ of the original mass. Data are obtained from Ridderinkhof.[19] For location of the basins see Figure 1.

a  HW = High water.

The mean annual tidal range in the Dutch Wadden Sea is not constant. The value slowly increases due to different processes — among them the relative sea level rise.[17] The mean tidal range also differs for the different parts of the Wadden Sea. Near Den Helder in the most westerly part, the tidal range is 1.34 m; at the island of Terschelling it is 1.85 m; and in the coastal zone near the border between the Dutch Wadden Sea and the Ems estuary, it is 2.09 m.

## III. CATCHMENT

### A. POPULATION AND INDUSTRY

The catchment of the Wadden Sea is mainly the drainage basin of the River Rhine (Figure 1), the most important river in western Europe. It receives water from the Federal Republic of Germany, Switzerland, France, Austria, Lichtenstein, Belgium, Luxembourg and The Netherlands.

The population density in these countries differs from less than 10 inhabitants per km$^2$ to over 200 per km$^2$. However, in the main part of the drainage basin the population density is over 200 per km$^2$.

Approximately 40–50% of the working part of the population is active in the industry while between 5–15% works in agriculture. The main industrial activities are chemicals, metals, cars and textiles. These activities affect the water quality of the River Rhine. Leaking of nutrients from agricultural activities also heavily influences the quality of the river water.

### B. FRESHWATER DISCHARGES

The River Rhine receives water from precipitation and the melting of snow in the Alps. The drainage basin comprises an area of ca. 185,000 km$^2$. The total length of the river is over 1000 km. The water discharge of the river varies strongly. The minimum (575 m$^3$ s$^{-1}$ in winter during ice cover) and maximum (13,000 m$^3$ s$^{-1}$) were recorded in 1926. The mean annual discharge is approximately 2200 m$^3$ s$^{-1}$. In Figure 4 the time series of the annual mean water discharge on the border between The Netherlands and the FRG at Spijk is given for the period for which nutrient measurements are also available.

Downstream of the border, the River Rhine splits into several branches. The two main branches (Rivers Rhine and Waal) flow westwards. Along the North Sea coast the river water is mixed with sea water coming from the Strait of Dover. Under the influence of the residual current and the wind, the water flows northeast, entering the Wadden Sea area near Den Helder.

The other branch, the River IJssel, flows northward into Lake IJssel, and is subsequently sluiced out during low tide from the lake into the Wadden Sea at two points. The amounts discharged annually at Den Oever and Kornwerderzand are also given in Figure 4.

Apart from these main rivers there are also local fresh water discharge points, but these are of minor importance.

As a result of the influx of fresh water, the coastal water in the Marsdiep tidal inlet (Figure 1) consists of approximately 13% of fresh water.[18,19] Approximately 60% of this amount is transported through Lake

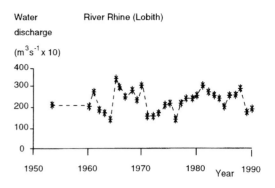

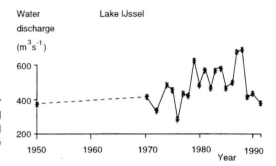

**Figure 4** Mean annual water discharge from the River Rhine near the border between The Netherlands and the Federal Republic of Germany and the mean annual amount of fresh water that is discharged from Lake IJssel.

IJssel, the rest via the Rivers Rhine and Waal[19] and the North Sea. In the eastern Wadden Sea only small amounts of water are sluiced out annually.

## C. NUTRIENTS IN THE FRESH WATER
### 1. Concentrations

The seasonal patterns of the nutrient concentrations in the water of the River Rhine at Spijk on the border with FRG are given for the years 1970/71, 1981/82 and 1990/91 (Figure 5). The curves only partly follow a common pattern.

Ammonium concentrations in the River Rhine were high in the early 1970s, then significantly decreased due to waste water treatment. The concentrations are relatively high in winter and low in summer. In Lake IJssel the seasonal variation in the low ammonium concentrations is not clear.

Due to the wastewater treatment, the concentrations of nitrate increased in the River Rhine since the early 1970s. The seasonal pattern in the River Rhine is not pronounced but in Lake IJssel is more marked. This pattern may be caused by local primary production.

The phosphate concentrations in the River Rhine do not show a clear seasonal pattern. In the water of Lake IJssel the seasonal curve is most obvious, showing very low concentrations in spring, except in the early 1980s when the phosphate load reached its maximum (see below).

The total nitrogen concentrations do show a seasonal curve in both the River Rhine and Lake IJssel during all years, something not observed for total phosphorus (Figure 5).

The changes in the seasonal curves of the nutrients in the River Rhine and Lake IJssel are caused by the combination of hydrodynamics and biology of this shallow lake.

Apart from the general seasonal picture there are differences in the mean annual levels of the nutrients for the different years. These long-term changes show low concentrations in the early 1950s. Concentrations of phosphate and total phosphorus peaked in the 1970s and early 1980s. Since then the concentrations have declined, except the total phosphorus concentrations in Lake IJssel. The mean annual nitrogen concentrations vary significantly; in the early 1970s the concentrations in the water of the River Rhine suddenly increased from approximately 250 $\mu$mol L$^{-1}$ to almost 400 $\mu$mol L$^{-1}$. Since 1970, the values have usually been above 300 $\mu$mol L$^{-1}$. For Lake IJssel a similar picture emerges, although data are only available since 1972. For total nitrogen there is no clear trend in the River Rhine, while the values in the water of Lake IJssel seem to have increased dramatically during the early 1970s but have shown no trend since then. Comparison of these values with data of Helder[20] and the long-term development suggest that for 1972 and 1973 both the dissolved and total nitrogen values in the files of Rijkswaterstaat may be too high.

There are some differences in the long-term trend of the mean annual nutrient concentrations of the River Rhine and Lake IJssel. The increase in mean annual phosphate concentrations in Lake IJssel starts later than the increase in the River Rhine. This is mainly thought to be attributed to buffering by the bottom sediments of Lake IJssel.[11] For the same reason, the concentrations of total phosphorus in the River Rhine increased more steeply than in the water of Lake IJssel. The decline of phosphate in the water of the River Rhine and Lake IJssel occurred during the same period. However, the decrease in total phosphorus in the River Rhine started in 1977 and in Lake IJssel water not before 1985.

## 2. Loads

In Figure 6 the nutrient loads are given for station Spijk over the period 1950 to 1991. The curves show the tremendous increase in phosphate and total phosphorus from 1950. Remarkably enough, the dissolved nitrogen loads only increased before the mid-1960s while phosphate has increased further since then. This was due to the use of polyphosphates in detergents and the use of artificial fertilizers in agriculture. Since the early 1980s the loads of total phosphorus and phosphate have declined, while the loads of dissolved nitrogen were roughly maintained. For total nitrogen there are not enough data to draw firm conclusions on the long-term trend including the period before 1970.

Figure 6 also shows the nutrient loads from Lake IJssel through the sluices at Den Oever and Kornwerderzand. The loads of phosphate and total phosphorus increased steeply from 1971 to 1984. Since then the loads have decreased again. These changes are partly caused by the variation in water discharge (Figure 4), and partly by changes in nutrient concentrations (Figure 5). The loads of total dissolved nitrogen and total nitrogen increased parallel with the phosphorus compounds. However, just as reported for the River Rhine station Spijk, the decrease in loads observed for phosphorus is less clearly demonstrable for nitrogen.

Nutrients dicharged from the Rivers Rhine and Waal reach the Wadden Sea by a mixture of direct water transport and sediment transport — either incorporated in organic particles or adsorbed to or included in sediment particles. The magnitude of this total import from the North Sea coastal zone to the Wadden Sea is not well quantified (Section V.D).

## D. NUTRIENTS IN THE NORTH SEA

The southern North Sea (Figure 1) receives water and nutrients from different sources, including the Dover Strait situated between France and the United Kingdom. Through this strait a mixture of water from the Atlantic Ocean and fresh water from the United Kingdom and France flows into the Southern Bight of the North Sea. Because the water supply through the Dover Strait is very high (155,000 m$^3$ s$^{-1}$ on average),[21] it determines the retention time and the base levels of nutrient concentrations in Dutch coastal waters.

The seasonal variation in nutrient concentrations in the English Channel and the Dover Strait are classical, with high values in winter, a decline in March and low values in early summer. In August the silicate values are above the growth-limiting concentrations. For phosphate, nitrate and nitrite this increase starts in October.[22]

The nutrient concentrations in the Dover Strait have increased two to three times from the early 1930s to the period 1986 to 1988.[22] However, firm evidence for the correctness of this factor is absent, due to lack of data and the use of different analyses in the past.

The most recent calculations of nutrient fluxes through the Dover Strait are given in Table 2. Comparison of these fluxes with those from the River Rhine and Lake IJssel indicates the important role of the Dover Strait on the nutrient levels of the southern North Sea.

## IV. WATER MOVEMENT AND TRANSPORT

### A. TIDAL CURRENTS

The average tidal curve for the western Wadden Sea shows a faster rise (approximately 5.75 h) than fall (approximately 6.7 h). Consequently, flood currents are faster than ebb currents. This phenomenon is one of the processes that governs the transport of sediment and small organisms from the North Sea to the Wadden Sea. Further eastward, the tidal curve changes so that the asymmetry becomes weaker.

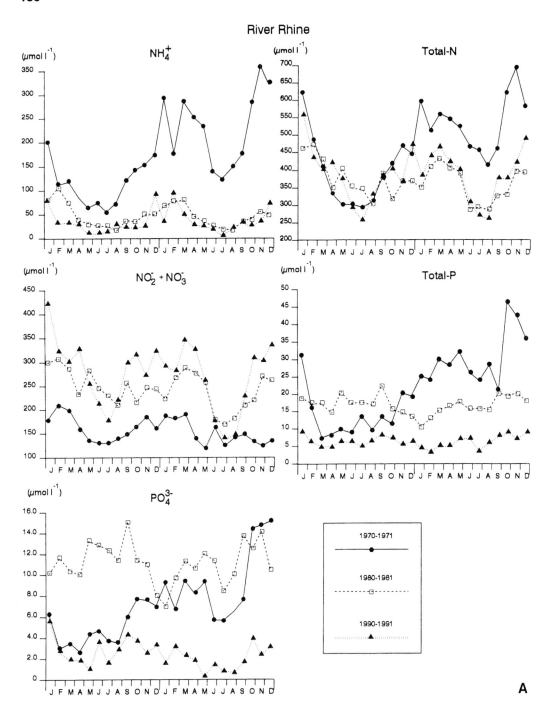

**Figure 5** Seasonal variation in nutrient concentrations of the fresh water of the River Rhine when entering The Netherlands and the fresh water of Lake IJssel near one of the discharge points.

The tidal wave on the North Sea moves from west to east. Consequently there is a time difference in high water of 4.5 hours between Den Helder and the border between the eastern Wadden Sea and the Ems estuary.

Tidal prisms (Table 1) are largest in the two westernmost basins, being 1140 and 1060 × 10⁶ m³, respectively. At the other inlets the tidal prisms range from $31 \times 10^6$ m³ between Rotumerplaat and Rottumeroog to $478 \times 10^6$ m³ in the Borndiep between Terschelling and Ameland.[19]

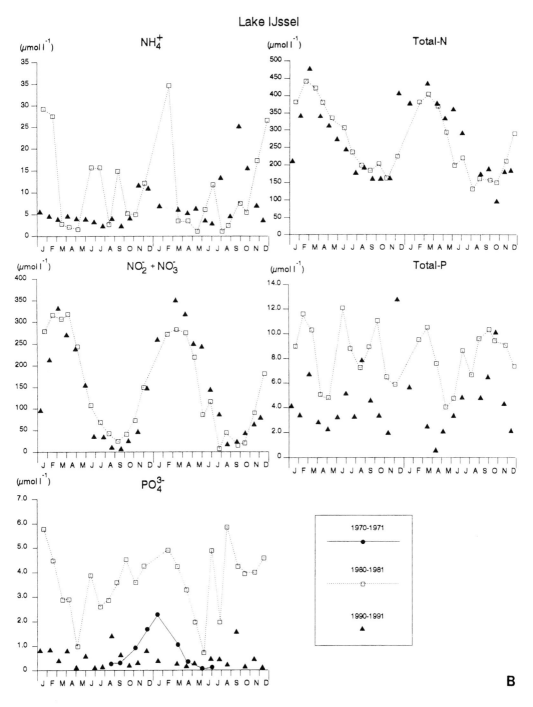

Figure 5 (B)

## B. WATER CIRCULATION

In the Wadden Sea the dominating mechanisms affecting water circulation are the tides and the wind. Only in the Marsdiep and Vlie basins are there substantial freshwater inputs, resulting in additional density-driven currents. Due to the tidal watersheds the water circulation is largely restricted to the individual basins. With strong winds, however, water is exchanged across the watersheds. Water circulation patterns are strongly influenced by the morphology of the basins, particularly by the large area of tidal flats and the presence of tidal channels. Current speeds in the center of these channels are larger than

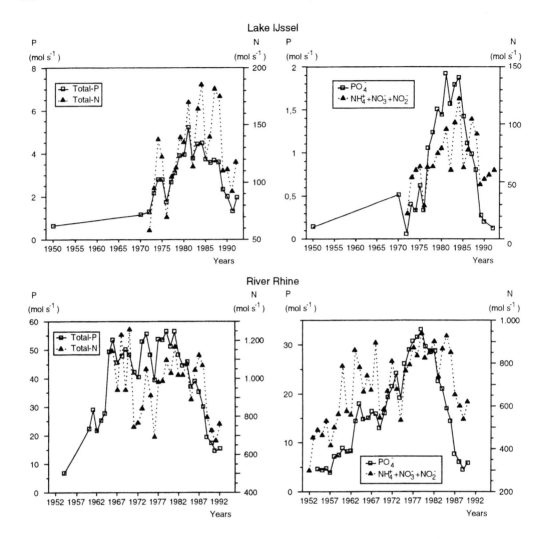

**Figure 6** Time series of annual nutrient loads (mol s⁻¹) of the River Rhine and Lake IJssel (total amount sluiced out at Kornwerderzand and Den Oever). "Total" is the sum of dissolved and particulate nutrients. (Data from Rijkswaterstaat.)

those in the shallower parts. This shear in current field is one of the basic mechanisms for mixing.[23] Another mechanism is the tidal random walk of water parcels due to residual eddies with a length scale comparable to the tidal excursion.[18] Most important, however, is the irregularity of the flow field which stems from the interaction of the tidal flow with the complex bathymetry of the system.[24] For a full physical description of the water movements and mixing processes in the Wadden Sea refer to the literature.[25]

## C. RESIDUAL WATER CIRCULATION

Changes in water level caused by the tidal wave generate residual currents. In a uniform periodic tidal current field, water parcels are carried only backwards and forwards with no net displacement. A net displacement or residual current can only exist if there is a continuous throughflow or if different water parcels exchange their positions during the tidal cycle. In the Marsdiep and Vlie basins, throughflow is caused by the input of freshwater from Lake IJssel. There is also a southward throughflow of the same order of magnitude from the Vlie basin towards the Marsdiep basin driven by the tide itself.[19] The total residual currents at the inlets of both basins are determined by the sum of these throughflows.

At the Marsdiep tidal inlet the total residual current is ca. $50 \times 10^6$ m³ per tide, and in the Vlie basin the throughflow is ca. $30 \times 10^6$ m³ per tide.[19] In both basins the residual transport is less than 5% of the

**Table 2** Nutrient Fluxes Through the Strait of Dover, the River Rhine on the Border Between FRG and The Netherlands and from Lake IJssel to the Western Dutch Wadden Sea

| | Total dissolved nitrogen | Total nitrogen | Orthophosphate | Total phosphorus | Dissolved silicate |
|---|---|---|---|---|---|
| Strait of Dover | 1,237 | 3,196 | 81 | 196 | 555 |
| River Rhine | 899 | 1,111 | 18 | 35 | |
| Lake IJssel | 89 | 165 | 1 | 3.5 | |

*Note:* Values are given in mol s$^{-1}$. Data from Laane et al.[22] and Rijkswaterstaat.

tidal prism. In the other basins no substantial input occurs from the mainland while the throughflow from one basin to another has not been quantified.

Detailed modeling of the water transport in the tidal channels[25] showed that the largest inflow of North Sea water occurs along the southern coast and the largest outflow of Wadden Sea water occurs along the northern coast of the inlets, indicating an anticlockwise residual circulation close to the inlets. Analysis of field measurements in the Ems estuary have shown that within this anticlockwise residual cell smaller residual cells are present.[26]

## D. TIME SCALES OF THE WATER

The flushing time scales of the Marsdiep tidal basin were first published in 1954.[8] More recent calculations of the freshwater flushing time[27,28] arrived at the same figure of ca. 15 tides for water discharged at Den Oever (Figure 1) and ca. 30 tides for water discharged at Kornwerderzand. These estimates did not, however, take into account the residual transport of North Sea water entering the Wadden Sea by the Vlie tidal inlet and leaving the area via the Marsdiep inlet. The most reliable results of mixing time scales as turnover times ($t_e$) were recently published.[19] The turnover times (tidal volume divided by the exchange coefficient with the open sea) are given in Table 1. Taking into account the southward residual currents, the flushing time of freshwater from Den Oever is approximately 20 tides and of water from Kornwerderzand ca. 24 tides.[19] These recent estimates are relevant in interpreting previously published exchange rates between the Wadden Sea and the North Sea,[8] as will be discussed in the next section.

## V. NUTRIENT CONCENTRATIONS

### A. CONCENTRATIONS IN THE WATER

Reliable data on nutrient concentrations in the entire Dutch Wadden Sea are available from 1970 onward. During that time phosphorus[9] and nitrogen[20] were measured in an extensive sampling program that covered the entire Dutch Wadden Sea. Since then the area has been monitored for all nutrients by Rijkswaterstaat. There are also some older data, but these are restricted to the western part.[8,27,29,30]

An overview of dissolved nutrient concentrations in the entire Dutch Wadden Sea during the early 1970s was presented by de Jonge and Postma[9] and Helder.[20] Their data show that the distribution in winter is more homogeneous than in summer, particularly for phosphate. In summer strong gradients in phosphate occur with highest concentrations near the freshwater sources and near the tidal watersheds where enhanced mineralization of organic matter occurs.[8,9] The pattern for ammonium and nitrate is less clear. Apart from the Ems estuary, inorganic nitrogen concentrations were highest in the western part of the Wadden Sea. This reflects the influence of the River Rhine, by the influx of nutrients from both Lake IJssel and the influx of particulate nutrients from the Dutch coast that originates from the River Rhine.

The early 1970s represent a period with relatively low inputs of both nitrogen and phosphorus from Lake IJssel (Figure 6); after these years inputs increased sharply, reaching maxima in the period 1980–1985. Due to this increase the concentrations in the Marsdiep and Vlie basins also increased strongly from 1972 onward (Figures 7, 8 and 9).

Some indications for the development of the nutrient levels and accompanying processes in the Marsdiep basin are shown in Figures 8 and 9 where components of nitrogen and phosphorus are given as functions of salinity for different years. During winter (December to February) nitrate behaved conservatively during all years. Both nitrate end-members at the freshwater side and at the seawater boundary were higher in 1985–1986 than in the early 1960s and 1970s. For the summer period (May to July) this increase was even more pronounced, particularly at the freshwater side. During summer, ammonium and nitrate concentrations in the coastal North Sea did not substantially change between 1960

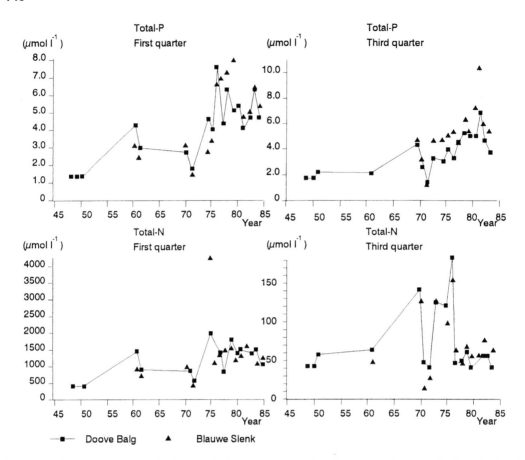

**Figure 7** Concentrations of total-P (A,B) and total-N (C,D) in the Marsdiep and Vlie basins in the first (A,C) and third quarter (B,D) of the year, respectively, during the period 1950–1985. All concentrations in µmol L⁻¹. Different lines represent different stations in the basins. (Obtained from van der Veer, H. W. et al., *Helgo. Meeresunters.*, 43, 501, 1989. With permission.)

and 1986. In the summer of 1986 nitrate was consumed in the western Wadden Sea (Figure 8), while a net production of ammonium and dissolved inorganic phosphorus (DIP) occurred (Figure 8 and 9). In 1972, a net consumption of ammonium occurred during the summer period, while in the early 1950s an uptake of DIP was observed. These data suggest that in the low salinity regions of the western Dutch Wadden Sea limitation of the summer production may have changed over past decades from DIP in 1950 and probably also in 1961 to ammonium in the early 1970s. The situation for 1986 is difficult to interpret. The data in Figure 8 suggest nitrate to be the limiting compound.

The curves in Figures 8 and 9 also show that the largest increase in nutrient concentrations in the western Wadden Sea occurred from the early 1970s onward, particularly due to the increase of nutrient concentrations in the northern part of Lake IJssel by that time.[11,30,31] This is most clear for nitrate, DIP and total phosphorus.

## B. PROCESSES REGULATING CONCENTRATIONS

Concentrations of nutrients in the different basins of the Wadden Sea are determined by the characteristics of the basins and the boundary conditions at the sea side and at the freshwater side. Nutrient-salinity plots often show nonconservative behavior (Figures 8 and 9), indicating that internal processes in the area are as important as the external inputs. The relative influence of external inputs and internal processes can be estimated from simple mass balances, in which the average concentration gradients in the basin are the sum of a background concentration determined by the open sea, and effects of external inputs, losses to the sediment and internal production and consumption processes.[31] Also important are the nondispersive tide-induced residual transports of particulate nutrients from the open sea into the Wadden Sea.[8,16,31] These mass balances indicate that the annual mean contents of total nitrogen and total phosphorus in the

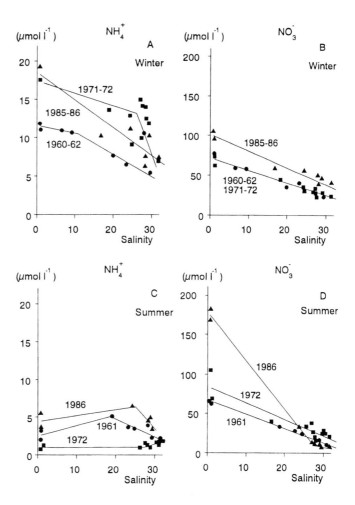

**Figure 8** Ammonium (A,C) and nitrate (B,D) as function of salinity in the Marsdiep basin during winter (December–February) and summer (May–July). Data for 1960–1962 were obtained from Postma,[29] for 1971–1972 from Helder,[20] and for 1985–1986 from the EON-Project Group,[36] Veldhuis et al.[35] and Rijkswaterstaat.

basins are largely controlled by the boundary conditions at the seaside and by the external inputs, particularly in the Marsdiep/Vlie and the Friesche zeegat basin. Approximately 30% of total nitrogen and 25% of total phosphorus seems to be determined by the freshwater inputs in these basins. Denitrification is responsible for the loss of ca. 110 mmol N m$^{-2}$ year$^{-1}$ in the Marsdiep/Vlie basins,[32] which would reduce total nitrogen in the Marsdiep/Vlie basins by less than 1 mmol m$^{-3}$. Net burial of nitrogen and phosphorus is not well quantified in the Wadden Sea, but tentative calculations[31] and measurements[33] indicate that it may be substantial.

Simple mass balance calculations are based on average concentrations in the entire basins. Obviously, lateral gradients occur in the basins and are related to salinity distribution, sediment composition, biological processes and chemical processes. Basically, two processes determine concentrations of dissolved nitrogen and phosphorus after they have been transported into the Wadden Sea. The first is uptake by primary producers, both phytoplankton and phytobenthos. The second is the mineralization of organic matter that is accumulated in the Wadden Sea. The net effect of the nutrient consumption by phytoplankton and the mineralization of organic matter is higher concentrations of dissolved inorganic nutrients in the inner parts of the basins compared with near the inlets (Figures 8 and 9).

## C. NUTRIENTS IN THE SEDIMENTS

Nutrient levels in the sediment layers are much less studied than those in the water column. Yet these levels may be of direct importance for benthic primary producers and, due to sediment-water exchanges, also for the phytoplankton. Seasonal patterns of inorganic nutrients in the interstitial waters of the upper 40 cm of the sediments of an exposed subtidal station near the island of Texel (Figure 1) were measured in 1978.[34] According to these data, concentrations of ammonium, phosphate and silicate in the upper 10 cm of this exposed station differed little from those in the overlying water, an indication for high apparent

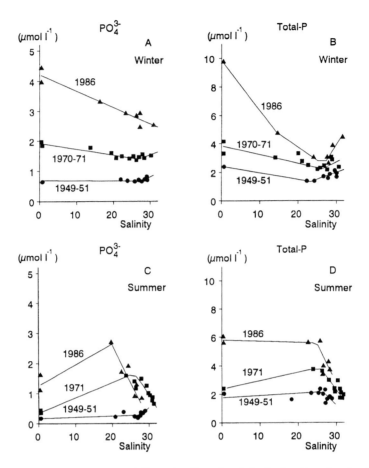

**Figure 9**  Dissolved inorganic P and total P as function of salinity in the Marsdiep basin during winter (December–February) and summer (May–July). Data for 1950 were obtained from Postma,[8] for 1970–1971 from de Jonge and Postma,[9] and for 1985–1986 from the EON-Project Group,[36] Veldhuis et al.[35] and Rijkswaterstaat.

vertical diffusion rates. Enhanced concentrations during periods of high mineralization rates were mainly observed in the layers between 10 and 25 cm depth. Nitrate rapidly decreases with depth due to nitrate reduction.[34,35] The absence of high concentrations in the upper layers indicates that nutrients mineralized in the sediment are quickly released to the water column or taken up by the phytobenthos. For phosphate this picture may be modified because of physical-chemical processes that favor sorption of phosphorus compounds to mainly iron(III) oxides.[32,33] Due to this, part of the phosphate produced by mineralization can be retained temporarily in the sediment. However, most of the sedimentary phosphorus in the Wadden Sea seems to be associated with Ca[33] and consequently is not directly available for primary producers. The limited amount of available data on this subject[8,32,33] indicate that a substantial internal loading of phosphorus from the sediment after reduction of external inputs is not to be expected in the Dutch Wadden Sea.

## D.  EXCHANGE WITH THE OPEN SEA

A main characteristic of estuaries is that they receive materials from both the sea and from the freshwater (river) side. These exchanges with the open sea are, however, difficult to quantify. For the Marsdiep basin, these exchanges with the North Sea have been a topic for decades. Recently, an analysis of the phosphorus budget of the Marsdiep basin in the period 1950 to 1985 was performed.[31] It was concluded that almost half of the phosphorus input of the Marsdiep basin originates from the adjacent Vlie basin, which in turn is fed by inputs from Lake IJssel and the North Sea coastal zone. Also, a clear effect in the exchange processes during the last decades was observed. Before 1970, loadings from Lake IJssel were low and exchange with the North Sea dominated the phosphorus budget of the Marsdiep basin. From approximately 1975 onwards, freshwater sources became more important until a maximum was reached in 1981.

Interestingly, the changes in external inputs seem to be correlated with changes in annual primary production (see below). Nowadays, dispersive export of particulate phosphorus seems of the same order of magnitude as the tide-induced residual inputs.

## E. RESPONSE TO CHANGING EXTERNAL NUTRIENT INPUTS

The response of the individual basins of the Wadden Sea to changing nutrient inputs from the River Rhine, either via the North Sea or via Lake IJssel, can be calculated. Increased concentrations in the North Sea directly result in an equally increased basin concentration in the Wadden Sea. The response to changed loadings by the freshwater sources are modified by the mixing time scales and the volumes of the basins. This simple picture is complicated by the effect of changed nutrient levels due to the other processes that also contribute to the actual concentrations. Data analysis[31] suggests that the residual phosphorus input has not been constant over the past decades but that it is linearly related to the particulate phosphorus concentration in the inlets. For the Marsdiep basin it was concluded that an increase of particulate phosphorus at the inlet of 1 mmol m$^{-3}$ increases total phosphorus content by almost 0.6 mmol m$^{-3}$ in the western Wadden Sea due to the residual transport. Responses of sedimentation fluxes are not known, but it can be hypothesized that nutrient contents of the sediments have increased due to the eutrophication processes of the last 40 years and that increased residual transports are at least partly counterbalanced by enhanced sedimentation.

The effects of the changed external inputs on the concentrations of dissolved inorganic nutrients cannot easily be analyzed because internal biological and physical-chemical transformations often dominate. For the Marsdiep and Vlie basin the effect of reduced or enhanced loadings from Lake IJssel have been studied with a complex ecosystem model (EMOWAD = Ecosystem Model Wadden Sea).[36,37] According to this model, effects on the concentrations of dissolved phosphate occur especially during the first and last months of the year, while in summer, phosphate remains low due to uptake by primary producers. For this period most pronounced effects were calculated for total pelagic primary production and chlorophyll contents.

# VI. SEDIMENT PROPERTIES

## A. COMPOSITION OF BOTTOM SEDIMENTS

Approximately 75% of the entire Dutch, German and Danish Wadden Sea consists of sand flats, 7% of mud flats and 18% of mixed sediments.[38]

Compared with the entire Wadden Sea, the Dutch part is a very sandy area. Approximately 93% of the intertidal zone in the western Dutch Wadden Sea consists of sand flats, approximately 1.5% is mud flats, and approximately 5% is mixed sediments. In the eastern Dutch Wadden Sea the values are similar to the overall mean values presented above (intertidal sand flats 82%, 5% mud flats, and 13% is mixed sediments).[38] In general, the coarse sand is found near the tidal inlets and at the weather side (usually the western part) of the tidal water sheds. The muddy sediments are found on sheltered localities south of the islands and in land reclamation works near the mainland.

## B. ACCUMULATION AND TRANSPORT OF SUSPENDED MATTER

The concentrations of suspended matter for each tidal basin show a longitudinal gradient which is typical for most coastal plain estuaries. Low concentrations occur near the tidal inlet. The highest concentrations are found near the mainland and at the tidal water sheds. The concentration gradient in the main channels is partly maintained by an accumulation process that is driven by the tide. Part of this material originates from the channel bed, another part originates from the intertidal flats (see below). The role of the tide-induced residual sediment transport was first recognized by Postma.[39] Later, the process was analyzed experimentally as well as theoretically.[40-42] A full description of the process consists of a complex of factors, each of them favoring the net upstream transport and accumulation or deposition of suspended sediment.

## C. RESUSPENSION OF SEDIMENTS AND MICROPHYTOBENTHOS

Resuspension processes are significant because they cause rapid changes in light conditions in the water column, supply food for the water column and facilitate the turnover of nutrients associated with the fine sediment fraction and microphytobenthos.[9,43,44]

Resuspension of bottom material is caused by tidal currents and wind waves. In the main channels the resuspension of sediments from the channel bed is mainly governed by the currents[16] while on the shallow tidal flats the resuspension is mainly caused by wind waves.[43] The transport of the resuspended material from the tidal flats is caused by the tidal currents.[43]

Direct quantitative information on the relation between wind speed and suspended mud concentrations is only available for the Ems estuary.[43] Insights for the other basins of the Wadden Sea can only be derived from these data. The actual functions will certainly differ because the percentage of intertidal flats in, for example, the Marsdiep tidal basin is extremely low (15%). Consequently, the resuspension of mud from the tidal flats probably is less important than in the Ems estuary, an area with approximately 50% tidal flats. Taking into account the low percentage of tidal flats in the Marsdiep basin (Table 1), the larger wind speed and the sandy sediments, the relative importance of the transports of mud and microphytobenthos between tidal flats and channels may be between 25% and 30% of the figures given for the Ems estuary. In the eastern basins, lateral transports of mud are more pronounced, mainly due to the relative increase in tidal flats and an increase in mud content. No firm data on the lateral transport of microphytobenthos are available for the Marsdiep basin.[45,46]

## VII. SYMPTOMS OF NUTRIENT ENRICHMENT

### A. PRIMARY PRODUCERS

#### 1. Primary Production

The primary production of the phytoplankton in the Marsdiep tidal basin varied strongly during the period 1950 to 1990 (Figure 10). The first primary production estimate dates back to the early 1950s[8] and was 20–40 g C $m^{-2}$ $a^{-1}$. In the mid-1960s the primary production was ca. 150 g C $m^{-2}$ $a^{-1}$.[48] From the mid-1960s to the mid-1980s the primary production at least doubled.[47] The highest annual primary production, measured in 1981/82 was 520 g C $m^{-2}$ $a^{-1}$.[47,49] A comparable trend is visible in the figures of the primary production of the microphytobenthos on the tidal flats along the Marsdiep tidal channel and the phytoplankton in the inner area of the Marsdiep tidal basin.

The rising trend has generally been ascribed to eutrophication. The annual primary production of both phytoplankton and microphytobenthos in and near the Marsdiep tidal channel are significantly correlated with the mean annual loads of orthophosphate from Lake IJssel[12,51] (Figure 11). This is a strong indication that eutrophication is indeed responsible for the year-to-year changes in primary production. Moreover, Figure 11 indicates that in particular the phosphate loads from Lake IJssel fuel the growth of microalgae in the Marsdiep basin, which agrees with the concentration distribution of phosphate in the western Dutch Wadden Sea and recently developed phosphorus budgets of the area.[11,31]

#### 2. Biomass

The biomass of the phytoplankton and microphytobenthos has also increased significantly ( $p = 0.01$ and 0.05, respectively)[12] in the Marsdiep tidal channel. This increase was not established for the inner area.[12] In the eastern Dutch Wadden Sea (Friesche zeegat) there is no statistically significant trend in phytoplankton chlorophyll $a$ in the tidal inlet or in the inner area, which is expected with the low direct input of nutrients there relative to the residual inputs.[12]

One of the phytoplankton species that occurs in large densities is *Phaeocystis pouchetii*. It has been established that the blooms of this nuisance species have increased tremendously during the last decades, although some data are available showing extended blooming periods at the end of last century (Figure 12).[52] Recent theoretical studies indicate that the occurrence of *Phaeocystis pouchetii* may be at least partly determined by N/P ratios in available nutrients, and thus by external and internal inputs of both nitrogen and phosphorus.

During the last 2 decades, the biomass (and consequently the growth) of macroalgae from the genus *Ulva* appears to have increased, but data to support this suggestion are not available. Before the closure of the former Zuiderzee, a large surface of the Marsdiep tidal basin (northern part of the former Zuiderzee) was also covered with eelgrass beds (*Zostera marina*). Today only some spots are left in the intertidal zone, but so far, the disappearance of eelgrass from the western Dutch Wadden Sea cannot be attributed to the eutrophication of the area.[53]

In the western Dutch Wadden Sea extensive mussel culturing has occurred since 1950. This activity may largely have affected the development of other macrofauna species during last decades.[13] Between 1950 and 1960 the mussel culturing may have increased food competition in the area, which may have

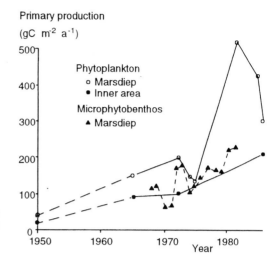

**Figure 10** Time series of annual primary production for phytoplankton and microphytobenthos at two localities in the Marsdiep tidal basin.[8,35,47-50]

decreased the stock of the intertidal macrofauna significantly (Figure 13). Since 1970, the eutrophication of the area could have improved food conditions, allowing both higher maximum yields of the mussel culture and higher biomasses of other macrofauna species (Figure 13).

## B. SECONDARY PRODUCERS

Following the response of the primary producers, the biomass of the secondary producers has also doubled in the western Dutch Wadden Sea since 1970 (Figure 13). This increase is statistically significant ($p = 0.0001$).[12,54,55] The data set of Beukema shows no decline, despite the decline in the loads of nutrients from the River Rhine and Lake IJssel. Unpublished data of Beukema show that this decline cannot yet be detected in 1992.

For the eastern part of the Dutch Wadden Sea no indication was found for such an increase in standing stock of macrozoobenthos.[12] It confirms the conclusion by Beukema and Cadée that eutrophication is the main factor responsible for the increased macrozoobenthos production in the western Dutch Wadden Sea,[54] although also other factors (climate) may also have played a role.

## C. FISH

The Wadden Sea is an important nursery area for several North Sea fishes among them the herring (*Clupea harengus*), sole (*Solea solea*) and plaice (*Pleuronectes platessa*). Since 1965 the juvenile fish have been dominated by plaice.[56] This is thought to be partly due to eutrophication and partly due to changes in the fish fauna.[3] Also, the length of juvenile plaice during the first winter increased by 1 cm over the period 1955 to 1973.[57] An increase in growth for sole has also been observed.[58,59] This may

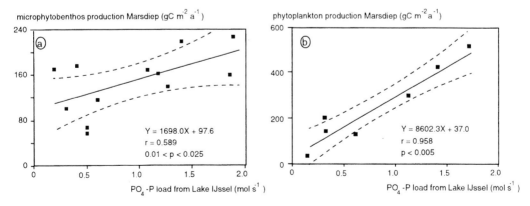

**Figure 11** Mean annual primary production of phytoplankton and microphytobenthos as a function of the mean annual orthophosphate loads from Lake IJssel. (Modified from de Jonge, V. N. and Essink, K., *Estuaries and Coasts: Spatial and Temporal Intercomparisons,* Elliot, M. and Ducrotoy, J.-P., Eds., Olsen & Olsen, Fredensborg, Denmark, 1991, 307; and de Jonge, V. N., *Hydrobiologia,* 195, 49, 1990. With permission.)

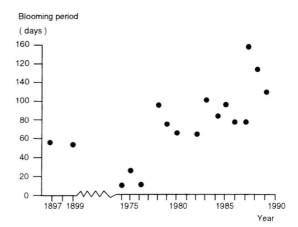

**Figure 12** Blooming period of the species *Phaeocystis pouchetii* (Haptophyceae) in the Marsdiep tidal channel. (From Cadée, G. C. and Hegeman, J., *Hydrobiol. Bull.*, 24, 111, 1991. With permission.)

indicate better feeding conditions due to eutrophication. However, firm evidence for the positive role of eutrophication on the growth of these species is lacking.

## VIII. CONSEQUENCES OF NUTRIENT ENRICHMENT

The productivity of the western Dutch Wadden Sea has increased. More primary production has increased food availability. This has improved feeding conditions for birds and man (mussel culturing, cockle fisheries, improved growth of juvenile flatfish, etc.) and can be considered a positive effect of eutrophication. This positive effect seems to be restricted because the productivity of the eastern basins of the Dutch Wadden Sea has remained more or less the same over time (Figure 13).

Increased production of organic matter has also had negative effects. By nature the Wadden Sea is an area where relatively low oxygen saturation values may occur. In particular fine particles and associated organic matter accumulate close to the tidal water sheds, and the mineralization of the organic material reduces oxygen levels.[61] During certain weather conditions and high densities of ciliates, oxygen may be depleted locally, leading to mortality of benthic organisms. Eutrophication may also enlarge the surface area where low oxygen values occur. To date, however, serious problems have not occurred.

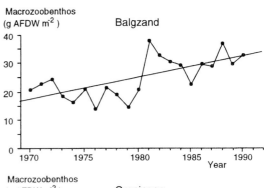

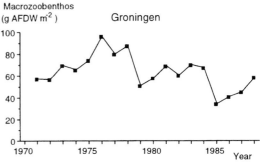

**Figure 13** Time series of the biomass of the macrozoobenthos in the western (Balgzand) and in the eastern (Groningen) Dutch Wadden Sea. At Groningen no data on *Arenicola marina* and large *Mya arenaria* were available. (Top figure from Beukema, J. J., *Mar. Biol.*, 111, 293, 1991. Bottom figure from de Jonge, V. N. and Essink, K., *Estuaries and Coasts: Spatial and Temporal Intercomparisons,* Elliot, M. and Ducrotoy, J.-P., Eds., Olsen & Olsen, Fredensborg, Denmark, 1991, 307. With permission.)

On the basis of available data, the species composition in the Wadden Sea has not become impoverished due to the high nutrient loads. However, changes have been observed in the succession of different groups of algae, which has resulted in local problems. The density of the populations of *Phaeocystis pouchetii* has temporarily led to foam on the beaches and coasts and offensive odors. Arguments have been advanced both for and against eutrophication as the cause of the change in succession of algal groups,[62,63] but concrete evidence is lacking.

The eutrophication of Dutch coastal waters has led to loading of the bottom sediments with phosphorus.[35] The amount buried is low and roughly 3% of the total annual load. Most of the buried phosphorus is calcium-associated and therefore not available for organisms in the short term. The biologically available amount is less than 20% of the total amount buried in the top 12 cm.[35] These values indicate that a reduction in nutrient loads may result in a quick response of the productivity of the system and not to a substantial internal loading.

## IX. MANAGEMENT OPTIONS FOR RESTORATION

Improvement of the water quality in the Wadden Sea depends on reduction of the nutrient loads from the River Rhine and the many smaller fresh water discharge points draining into the North Sea or directly into the Wadden Sea. Reduction of the nutrient loads from the North Sea countries by 50% before 1995 is implemented by the International Conferences on the Protection of the North Sea held in 1984, 1987 and 1990.

In the longer term, a fundamental reorientation is required in the way policy makers and planners approach the development of natural systems. Central to this new approach are

- The "wise use principle" — in which sustainable development of the ecosystem is the main goal.
- Treatment of the estuary as one management unit.
- Maintenance of natural biological and physical linkages among the system and surrounding land, including the catchment area.
- Subjection of the development of activities to a comprehensive Environmental Impact Assessment (EIA) procedure.
- Application of the Precautionary Principle.

The effects of reducing nutrient discharges can be studied in several ways. One of the possible techniques is the application of computer simulation models. A serious problem in all available ecosystem models, however, is their treatment of the transport of mud, sand and particulate organic matter, and the nutrients therein. Incorporation of correct particle transport subroutines offers the possibility for answering questions about effects of increased as well as decreased nutrient loads more accurately. More research is needed in this area.

## REFERENCES

1. **Dijkema, K. S.,** Large scale geomorphological pattern of the Wadden Sea area, in *Geomorphology of the Wadden Sea Area, Report 1 of the Wadden Sea Working Group,* Dijkema, K. S., Reineck, H.-E., and Wolff, W. J., Eds., Stichting Veth tot Steun aan Waddenonderzoek, Leiden, 1980, 72.
2. **Veenstra, H. J.,** Introduction to the geomorphology of the Wadden Sea area, in *Geomorphology of the Wadden Sea Area, Report 1 of the Wadden Sea Working Group,* Dijkema, K. S., Reineck, H.-E. and Wolff, W. J., Eds., Stichting Veth tot Steun aan Waddenonderzoek, Leiden, 1980, 8.
3. **de Jonge, V. N., Essink, K., and Boddeke R.,** The Dutch Wadden Sea: a changed ecosystem, *Hydrobiologia,* 265, 45, 1993.
4. **Thijsse, J. Th.,** Een halve eeuw Zuiderzeewerken 1920–1970, in *Tjeenk Willink,* Groningen, 1972, 469.
5. **van der Hoeven, P. C. T.,** Observations of surface water temperature and salinity, *State Office of Fishery Research (RIVO): 1860–1981, Scientific Report W.R. 82-2,* De Bilt, 1982, 118.
6. **Glim, G. W., Kool, G., Lieshout, M. F., and de Boer, M.,** Erosie en sedimentatie in de binnendelta van het zeegat van Texel 1932–1982, in *Rijkswaterstaat Directie Noord-Holland,* Rep. ANWX 87.H201, 1987, 33.
7. **Wolff, W. J. and Binsbergen, M.,** Het beheer van de Wadden; de visie van de Werkgroep Waddengebied, Stichting Veth tot Steun aan Waddenonderzoek, Arnhem, 1985.
8. **Postma, H.,** Hydrography of the Dutch Wadden Sea, *Arch. Néerl. Zool.,* 10, 405, 1954.

9. **de Jonge, V. N. and Postma, H.,** Phosphorus compounds in the Dutch Wadden Sea, *Neth. J. Sea Res.*, 8, 139, 1974.

10. **Cadée, G. C.,** Has input of organic matter into the western part of the Wadden Sea increased during the last decades?, Laane, R. W. P. M. and Wolff, W. J., Eds., The role of organic matter in the Wadden Sea, *Neth. Inst. Sea Res. Publ. Ser.*, 10, 71, 1984.

11. **van der Veer, H. W., Van Raaphorst, W., and Bergman, M. J. N.,** Eutrophication of the Dutch Wadden Sea: external nutrient loading of the Marsdiep and Vliestroom basin, *Helgo. Meeresunters.*, 43, 501, 1989.

12. **de Jonge, V. N. and Essink, K.,** Long-term changes in nutrient loads and primary and secondary producers in the Dutch Wadden Sea, in *Estuaries and Coasts: Spatial and Temporal Intercomparisons,* Elliott, M. and Ducrotoy, J.-P., Eds., Olsen & Olsen, Fredensborg, Denmark, 1991, 307.

13. **van der Veer, H. W.,** Eutrophication and mussel culture in the western Dutch Wadden Sea: impact on the benthic ecosystem; a hypothesis. *Helgo. Meeresunters.*, 43, 517, 1989.

14. **de Jonge, V. N.,** The Ems Estuary, The Netherlands, in *Eutrophic Shallow Estuaries and Lagoons,* McComb, A. J., Ed., CRC Press, Boca Raton, FL, chap. 7, 1995.

15. *Klimaatatlas van Nederland,* KNMI, Staatsuitgeverij, The Hague, 1972.

16. **Postma, H.,** Hydrography of the Wadden Sea: movements and properties of water and particulate matter, Wadden Sea Working Group, Report 2, Leiden, The Netherlands, 1982.

17. **Führböter, A.,** Changes of the tidal water levels at the German North Sea coast, *Helgo. Meeresunters.*, 43, 325, 1989.

18. **Zimmerman, J. T. F.,** Mixing and flushing of tidal embayments in the western Dutch Wadden Sea, Part II. Analysis of mixing processes, *Neth. J. Sea Res.*, 10, 397, 1976.

19. **Ridderinkhof, H., Zimmerman, J. T. F., and Philippart, M. E.,** Tidal exchange between the North Sea and Dutch Wadden Sea and mixing time scales of the tidal basins, *Neth. J. Sea Res.*, 25, 331, 1990.

20. **Helder, W.,** The cycle of dissolved inorganic nitrogen compounds in the Dutch Wadden Sea, *Neth. J. Sea Res.*, 8, 154, 1974.

21. **Prandle, D.,** Monthly mean residual flows through the Dover Strait 1949–1980, *Mar. Biol. Assoc.*, 64, 722, 1984.

22. **Laane, R. W. P. M., Groeneveld, G., de Vries, A., van Bennekom, A. J., and Sydow, J. S.,** Nutrients (P, N, Si) in *The Channel and the Strait of Dover: Seasonal and Year to Year Variation and Fluxes to the North Sea,* Rijkswaterstaat, 1990, 18.

23. **Okubo, A.,** The effect of shear in an oscillatory current on horizontal diffusion from an instantaneous source, *Int. J. Oceanol. Limnol.*, 1, 194, 1967.

24. **Ridderinkhof, H. and Zimmerman, J. T. F.,** Mixing processes in a numerical model of the western Dutch Wadden sea, in *Residual Currents and Long-term Transport,* Cheng, R. T., Ed., *Lecture Notes on Coastal and Estuarine Studies,* Springer-Verlag, Berlin, 38, 1990, 194.

25. **Ridderinkhof, H.,** Residual Currents and Mixing in the Wadden Sea, Ph.D. thesis, State University of Utrecht, Utrecht, 1990.

26. **de Jonge, V. N.,** Tidal flow and residual flow in the Ems estuary, *Estuarine Coastal Shelf Sci.*, 34, 1, 1992.

27. **van Bennekom, A. J., Krijgsman-van Hartingsveld, E., van der Veer, G. C. M., and van Voorst, H. F. J.,** The seasonal cycles of reactive silicate and suspended diatoms in the Dutch Wadden Sea, *Neth. J. Sea Res.*, 8, 174, 1974.

28. **Zimmerman, J. T. F.,** Mixing and flushing of tidal embayments in the western Dutch Wadden Sea, Part I. Distribution of salinity and calculation of mixing time scales, *Neth. J. Sea Res.*, 10, 149, 1976.

29. **Postma, H.,** The cycle of nitrogen in the Wadden Sea and adjacent areas, *Neth. J. Sea Res.*, 3, 186, 1966.

30. **Duursma, E. K.,** Dissolved organic carbon, nitrogen and phosphorus in the sea, *Neth. J. Sea Res.*, 1, 1, 1960.

31. **van Raaphorst, W. and van der Veer, H. W.,** The phosphorus budget of the Marsdiep tidal basin (Dutch Wadden Sea) in the period 1959–1985: Importance of the exchange with the North Sea, *Hydrobiologia,* 195, 21, 1990.

32. **van Raaphorst, W., Ruardij, P. and Brinkman, A. G.,** The assessment of benthic phosphorus regeneration in an estuarine ecosystem model, *Neth. J. Sea Res.*, 22, 23, 1988.

33. **de Jonge, V. N., Engelkes, M. M., and Bakker, J. F.,** Bio-availability of phosphorus in sediments of the western Dutch Wadden Sea, *Hydrobiologia,* 253, 151, 1993.

34. **Rutgers van der Loeff, M. M.,** Time variation in interstitial nutrient concentrations at an exposed subtidal station in the Dutch Wadden Sea, *Neth. J. Sea Res.*, 14, 123, 1980.

35. **Veldhuis, M. J. W., Colijn, F., and Venekamp, L. A. H.,** The spring bloom of *Phaeocystis pouchetii* (Haptophyceae) in Dutch coastal waters, *Neth. J. Sea Res.*, 20, 37, 1986.

36. **EON-Project Group,** Ecosysteemmodel van de westelijke Wadden Zee, Netherlands Institute for Sea Research, *NIOZ Report* 1988–1, Texel, The Netherlands, 1988.

37. **Lindeboom, H. J., van Raaphorst, W., Ridderinkhof, H., and van der Veer, H. W.,** Ecosystem model of the western Wadden sea: a bridge between science and management?, *Helgo. Meeresunters.*, 43, 549, 1989.

38. **Philippart, C. J. M., Dijkema, K. S. and van der Meer, J.,** Wadden Sea seagrasses: where and why?, *Neth. Inst. Sea Res. Publ. Ser.*, 20, 177, 1992.

39. **Postma, H.,** Sediment transport and sedimentation in the estuarine environment, in Estuaries, Lauff, G. H., Ed., *American Association for the Advancement of Science,* 83, 158, 1967.

40. **Postma, H.,** Transport and accumulation of suspended matter in the Dutch Wadden Sea, *Neth. J. Sea Res.*, 1, 148, 1961.

41. **Straaten, L. M. J. U. and van, Kuenen, P. H.,** Tidal action as a cause for clay accumulation, *J. Sediment. Petrol.*, 28, 406, 1958.
42. **Groen, P.,** On the residual transport of suspended matter, by an alternating tidal current. *Neth. J. Sea Res.*, 3/4, 564, 1967.
43. **de Jonge, V. N.,** Wind Driven Tidal and Annual Gross Transports of Mud and Microphytobenthos in the Ems Estuary and its Importance for the Ecosystem, Ph.D. thesis, State University of Groningen, Groningen, 1992.
44. **de Jonge, V. N. and Villerius, L. A.,** Possible role of estuarine processes on phosphate-mineral interactions, *Limnol. Oceanogr.*, 34, 330, 1989.
45. **Kamermans, P.,** Similarity in food source and timing of feeding in deposit and suspension-feeding bivalves, *Mar. Ecol. Prog. Ser.*, 104, 63, 1994.
46. **Hummel, H.,** Food intake of *Macoma balthica* (Mollusca) in relation to seasonal changes in its potential food on a tidal flat in the Dutch Wadden Sea. *Neth. J. Sea Res.*, 19, 52, 1985.
47. **Cadée, G. C.,** Increased phytoplankton primary production in the Marsdiep area (western Dutch Wadden Sea), *Neth. J. Sea Res.*, 20, 285, 1986.
48. **Postma, H. and Rommets, J. W.,** Primary production in the Wadden Sea, *Neth. J. Sea Res.*, 4, 470, 1970.
49. **Cadée, G. C. and Hegeman, J.,** Primary production of phytoplankton in the Dutch Wadden Sea, *Neth. J. Sea Res.*, 8, 240, 1974.
50. **Cadée, G. C. and Hegeman J.,** Phytoplankton primary production, chlorophyll and composition in an inlet of the western Wadden Sea (Marsdiep), *Neth. J. Sea Res.*, 13, 224, 1979.
51. **de Jonge, V. N.,** Response of the Dutch Wadden Sea ecosystem to phosphorus discharges from the River Rhine, *Hydrobiologia,* 195, 49, 1990.
52. **Cadée, G. C. and Hegeman, J.,** Historical phytoplankton data of the Marsdiep, *Hydrobiol. Bull.*, 24, 111, 1991.
53. **de Jonge, V. N. and de Jong, D. J.,** Role of tide, light and fisheries in the decline of *Zostera marina* L. in the Dutch Wadden Sea, *Neth. Inst. Sea Res. Publ. Ser.*, 20, 161, 1992.
54. **Beukema, J. J. and Cadée, G. C.,** Zoobenthos responses to eutrophication of the Dutch Wadden Sea, *Ophelia*, 26, 55, 1986.
55. **Beukema, J. J.,** Long-term changes in macrozoobenthic abundance on the tidal flats of the western Wadden Sea, *Neth. J., Sea Res.*, 43, 405, 1989.
56. **Boddeke, R.,** Visserij-biologische veranderingen in de westelijke Waddenzee, *Visserij*, 20, 213, 1967.
57. **Zijlstra, J. J.,** On the importance of the Wadden Sea as a nursery area in relation to the conservation of the southern North Sea fishery resources, *Symp. Zool. Soc. Lond.*, 29, 233, 1972.
58. **de Veen, J. F.,** On changes in some biological parameters in the North Sea sole (*Solea solea* L.), *J. Cons. Int. Explor. Mer*, 37, 60, 1976.
59. **van Beek, F. A.,** On the growth of sole in the North Sea, *I.C.E.S.,* Copenhagen, Report, *C.M./G,* 1988, 24.
60. **Beukema, J. J.,** Changes in composition of bottom fauna of a tidal-flat area during a period of eutrophication, *Mar. Biol.*, 111, 293, 1991.
61. **de Groot, S. J. and Postma, H.,** The oxygen content of the Wadden Sea, *Neth. J. Sea Res.*, 4, 1, 1968.
62. **Riegman, R., Noordeloos, A. A. M., and Cadée, G. C.,** Phaeocystis blooms and eutrophication of continental coastal zones of the North Sea, *Mar. Biol.*, 112, 144, 1992.
63. **Smayda, T. J.,** Novel and nuisance phytoplankton blooms in the sea: evidence for a global epidemic, in *Toxic Marine Phytoplankton,* Graneli, E., Sundström, B., Edler, L.. and Anderson, D. M., Eds., Elsevier, New York, 1990, 29.

Chapter 10

# Water Exchange Between Shallow Estuaries and the Ocean

*Clifford J. Hearn*

## CONTENTS

I. Introduction ................................................................................................................. 151
II. The Channels ............................................................................................................. 153
III. Barotropic Motion ..................................................................................................... 153
IV. Baroclinic Motion ...................................................................................................... 153
V. Two-Layer Flow through a Channel ......................................................................... 155
VI. Relation between Baroclinic and Barotropic Flow ................................................... 159
VII. Baroclinic Ocean Exchange ..................................................................................... 163
VIII. Barotropic Ocean Flushing ...................................................................................... 163
IX. Observations of the Retention Coefficient .............................................................. 164
X. Low Frequency Tides ................................................................................................ 166
XI. Conclusion and Discussion ...................................................................................... 167
Acknowledgments .............................................................................................................. 168
Appendices ......................................................................................................................... 168
References .......................................................................................................................... 171

## I. INTRODUCTION

This chapter treats the ocean exchange for estuaries consisting of very shallow basins joined to the ocean through narrow channels. They are a type of bar-built, or lagoonal estuary[1-3] with the important characteristic features of being both very shallow (of depth 1 or 2 meters or less) and having a long and narrow channel to the ocean. Such systems are found worldwide and some good examples are shown in Figure 1 from southwestern Australia.

The isolation of these estuaries from the ocean is due to bar building by littoral drift which has usually followed a fall in sea level over the last few thousand years; this has often been accentuated by a drop in riverflow. In some cases, neighboring estuaries have become joined as a result of the ocean outlet of one of them being totally blocked by littoral drift; floods then cut an interconnecting channel between the estuaries. For example, the Peel-Harvey Estuary[4] (Figure 1 and in more detail in Figure 10) is formed by the joining of Harvey Estuary (which originally flowed to the ocean) and Peel Inlet. This is a common cause of the multibasin structure of many of these estuaries, but in other cases the inner channels and basins may have developed from deposition of sediment in a single estuary.

The flushing of an estuary is concerned with the residence time of a parcel of water within one of the basins. There are two distinct processes by which water is expelled from the estuary into the ocean. The first is simply a consequence of river flow, and in the very shallow basins that are being discussed river water mixes efficiently with basin water, and the resultant flushing time is simply determined by the time for riverflow to fill the estuarine basins. This time is often very seasonal and many estuaries (such as those in southwestern Australia) experience a wet/dry annual cycle with long periods without riverflow. Although riverflow induces a true flushing of the estuary it may also introduce a contaminant (such as phosphorus) which contributes to eutrophication. The second process by which water leaves the estuary is exchange with ocean water through the channel. This process is often rather weak, and the estuaries, which have become fresh under the influence of riverflow, take many months of drought to reach marine salinity. The process of ocean exchange has two components. The first of these is barotropic and is mainly a consequence of tidal motion. The other mechanism for ocean exchange is driven by the density difference between the fresher estuarine and more saline ocean water. This takes the form of a baroclinic two-layer flow through the channel with the denser ocean water flowing inward beneath the outward

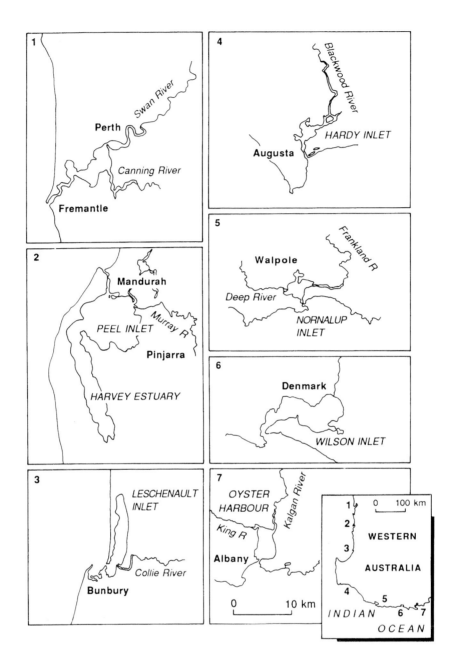

**Figure 1** The estuaries of southwestern Australia: 1, Swan; 2, Peel-Harvey; 3, Leschenault; 4, Hardy; 5, Nornalup; 6, Wilson; 7, Oyster Harbour.

flowing basin water. Such baroclinic exchange is strongest at the end of the wet season when density differences are greatest and declines in importance as marine salinity is approached in the estuary. However, baroclinic exchange tends to be blocked by the barotropic tidal currents and this process limits the ocean exchange in these estuaries.

This chapter will be concerned with these two processes of barotropic and baroclinic exchange through the channel. The physics of water movements in the basins themselves involves wind forcing, wind mixing, and convection;[5] it has been reviewed recently[6] and will not be discussed here. Understanding of the total behavior of an estuary requires a complete estuarine model which includes ocean exchange, basin dynamics, and riverflows; ecological models[7] also incorporate chemical, biological, and sediment exchange.

## II. THE CHANNELS

The channels are usually a few meters deep and a hundred or more meters wide. Typical tidal currents through the narrow channels are between 0.1 and 0.3 m s$^{-1}$ with maximal channel speeds during spring tides; speeds as high as 1 m s$^{-1}$ are encountered in artificial channels, or parts of natural channels which have been constricted and supporting walls built (such as under a road bridge). Speeds in the channels are often controlled by the threshold speed for the erosion of sand which is about 0.1 to 0.2 m s$^{-1}$; only the very finest silts and clay are stable for speeds up to 1 m s$^{-1}$. The tidal currents in the channels happen to be rather similar to the speed of internal waves propagating on the interface between ocean and brackish water in the channels and this will be shown to be a very important aspect of the ocean exchange problem. The length of the channels is very variable and it is this quantity that has greatest effect on the ocean exchange through its influence on the tidal range in the basins and also because of the importance of the channel volume.

Water is forced through a channel by pressure gradients due to differences in water height and density (associated with the variations of salinity and temperature between the various basins and the ocean). It is useful to separate the flow into its barotropic and baroclinic components, and these are discussed in the following sections.

## III. BAROTROPIC MOTION

The barotropic component provides the net flux of fluid and controls the changes in the volume of water in the basins. It is ultimately driven by either changes in sea level, river flow, or wind set-up. If, for example, the ocean level is raised a sequence of changes in water level will propagate through the estuary and produce barotropic flow through the channels; similarly, a head of water is created by river flow into one of the basins.

The barotropic flow through the channel-basin system is characterized by a set of channel exchange times, which are discussed in some detail by Hearn and Lukatelich,[6] and wave propagation along the basins. These channel exchange times are derived from the hydraulic properties of the channels, and the volumes of the basins, and represent the characteristic times for differences in water level between the basins to decay away as equilibrium is approached. For the shallow estuaries under discussion, the tidal slope of the water surface in the basins is controlled by the ratio of the typical length-scale of the basins to half the basin wavelength of the highest frequency (significant) astronomical tidal component. The speed of propagation of a long wave in 2–10 m of water is 5–10 m s$^{-1}$ so that for a semidiurnal tide the wavelength is about 200–400 km and twice as large for a diurnal tide. Most of the estuaries in southwest Australia and on many other coasts[8] have diurnal tides and lengths of a few tens of kilometers and so the dominant tidal ratios are below 0.1. Consequently, the water in the basins moves as a level surface except close to the channels where geometric effects are important. The basins of Lake Songkhla in Thailand[9] are larger but the ratio is still much less than unity while the large Patos Lagoon on the South Atlantic seaboard of Brazil is about 250 km long so that the tidal ratio is of order 1, and standing wave effects could be important.

The ability of periodic barotropic components to flush the estuary depends on mixing within the basins in order that the water mass that enters a basin on a flood tide does not simply leave on the following ebb. The ratio of the volume of the estuary to the spring tidal prism multiplied by the tidal period is referred to as the volumetric tidal exchange time. It is the (minimum) barotropic flushing time which would occur if there were total mixing between the tidal prism and basin water and the channels were of negligible volume.

## IV. BAROCLINIC MOTION

Density differences between the basins tends to form a two-layer structure in the channels. For purely baroclinic motion, equal volumes of the two layers are exchanged but generally the flow will contain a net barotropic component. This baroclinic exchange is forced by the density difference and is limited by friction, bottom topography (since ridges or holes can block the lower layer) and two-layer hydrau... control in the channel.

The frictional stress scales as $C_D v^2$ where $v$ is the water velocity, and $C_D$ is a drag coefficient (which can be derived from the Manning or Chezy formulae), while the inertial forces (which dominate the hydraulic control) scale as $hv^2/2L$ where L is the horizontal length scale of the channel (in the direction of flow) and h the depth. Therefore, the importance of inertial as compared to frictional forces can be quantified by a Reynolds number R,

$$R \equiv \frac{h}{2C_D L} \tag{1}$$

The channels of interest here are usually several meters deep so R is usually greater than unity, but friction could become important in some shallower, and longer, channels.

Irregularities in the floor of the channel can be an effective block to the spread of a very thin bottom layer. This is often the case for flow through channels leading to secondary basins of the estuary. A hole in the channel floor can act as a sink into which the very thin bottom layer, sometimes found in secondary channels, flows without further progress. The secondary channels can also show complicated strata involving two basin layers beneath which may be a layer of high salinity marine water (compare to Figure 17); the flow of this marine layer is very susceptible to holes and sills in the channel floor (Appendix 1).

Calculation of the baroclinic flow through a channel requires a knowledge of the hydrodynamic conditions at the ends of the channel which in turn necessitates a model of the complete estuary. However, for given density difference in the channel, there is a maximum (hydraulically controlled) baroclinic flow which can be achieved and it is assumed in most studies of estuarine flow through channels (a good general discussion is given by Officer[10]) that there is sufficient forcing between the basins for this maximum to be reached. This assumption decouples the channel and basin dynamics and makes the ocean exchange problem tractable. Hydraulic control limits the speed of the two layers to a value which is of similar magnitude to the speed of an internal wave propagating along the interface between the two layers. For convenience a speed c that governs the internal dynamics is defined here as,

$$c = \sqrt{g'd} \tag{2}$$

where d is the depth of the channel and $g'$ is the reduced gravity between the layers,

$$g' \equiv \frac{\Delta\rho}{\rho} g \tag{3}$$

in which $\Delta\rho$ is their density difference, $\rho$ the mean density and g the acceleration due to gravity; the actual internal wave speed depends on the thickness of the layers. The maximum salinity difference encountered in estuaries is about 35‰, giving a fractional density difference of $25 \times 10^{-3}$ and c approximately 1 m s$^{-1}$ for a representative channel of depth 5 m. For a typical channel of width 100 m, this produces a flux of about 125 m$^3$ s$^{-1}$ using c/4 as the maximum speed of baroclinic exchange (which will be justified later) so that for a characteristic basin volume of $100 \times 10^6$ m$^3$, the flux would take about 10 days to flush the basin.

An interesting aspect of purely baroclinic processes is that the flushing time increases as the density difference decreases since it is inversely proportional to the speed c in Equation 2, and therefore by Equation 3, varies inversely as the square root of the density difference. Assuming a linear relation between salinity and density, the rate of change of the salinity difference between the ocean and basin will have the form,

$$\frac{d\Delta S}{dt} = -\frac{\Delta S}{\tau(\Delta S)} \tag{4}$$

where $\tau(\Delta S) \propto 1/\sqrt{\Delta S}$ is the flushing time, so that

$$\Delta S = \frac{(\Delta S)_0}{\left[1 + (t/2\tau_0)\right]^2}$$

(5)

with the suffix indicating the initial salinity difference (at time $t = 0$) and $\tau_0$ being the initial flushing time. Figure 2 shows the predicted recovery to marine salinity by the ocean flushing (Equation 5) for a basin with an initial salinity of 5‰, and $\tau_0 = 10$ days (as derived above). Marine salinities are achieved after about 60 days and the baroclinic flushing becomes less effective with time; for comparison the dashed curve shows the salinity variation when the flushing time $\tau_0$ is held constant at its initial value. The internal speed c will decrease from the value quoted above as the density difference relaxes and a more typical value would be in the range 0.1 to 0.3 m s⁻¹. The barotropic speeds through the channels, which were discussed in Section III, are comparable to or greater than this internal speed so that the barotropic flow affects the baroclinic component. The next section discusses the theory of two-layer flow through a channel which is central to an understanding of the effect of barotropic flow on baroclinic flushing.

## V. TWO-LAYER FLOW THROUGH A CHANNEL

This problem was originally studied by Stommel and Farmer[11] and has been the subject of a number of theoretical and experimental studies.[12-19] The channel contains two layers flowing, without mixing, under steady state conditions between the two basins. The depth of the upper layer is denoted by $d_1$ and that of the lower layer by $d_2$ with h representing the height of the bottom of the channel above a datum which is chosen so that $h = 0$ at the minimum cross section of the channel; for a flat-bottom channel $h = 0$ everywhere. Dimensionless depths are used in which the (unperturbed) height of the fluid above the datum is normalized, that is

$$d_1 + d_2 + h = 1$$

(6)

Although a difference of surface height along the channel is required to drive the barotropic flow, this is small compared to the variation of the layer thicknesses and it is ignored here so that Equation 6, is always valid; this formally assumes that the barotropic speed is much less than that of long surface waves (approximately 10 m s⁻¹), i.e., the barotropic Froude number is small. Because there is no mixing of fluid, the volume flux in each layer is the same at all points of the channel so that the particle speed in each layer only varies as a result of changes in the cross section of the channel and the fractional depth of that

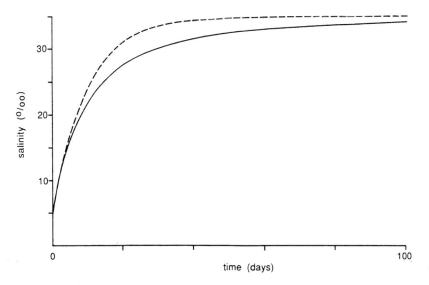

**Figure 2** The variation of salinity in a basin connected to the ocean by a narrow channel with a purely baroclinic hydraulically controlled exchange (solid line) compared with a constant flushing time (dashed). The basin is assumed to have been flooded with river water at time $t = 0$ to produce an initial salinity of 5‰.

layer. If the volume flux of each layer is expressed in terms of the minimum channel cross section this flux becomes a speed which will be denoted by $\tilde{U}_i$ where $i = 1, 2$; all speeds are expressed dimensionlessly by scaling to the speed c defined in Equation 2 and depths are also dimensionless and normalized in Equation 6. The flux per unit channel cross section $U_i(x)$ at any point along the length of the channel is then

$$U_i(x) = \frac{\tilde{U}_i}{b(x)[1 - h(x)]}$$

(7)

where x is the horizontal spatial coordinate along the channel, and b(x) is the channel width divided by its minimum value. The functions b(x), h(x), and hence, $U_i(x)$, are specified and the latter reaches a maximum value of $\tilde{U}_i$ at the minimum cross section. The particle speed within each layer is

$$u_i = \frac{U_i}{d_i}$$

(8)

and the problem is to find the particle speeds $u_i$ (or alternatively the layer depths $d_i$) as functions of x.

The baroclinic motion is controlled by Bernoulli's equation and is equivalent to the difference between the total energies of the layers being spatially constant; their sum determines the barotropic motion. The internal Froude numbers $F_i$ (which enter the analysis naturally) are defined in terms of dimensionless particle speeds and depths as

$$F_i^2 \equiv \frac{u_i^2}{d_i}$$

(9)

Armi[17] analyzed this problem by expressing the difference of total energies in terms of the Froude numbers $F_i$ using Equations 6 to 9, with the condition (Equation 7) serving as a relation between $F_i$ and x. The present analysis uses the variables,

$$z \equiv F_1^2, \; y \equiv \frac{d_2}{d_1}$$

(10)

so that the difference in total energy between the layers divided by $\rho c^2$, which will be denoted by $\gamma$,

$$\gamma = \frac{(u_2^2 - u_1^2)/2 + d_2}{d_1 + d_2}$$

(11)

can be expressed in terms of y and z using Equations 7 to 10; this gives the loci of constant energy difference $\gamma$ in the (y,z) plane. The following analysis treats a flat-bottomed channel (h = 0); the case of h varying to form a sill (with the width taken as constant) is presented in Appendix 1.

For a flat-bottomed channel the energy loci are given by

$$\gamma \neq \frac{\beta}{1 + \beta} \quad z = 2y^2 \frac{\gamma - (1 - \gamma)y}{\beta^2 - y^2}$$

(12)

where $\beta$ is the ratio of flux in the lower and upper layers (which is constant), that is

$$\beta = \frac{\tilde{U}_2}{\tilde{U}_1} = \frac{U_2}{U_1}$$

(13)

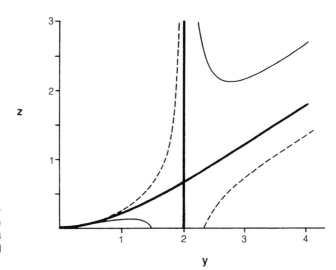

**Figure 3** Loci of constant energy difference $\gamma$ in the (y,z) plane for $\beta = 2$. The solid black line is the locus of $\gamma = 2/3$ (lock-exchange), the finer line $\gamma = 0.6$, and the dashed line $\gamma = 0.7$.

Loci are illustrated in Figure 3 for the case $\beta = 2$ and three different values of $\gamma$. For any value of $\beta$ there is a special value of $\gamma$ given by $\beta/(1 + \beta)$ in which the locus is not given by Equation 12, but for other values of $\gamma$ the loci diverge at $y = \beta$. At this special value of $\gamma$ the loci converge to a saddle,

$$\gamma = \frac{\beta}{1+\beta} \begin{cases} z = \dfrac{2y^2}{(1+\beta)(\beta+y)} & \text{for } y \neq \beta \\ \text{or } y = \beta \text{ for all } z \end{cases}$$

(14)

as shown in Figure 3 (where $\beta = 2$) for $\gamma = 2/3$.

For the flat-bottomed basin under consideration, the layer thicknesses are functions of y only; using Equations 6 and 10,

$$d_1 = \frac{1}{1+y} \, , \, d_2 = \frac{y}{1+y}$$

(15)

and so the particle speeds in Equation 8 become

$$u_1 = U_1(1+y) \qquad u_2 = U_2 \frac{1+y}{y}$$

(16)

which are equal for $y = \beta \, (\neq 0)$. The energy difference $\gamma$ is independent of the direction of the flows (since it merely depends on kinetic energy); however, a sign convention becomes necessary in order to determine the net particle speed $u_0$ through the channel; positive $\beta$ will denote a bottom layer flowing in the opposite direction to the upper layer so that,

$$u_0 = U_1 - U_2 = U_1 (1 - \beta)$$

(17)

The net particle speed is a function of x and its maximal value, at the narrowest point, is

$$u_0 = \tilde{u}_0 \equiv \tilde{U}_1(1-\beta)$$

(18)

The locus of $\gamma$ gives z as a function of y; $U_1$ can then be determined at each point on the $\gamma$ locus since Equations 8 to 10 and 15 give

$$z = U_1^2(1+y)^3 \qquad (19)$$

which allows x to be found from Equation 7 (since $\tilde{U}_1$ is specified). This depends on the shape of the channel, and a suitable form is

$$b(x) = 1 + |x|^{\frac{1}{n}} \qquad (20)$$

where x has its origin at the narrowest point and is scaled by the characteristic length over which the channel doubles in width. The exponent $\frac{1}{n}$ (with $0 < n < 1$) controls the shape of the channel, which may be considered as two half-channels (for positive and negative x) with different exponents; for $n \ll 1$ the channel broadens very suddenly near $x = 1$ while $n = 1$ gives a linear channel.

The special $\gamma$, which was discussed above, is referred to the lock-exchange value,

$$\gamma_{l.e.} \equiv \frac{\beta}{1+\beta} \qquad (21)$$

and Equation 14 is called the lock-exchange solution (compare the locus of $\gamma = 2/3$ in Figure 3). The loci can be considered as trajectories along which the system passes (with some constant $\gamma$) from one basin to the next; each $\gamma$ corresponds to a different solution of the channel problem. The special case of $\gamma = \gamma_{l.e.}$ forms a saddle-point on which the loci converge as $\gamma$ increases or decreases toward $\gamma_{l.e.}$.

Figure 4 shows the same loci of $\gamma$ as Figure 3 together with (in broken line) three members of the family of loci in the (y,z) plane of constant $U_1$ as given by Equation 19. Moving along the $\gamma$ loci involves cutting across the $U_1$ loci so that the value of $U_1$ changes as progress is made along the channel from one basin to the other. As the system moves from one basin towards the narrowest section of the channel, the value of $U_1$ must increase in inverse proportion to the channel cross-sectional area, as is shown by Equation 7, and then decrease toward the other basin. Figures 3 and 4 show that each locus divides into an upper and lower trajectory (on either side of the line $y = \beta$) and that $U_1$ passes through a maximum only on the lower trajectory; Figure 5 shows the lower trajectory of one of the loci (with $y > \beta$).

One particular $U_1$ locus just touches the $\gamma$ locus (at the narrowest point) as illustrated in Figures 5 and 6. This locus corresponds to $\tilde{U}_1$ and it is simple to show that the contact point lies on the curve

$$z = \frac{y^3}{\beta^2 + y^3} \Rightarrow F_1^2 + F_2^2 = 1 \qquad (22)$$

where the definition (Equation 9) has been used; $F_1^2 + F_2^2$ is referred to as the composite internal Froude number by Armi[17] and Equation 22 is called the control curve.

When a solution passes through the control curve the flow is said to be hydraulically controlled in that $U_1$ is maximal for that value of $\gamma$ and $\beta$. The existence of control is dependent on the conditions at the end of the channel. A simple choice of end conditions could be that both internal Froude numbers are small so that z is small and y is finite. In that case the system can behave as indicated in Figure 6 (with control point P), which shows one solution with control and another (ABA) which stops at B (on the low $U_1$ side of P) and returns to A through previous points on the locus. The solution with control has a portion PC that lies above the control curve and is said to be supercritical; it may therefore make an internal jump to a subcritical state as illustrated by the arrowed line in which y (and hence, the layer thicknesses) changes rapidly at constant z and $\gamma$.

The flow which is of most interest to the present estuaries corresponds to well-mixed basins so that in the basin with lower density y is close to zero ($d_2 = 0$) and in the denser basin $y \gg 1$ ($d_2 \gg d_1$). The only trajectory which passes continuously across $y = \beta$ to satisfy both these end conditions is the lock-exchange locus $\gamma = \gamma_{l.e.}$; the value of z, is small as y tends to zero (in the less dense basin) and large in the denser basin.

The vertical arm of the lock exchange locus is also important for end conditions in which z is small but y is close to $\beta$. The center of the saddle, which is an intersection with the control curve, is called a

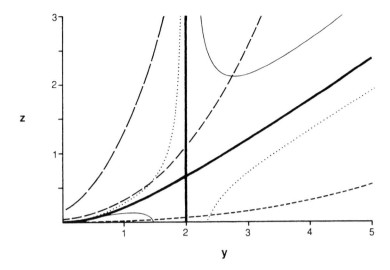

**Figure 4** The loci of constant energy difference repeated from Figure 3 together with three members of the family of loci of constant $U_1$ (broken lines) with $U_1 = 0.05$ (short dashes), 0.2 (longer dashes), 0.4 (longest dashes).

virtual control point and was first discovered by Wood;[12,13] along this vertical line the layer thicknesses are constant; for $\beta \neq 0$ Equations 15 and 16 show the particle velocities to be equal and increase as $\sqrt{z}$ (the vertical arm for $\beta = 0$ is considered in the next section).

## VI. RELATION BETWEEN BAROCLINIC AND BAROTROPIC FLOW

The two-layer flows contain a mean or barotropic component (which only vanishes when $\beta = 1$) that is denoted by $\tilde{u}_0$ in Equations 17 and 18. The lock-exchange trajectory (the solution of central interest here) intersects the control curve at $y = \beta$, and a second point $y_1$ which corresponds to the maximum of $U_1$, and is given by Equations 14 and 22 as,

$$y_1 = \frac{1-\beta}{4} + \sqrt{\beta + \left[\frac{1-\beta}{4}\right]^2}$$

(23)

$\tilde{U}_i$ can then be found from Equations 14 and 19, i.e.,

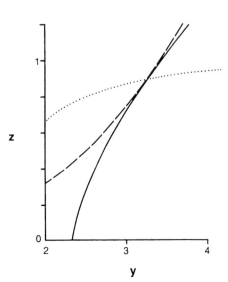

**Figure 5** Dashed line is locus of $U_1 = 0.11$ chosen to just touch the locus for $\gamma = 0.7$ taken from Figures 3 and 4 and the dotted line is the control curve.

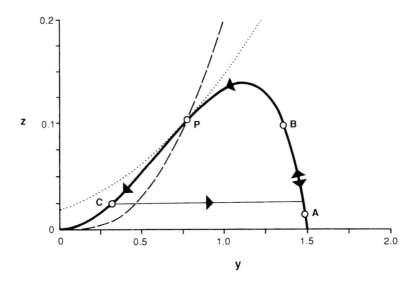

**Figure 6** The (y,z) plane for $\beta = 2$ showing the energy locus $\gamma = 0.8$ (heavy line), the control curve (dashed), the locus (dotted) of $\tilde{U}_1$ (0.135) and control point P. ABA is a solution for channel flow without control and ABPCA is a controlled solution.

$$\tilde{U}_1^2(1+y_1)^3 = \frac{2y_1^2}{(1+\beta)(\beta+y_1)} \tag{24}$$

from which $\tilde{u}_0$ as a function of $\beta$ is derived from Equation 17 and displayed in Figure 7. If $u_T$ is defined as[18]

$$u_T \equiv \left(\frac{2}{3}\right)^{\frac{3}{2}} \tag{25}$$

it can be shown that

$$\lim_{\beta \to 0} \tilde{u}_0 = u_T, \quad \lim_{\beta \to \infty} \tilde{u}_0 = -u_T \tag{26}$$

and these limits are evident in Figure 7. If the magnitude of $\tilde{u}_0$ does not exceed $u_T$ the flow can be represented by a lock-exchange locus with $\beta$ determined from Figure 7. The variation of the height of the bottom layer with distance x along the channel is shown in Figure 8 for various values of $\tilde{u}_0$ using a linear channel profile based on (20) (with n = 1). The layer depths for $\tilde{u}_0 = 0$ ($\beta = 1$) are shown in Figure 8a, while 8b shows the case $\tilde{u}_0 = u_T$ ($\beta = 0$); note the thinning of the lower layer by the barotropic motion in Figure 8b.

It is also possible to represent the flow on the $\beta = 0$ plane if $\tilde{u}_0$ exceeds $u_T$. The limit $\beta \to 0$ involves the lower layer becoming stationary; the lock-exchange locus has $\gamma = 0$ with the vertical arm degenerating into the z axis, by Equation 23 $y_1 = 1/2$, and by Equation 15, $d_2 = 1/3$ at the minimum cross section of the channel. The energy loci for $\beta = 0$ are shown in Figure 9 and given by Equations 12 and 14 as

$$\gamma \neq 0 \quad z = 2[(1-\gamma)y - \gamma]$$

$$\gamma = 0 \begin{cases} y = 0 \\ y \neq 0 \quad z = 2y \end{cases} \tag{27}$$

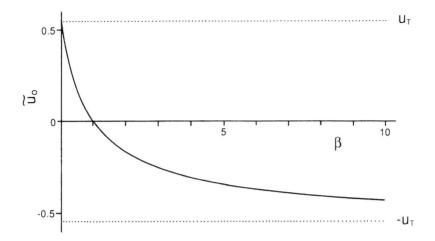

**Figure 7**  The variation of $\tilde{u}_0$, the barotropic current through the channel at the point of minimum cross section, with $\beta$ the ratio of volume flux in the lower to upper layer.

so that the $\beta = 0$ plane is a rather special case in that all the energy loci degenerate into the straight lines shown in Figure 9 which run continuously across the (y,z) plane. Only the lock-exchange locus $\gamma = 0$, which corresponds to $\tilde{u}_0 = u_T$, passes through the origin; the thickness of the lower layer for this case is shown in Figure 8b, with the flow moving from A to B in going from the less to more dense basin (left to right in Figure 8b).

Negative energy difference $\gamma$ gives pseudo lock-exchange loci with positive intersects on the z axis forming termination points for two-layer flow (the thickness of the stationary lower layer becomes zero at y = 0). On the other side of the termination point, only the lighter layer exists and the trajectory is simply the z axis; there is no requirement of constant energy difference (flow is totally controlled by the energy of the single layer). Starting from one end of the channel, z can increase along the z axis from zero to some termination point C as shown in Figure 9 and then (y,z) can proceed along the trajectory CD of the two-layer energy locus $\gamma < 0$. The value of $U_1$ increases to the extreme value of $\tilde{U}_1$ (which is identical to $\tilde{u}_0$ since $\beta = 0$) at the intersection of this energy locus with the z = 1 line (which is the control curve for $\beta = 0$), and this must coincide with the minimum cross section of the channel. The layer thickness for $\tilde{u}_0 = 0.64$ is shown in Figure 8c; the lower layer commences a distance of 0.4 to the left of the minimum cross section.

When $\tilde{u}_0 = 1$, the termination point of the locus is E in Figure 9 with $\gamma = -1/2$ and this is also the control point so that the two-layer flow terminates at the minimum cross section; the layer thickness is shown in Figure 8d. The solution for $\tilde{u}_0 > 1$ is similar to that for $\tilde{u}_0 = 1$ except that z increases beyond the point E to some general point denoted by G and then decreases to E (the control point) where it joins the energy locus with $\gamma = -1/2$. This point G has maximal $U_1$ (with $z = \tilde{u}_0^2$ from Equation 19), and therefore, corresponds to the minimum cross section of the channel; the two-layer flow starts further downstream. Two examples, with $\tilde{u}_0 = 1.5$, and 3.0, are shown in Figure 8e,f with the lower layer commencing at distances of 0.5, and 2.0, respectively, to the right of the minimum cross section. It is clear that $U_1$ is stationary with respect to y at E (the $U_1$ and $\gamma$ loci touch on the control curve) but this does not correspond to the maximum of $U_1$ as a function of x (which is located at the minimum channel cross section) and so

$$\frac{dy}{dx} \to \infty$$

(28)

at E. Therefore, the lower layer forms an abrupt front, with the interface becoming vertical at the bottom of the channel (as shown by Figure 8d,e,f); in contrast, the slope of the terminating lower layer in 8c is finite.

This analysis has described the blocking of baroclinic flow by a barotropic current from the less to more dense basin. The same type of behavior occurs for a current in the opposite direction and if

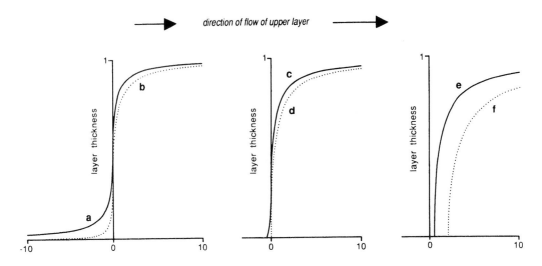

direction of flow of upper layer

distance from minimum channel cross-section in dimensionless length

**Figure 8** The variation of the height of the bottom layer in a two layer flow in which the upper layer is moving to the right for the following values of net barotropic flow $\tilde{u}_0$: (a) 0, (b) 0.54, (c) 0.62, (d) 1.0, (e) 1.5, (f) 3.0.

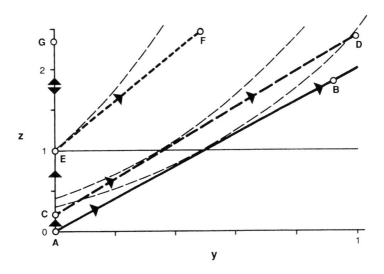

**Figure 9** Loci (heavy lines) of $\gamma = 0$ (solid), $-0.1$ (long dash), and $-0.5$ (short dash) for $\beta = 0$. The lighter broken lines are the loci of $U_1 = 0.55$, 0.64 and 1.0, the solid horizontal line at $z = 1$ is the control curve. Referring to Figure 8 curve (b) is AB, (c) is ACD, (d) is AEF, (e) and (f) are AEGEF with G at $z = 1.5^2$ and $3.0^2$, respectively.

$\tilde{u}_0 < -u_T$ the upper layer becomes stationary and extends only a finite distance into the channel; the analysis is identical to the above if $\beta$ is mapped to $1/\beta$ and the roles of the two layers interchanged.

The two-layer flow inside a channel in general carries both a barotropic current and a mass (or salt) flux. If the latter is specified, the internal wave speed c, defined in Equations 2 and 3, becomes a function of $\beta$. The case of zero net salt flow is relevant to a steady state estuary, and the analysis, which was developed by Stommel and Farmer,[11] is discussed in Appendix 2.

The analysis presented here ignores the Coriolis term, that is, the system is irrotational. This is justified[20] if the channel is much narrower than the internal Rossby radius of deformation L ~ c/f where f is the coriolis parameter (~ $7 \times 10^{-5}$ s$^{-1}$). The minimum speed of interest here is c ~ 0.05 ms$^{-1}$, so L >500 m and rotation is justifiably neglected in the estuarine channels.

## VII. BAROCLINIC OCEAN EXCHANGE

The analysis of the preceding section establishes that in the absence of barotropic flow ($\beta = 1$) through the channel $\tilde{U}_1 = 1/4$ so the maximum baroclinic flow is $c/4$ times the minimum cross-sectional area of the channel. This is the assumption that was made in Section IV and upon which Figure 2 is based.

The baroclinic exchange ceases when the barotropic current through the channel reaches $0.54c$. For Peel Inlet in southwestern Australia (Figure 10) the mean tidal speed through the ocean channel is about $0.3$ m s$^{-1}$. Section III estimated a maximum c of 1 m s$^{-1}$ and this would be blocked by a barotropic current of $0.5$ m s$^{-1}$ so that the baroclinic two-layer exchange would be replaced by single-layer barotropic flow for a significant part of the spring tidal cycle. As the salinity difference, and hence the internal wave speed c, decreases the barotropic flow becomes even more effective in blocking baroclinic exchange. The simple model of baroclinic flushing presented in Section IV can be modified to include this barotropic blocking. If the tidal prism from the channel does not mix with basin water the barotropic flow does not flush the basin but serves to partially block the baroclinic exchange. Figure 11 uses three volumetric tidal exchange times (Section III) to show the effect of barotropic blocking on the baroclinic flushing originally presented in Figure 2 (for a spring-neaps cycle in which $u_{neap}/u_{spring} = 0.1$). The longest tidal exchange time gives a salinity variation equivalent to Figure 2 and it is observed that tidal exchange times of some tens of days (typical of these estuaries) do significantly reduce the baroclinic flushing; note the modulation by the spring-neaps cycle.

Tidal distortion in shallow channels tends to shorten one of the phases of the tide producing stronger currents in the phase (referred to as flood- or ebb-dominated systems[21]) so that the blocking of baroclinic currents may be asymmetric between the two phases of the tide.

## VIII. BAROTROPIC OCEAN FLUSHING

Barotropic flow is marked by a nearly vertical front between the water advecting through the channel and the water of the receiving basin (with a thin layer of moving channel water downstream of the front). A very good example is the Mandurah Front[6] observed in Peel-Harvey Estuary[4,22-25] (Figure 10) downstream of the ocean channel during spring flood tides after strong riverflow (winter and early spring).

The dynamics of barotropic flushing depend on the fraction of the tidal prism that remains in the basin and does not simply exit on the next ebb tide. If mixing in the basin is very strong, the dynamics are largely controlled by the ratio of the volumes of the channel and tidal prism. It is useful to define the fraction of the prism that remains in the basin on one tidal cycle as a retention coefficient $\mu$ and this controls the extent of barotropic flushing; the curves in Figure 11 are based on $\mu = 0$ so that there is no barotropic flushing. For complete mixing,

$$\mu = \begin{cases} 0 & V_{prism} < V_{channel} \\ 1 - \dfrac{V_{channel}}{V_{prism}} & V_{prism} \geq V_{channel} \end{cases} \tag{29}$$

Incomplete mixing around the ends of the channel will reduce the actual retention coefficient below the value given by Equation 29. As an example, consider again the Mandurah Channel to Peel Inlet (Figure 10) which has a volume of about $6 \times 10^6$ m$^3$. The maximum spring tidal prism for the total estuary was $10 \times 10^6$ m$^3$ before dredging work on the channel in 1987, after which it increased to $15 \times 10^6$ m$^3$; the volume of the channel was also slightly changed. The mean tidal prism (averaged over a year) increased from 5.6 to $8.6 \times 10^6$ m$^3$ so that the mean value of $\mu$ crossed the threshold in Equation 29 from near zero to about 0.3.

Figure 12 shows barotropic ocean tidal flushing for a volumetric flushing time of 10 days with ratios of channel volume to spring tidal prism of 1, 0.5 and 0.1 (the neaps/springs tidal ratio is 0.1). The barotropic flushing vanishes at neap tide for the smallest channel (which has a volume equal to the neap tidal prism). For the next larger channel, the salinity is observed to remain static over about half of the fortnightly spring-neaps cycle and for the largest channel the retention is zero even at spring tide and there is no barotropic flushing. The combined effect of barotropic and blocked baroclinic flushing will be discussed in Section XI (Figure 14).

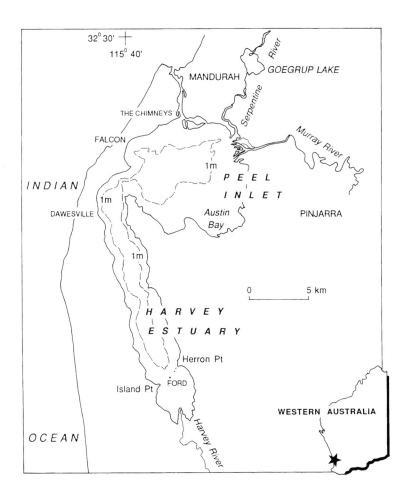

**Figure 10**   The Peel-Harvey Estuary.

In shallow systems the dynamics of the baroclinic and barotropic flows are clearly very different and have been treated separately in this chapter. However, in deep systems (for example the flow across the sill of a fjord[26]) it may be simpler to merely combine the effects of baroclinic and barotropic flow and consider the effect of tidal range on the *total* ocean exchange.

## IX. OBSERVATIONS OF THE RETENTION COEFFICIENT

During late spring and early summer the Harvey River ceases to flow into the Peel-Harvey Estuary (Figure 10) and the salinity of the Harvey basin slowly returns to marine salinity as higher salinity water flows through the channel from Peel Inlet.[6] Figure 13 shows a 36-hour time series of top and bottom salinities at a station inside this channel near spring tide in early summer when salinities were close to 25‰ at the Harvey basin end of the channel and about 29‰ at the Peel end; the figure also shows the velocity through the channel. The time series can be divided into the end of a flood tide, followed by ebb and flood tides, and the start of another ebb. At the end of each ebb and flood, the top and bottom salinities are about steady in time and representative of water at the northern end of Harvey basin and southern part of Peel Inlet, respectively. As the tidal current changes direction there is a transition region (of duration about 3 hours) in which a mixture of water from the two basins is observed at the channel station. The spring tidal excursion in the Peel to Harvey channel is about 10 km while the channel is some 1500 m long so that according to Equation 29 the retention coefficient at spring tide should be close to unity. The total salt flow through the channel is positive (i.e., toward Harvey basin) and can be estimated from Figure 13 to be roughly half the tidal prism times the salinity difference between the Peel and Harvey basins.[27]

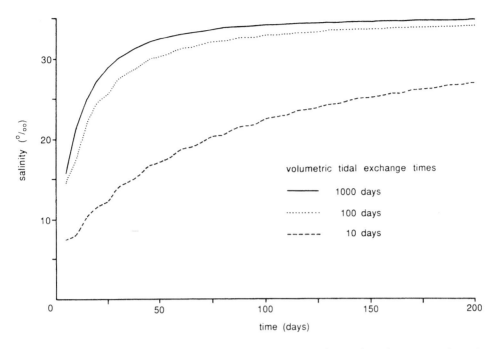

**Figure 11** Variation of basin salinity due to controlled baroclinic exchange through a narrow channel to the ocean with barotropic blocking by three different superimposed volumetric tidal flushing times (with zero retention coefficient) and an initial salinity of 5‰.

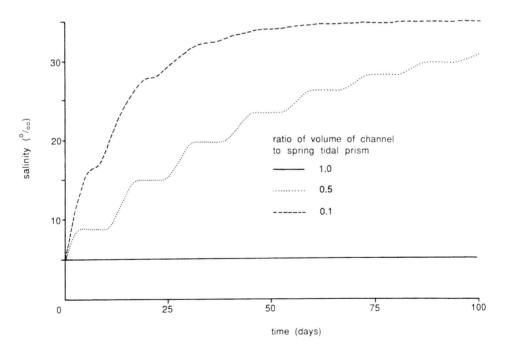

**Figure 12** Variation of basin salinity due to barotropic tidal flushing only with a volumetric tidal flushing time of 10 days and three values of the ratio of channel volume to spring tidal prism; the neaps/springs tidal ratio is 0.1.

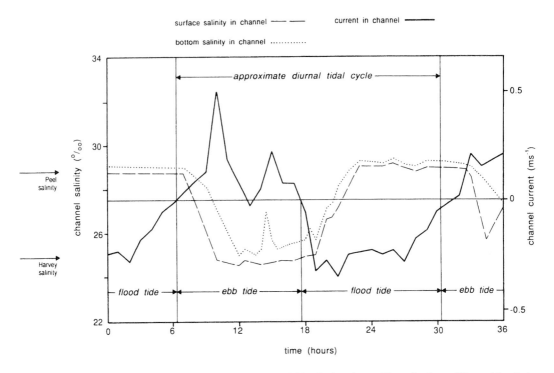

**Figure 13** Time series of current and top and bottom salinities in the channel from the inner (Harvey) basin to the outer basin (Peel Inlet), of the Peel-Harvey Estuary in southwestern Australia, during a 36-hour period in early summer (December 1984).

The retention coefficient $\mu$ is therefore about 0.5 but can only be accurately estimated from long time series of salinity and current in the channel; individual tidal cycles rarely give zero net transport because of residual riverflows and low frequency changes in sealevel (discussed in the next section). The reduction in retention coefficient below unity is due to incomplete mixing in the basins as manifested by the gradual salinity transition observed in Figure 13 after slack tide. The characteristic internal wave speed c for Figure 13 is of order $0.1 \text{ m s}^{-1}$ so that baroclinic processes are blocked except near to slack water. At the start of the ebb the surface salinity drops much more quickly than the bottom salinity and this is evidently due to baroclinic motion during the short slack period; the water is then tidally advected through the channel.

Black and Hatton[28,29] in their study of the tidal flow through McLennan Strait in the Gippsland Lakes of southeastern Australia found that the tidal excursion was approximately equal to the length of the channel (10 km) so that the retention coefficient is small. They determined a model retention coefficient numerically by tracking particles moving with the measured current in the channel over a 2-month period. Particles were released into the model every 10 minutes and removed after they passed out of the channel. This provides a valuable spectrum of particle ages in the channel which depends on the retention coefficient which is determined by Equation 29, the spring to neaps ratio, and additional flows due to low frequency tides, river-discharge, etc. With the inclusion of the finite basin mixing, this model would represent the essence of the exchange process.

## X. LOW FREQUENCY TIDES

Low frequency variations in sea level have high penetration through the narrow channels into the estuarine basins. These tides are created mainly by meteorological variations with periods from a few days to weeks and at mid-latitudes have ranges of order several tenths of a meter; this is a significant fraction of the depth of a very shallow basin and so they provide a major part of the ocean flushing. Estimating the average volumetric exchange time of meteorological tides requires a knowledge of the channel exchange time[6] and a long time series of ocean water level. For the Peel-Harvey Estuary, Hearn and Lukatelich[6] have shown that a channel exchange time of $\tau = 2.2$ days for the Mandurah Channel (prior to dredging in 1987) gives an exchange of 8.3 estuarine volumes per year, which was significantly increased by dredging to 20.2 volumes by the reduction of $\tau$ to 1.4 days; this is to be compared with the astronomical annual volumetric tidal exchange of 11.8, and 18.1 volumes, respectively. Because of their low frequency, the meteorological tidal prisms have retention coefficients close to unity so that (at the very least) their effect is comparable to the astronomical tides. By contrast, Lake Songkhla in Thailand[30] which is a low-latitude estuary has negligible meteorological tides.

A spring-neaps modulation of the mean water level is present in very shallow basins due to frictional effects being comparatively greater during the low water phase of a tidal cycle and therefore tending to elevate the mean water level relative to that of the ocean. The elevation of the mean water level increases with the ratio of the amplitude of the tide to the water depth in the estuary. In some estuaries[31,32] the effect is so extreme that low water springs are higher than low water neaps. In the present estuaries the major effect comes from the channels, which show longer channel exchange times[6] at low water levels because of the effects of friction and the reduction in the cross-sectional area of the channel. The latter effect is much more pronounced in natural channels (which have strongly sloping sides) than in artificial or modified channels. It will be sensitive to the predicted long-term rise in global sea level and this will affect near-bed sediment transport patterns.[33] Very shallow basins also develop their own local 'wind-induced' tides due to the large set-up of the surface by wind stress.[34]

## XI. CONCLUSION AND DISCUSSION

The exchange of water between the ocean and multibasin shallow estuaries through long narrow channels has been reviewed in this chapter. The flow consists of barotropic and baroclinic motion. For maximum density differences and typical channel cross sections and basin areas the baroclinic flushing time determined by hydraulic control in a typical channel is of the order 10 days; this time increases as the density difference is reduced. The barotropic motion is induced by astronomical and meteorological tides and usually produces maximum currents $\tilde{u}_0$ (at the narrowest part of the channel) of 0.1–0.3 m s$^{-1}$ in natural channels and up to 1 m s$^{-1}$ in artificial channels. Baroclinic motion is blocked by these currents (so that one of the layers becomes stationary) once $\tilde{u}_0$ passes a threshold of about 0.54c where c is a characteristic speed of the baroclinic motion defined in Equation 2. When $\tilde{u}_0$ reaches c the stationary layer terminates at the narrowest part of the channel. For greater values of $\tilde{u}_0$ the edge of the stationary layer forms a vertical front and with increasing $\tilde{u}_0$ this is progressively expelled to the downstream side of the channel. Barotropic flows block the baroclinic motion so that it is only operative during periods of slack tide or at neaps and this typically increases the flushing times mentioned above by a factor of at least two.

Barotropic flow is characterized by unidirectional single fluid flow through the channel with a stationary front downstream of the channel. These fronts have surface manifestation just inside the basins of the estuaries under conditions of strong salinity differences at peak spring tidal flood. The barotropic exchange is controlled by the volumetric exchange time (the time that it would take for the inward phase of the spring tidal flow to fill the estuary) and the retention coefficient $\mu$ (the fraction of the flood tidal prism that remains in the estuary and does not exit on the next ebb tide). The magnitude of the volumetric exchange time is typically 10 days at spring tides. The magnitude of $\mu$ is controlled by the ratio of the volumes of the channel and tidal prism and mixing within the basins. The retention coefficient, therefore, varies through the spring-neaps cycle and is maximal at springs. At spring tides the estuary tends to flush by barotropic exchange while the baroclinic process may become significant at neaps.

These processes can be included in a single model which has been implemented for the Peel-Harvey Estuary in southwestern Australia (Figure 10) and the results are presented in Figure 14. It shows the effect of ocean exchange on salinities in Peel Inlet and Harvey basin starting from an initial salinity of 5‰ at the end of winter. The parameters that determine the salinity recovery are the volumetric exchange time (10 days), the baroclinic flushing time at the maximum salinity difference (10 days), the ratio of the channel volume to the volume of the spring tidal prism (0.5 for Mandurah Channel), and the neaps/spring tidal ratio (0.1). It is evident in Figure 14 that flushing occurs over a seasonal time scale and that there are two types of exchange which vary through the spring-neaps cycle. Barotropic flushing has a nonlinear dependence on the magnitude of the tidal prism through the retention coefficient and this has been demonstrated by the dredging of the ocean channel to Peel Inlet; the predicted surface fronts at spring tides have become more pronounced since this dredging. Meteorological tides would also influence the salinity variation in Figure 14 and are an important aspect of the ocean exchange for this type of estuary at midlatitudes.

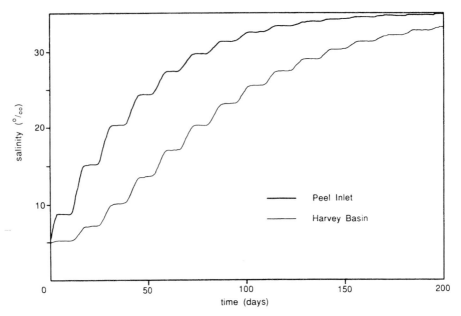

**Figure 14** The variation of salinity in Peel Inlet (heavy line) and Harvey basin (lighter line) due to ocean flushing from an initial salinity of 5‰ for barotropic tidal and baroclinic exchange.

## ACKNOWLEDGMENTS

The author acknowledges support from the Australian Research Council, and the Western Australian Waterways Commission, and is grateful to Prof. A. J. McComb and Dr. R. J. Lukatelich for collaboration on the studies of Peel-Harvey Estuary.

## APPENDIX 1: THE EFFECT OF A SILL IN THE CHANNEL

The analysis presented in Section V is based on the energy loci for a flat-bottomed channel. The natural channels joining the basins also contain sills. To consider both narrowing and depth variations simultaneously is both difficult and unnecessary because the spatial scales of the sills are much smaller than those of the channel narrowing and so in the subsequent analysis the width b is considered constant and only h varied. The datum used for h (which is measured vertically upwards) is taken to be the crest of the sill so that h(x) is negative. The earlier analysis continues to apply but the form of the locus of energy difference is changed. This is now conveniently expressed in terms of a quantity

$$\alpha \equiv \gamma \, \tilde{U}_1^{-\frac{2}{3}} \tag{A1}$$

which gives the energy loci as

$$z = \frac{2(y/\beta)^2 \left(1 - \alpha \, z^{\frac{1}{3}}\right)}{1 - (y/\beta)^2} \tag{A2}$$

Figure 15 shows that all of the loci start from the origin of the $(y,z)$ plane and diverge at the $y = \beta$ singularity which was encountered in the earlier analysis (it corresponds to the particle speeds of the two layers becoming equal). Loci with $\alpha < 3/2$ display a maximum value for y but for $\alpha > 3/2$, y diverges with a gap in the allowed values of z. The special case $\alpha = 3/2$ provides a unique solution which satisfies the lock-exchange conditions with $z = 1$. $U_1$ is maximal at the intersection of the locus with the control curve.

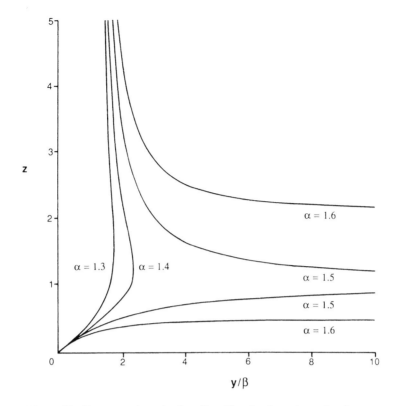

**Figure 15**  Energy contours for the sill problem for the values of $\alpha$ shown.

The position of this intersection is dependent on the value of $\beta$ and occurs at y = 0.6 for $\beta$ = 1 (which is to be compared with y = 1 for the simple contraction considered earlier) and the maximum of $U_1$ at this point is 0.208 (which is slightly less than the corresponding value of 0.25 for a simple contraction).

The limit z → 1 for the sill problem, which is referred to as exit control by Farmer and Armi,[19] is necessary to achieve maximum flow over the sill and the solution must also pass along the trajectory to reach the control point at the sill. Because the sill is localized, the flow moves only a finite distance along the locus in the (y,z) plane; it starts from a value of y associated with the layer heights in one basin and proceeds to that for the other basin.

As $\beta$ is reduced, so $\bar{U}_1$ tends towards unity because the energy loci in the (y,z) plane reduce to

$$\alpha < 3/2, \ y = 0$$

$$\alpha \geq 3/2, \ z^3 + 6z^2 + 4z(3 - 2\alpha^3) + 8 = 0 \tag{A3}$$

Hence, $\bar{u}_0$ reaches a maximum of unity at $\beta$ = 0 and the lower layer becomes stationary. The value of y at the control drops from 0.6 (at $\beta$ =1) to zero when $\beta$ = 0 so that the lower layer just fails to be able to spread over the top of the sill (recall that the water column has unit depth at the crest of the sill so that h is negative). To achieve $u_0$ > 1 the system must ascend the z-axis and then move onto an energy locus; if the flow is directed in the opposite direction the upper layer will become stationary as the lower layer is forced beneath it.

Figure 16 shows a front associated with dominant barotropic flow just inside the Harvey basin into which water flows from Peel Inlet (Figure 10). The higher salinity marine water which has crossed Peel Inlet from Mandurah Channel[6] drops below the slightly fresher water at the Harvey end of the channel and a front marked by a surface foam line is visible during major flood tides. There is a sill and hole[25] at the Peel end of the channel and it is evident in Figure 16 that this bathymetry is modifying the flow within the channel. Based on the measured salinities in Figure 16, the internal speed c defined in Equation

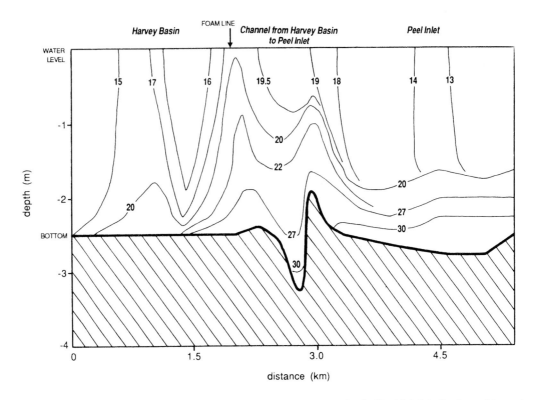

**Figure 16** Contours of salinity (‰) through the channel from the outer basin (Peel Inlet) to the inner (Harvey) basin, in southwestern Australia, in spring (October 1988) showing higher salinity bottom water in the outer basin passing through the channel; note the influence of the sill at the outer end of the channel.

2 is 0.3 m s$^{-1}$. The marine water at the bottom of Peel Inlet is diluted in its passage through the channel with salinity decreasing from 30 to 27‰. This water overflows the sill and it is evidence that upward mixing increases the surface salinities to exceed those in Peel Inlet. This channel surface salinity maximum has been a common feature since the dredging of Mandurah Channel in 1987 (mentioned above) and is apparently due to upwelling over the sill. Maximum salinities are observed just above the sill which apparently increases the depth of the lower layer. This corresponds to an increase of y in Figure 15 which is possible with an increase in $U_2$. The channel width continues to contract at the end of the sill, so $U_2$ must increase so that the flow is not critical.

## APPENDIX 2: STOMMEL-FARMER STEADY STATE ESTUARINE MODEL

In Section VI it was shown that control imposes a relation between the barotropic flow and β. Stommel and Farmer[35] modeled an estuary which consists of a basin fed by a river and joined to the ocean through a channel. In the steady state the barotropic outflow through the channel must be equal to the riverflow and there must also be zero net salt transport. The density difference, Δρ between the two layers in the basin is related to β by this latter condition,

$$1 - \beta = \frac{\Delta\rho}{(\Delta\rho)_0} \tag{A4}$$

where $(\Delta\rho)_0$ is the density difference between fresh and seawater; Equation A4 assumes a linear relation between density and salinity. If there is no mixing in the basin, $\Delta\rho = (\Delta\rho)_0$ so by Equation A4 β = 0, and by Equation 13 the bottom layer is stationary; as mixing increases Δρ tends to zero and β towards unity so that the layers develop equal and opposite fluxes.

The barotropic flow through the channel is equal to the riverflow so that if this flow is hydraulically controlled the strength of the riverflow will determine $\beta$ (compare Figure 7), and hence, by Equation A4, the strength of the mixing in the basin. The net flow $u_0$ introduced previously was normalized in terms of the internal speed c defined in Equation 2, but since this is not a fixed quantity in the Stommel and Farmer[35] problem, it is more convenient to normalize the river flow $\tilde{U}_{river}$ using an internal wave speed based on the maximum density difference $(\Delta\rho)_0$; the tilde indicates that the river flow is represented by a speed averaged over the cross section at the narrowest point of the channel. Hence, using Equations 3 and 17,

$$\tilde{u}_0 = \frac{\tilde{U}_{river}}{(1-\beta)^{\frac{1}{2}}} \tag{A5}$$

$$\tilde{U}_1 = \frac{\tilde{U}_{river}}{(1-\beta)^{\frac{3}{2}}} \tag{A6}$$

so that at the minimum cross section of the channel Equations 19 and A6 give

$$z = \frac{\tilde{U}_{river}^2}{(1-\beta)^3}(1+y)^3 \tag{A7}$$

and using the control condition Equation 22

$$\tilde{U}_{river} = \frac{(1-\beta)^{\frac{3}{2}}}{(1+\sqrt{\beta})^2} \tag{A8}$$

with control occurring at $y = \sqrt{\beta}$ so that complete mixing ($\beta = 1$) involves zero riverflow $\tilde{U}_{river}$ and y = 1 (layers are of equal thickness); while for $\tilde{U}_{river} \to 1$ ($\beta \to 0$) the mixing decreases with $y \to 0$ (thickness of lower layer vanishes).

The maximal value of $\tilde{U}_{river}$ always occurs on the right hand side of the saddle at $y = \beta$ so the energy locus tends to a vertically mixed ocean on one side of the channel and a two layer structure in the basin. The Stommel and Farmer[35] criterion is a statement of the minimum amount of mixing necessary to sustain the riverflow with control. If mixing is increased beyond this value the height of the bottom layer will be reduced and in the limiting case of total mixing the lock-exchange solution will be reached; Stommel and Farmer[35] refer to this as overmixing. If $\beta \sim 1$ so that $y \to 1$ it is possible to expand y as a function of small $\tilde{U}_{river}$

$$y = 1 - 3\sqrt{2\tilde{U}_{river}^2} \tag{A9}$$

## REFERENCES

1. **Dyer, K. R.**, *Estuaries: A Physical Introduction,* John Wiley & Sons, London, 1973.
2. **Kennish, M. J.**, *Ecology of Estuaries,* Vol. 1, *Physical and Chemical Aspects,* CRC Press, Boca Raton, FL, 1986.
3. **Fairbridge, R. W.**, *Chemistry and Biogeochemistry of Estuaries,* Olausson, E. and Cato, I., Eds., John Wiley & Sons, Chichester, 1980.
4. **Hodgkin, E. P., Birch, P. B., Black, R. E., and Humphries, R. B.**, The Peel-Harvey estuarine system study (1976–1980), Rep. No. 9, Department of Conservation and Environment, Perth, Western Australia, 1980.
5. **Smith, N. P.**, Wind domination of residual tidal transport in a coastal lagoon, in *Coastal and Estuarine Studies 38: Residual Currents and Long Term Transport,* Cheng, R., Ed., Springer-Verlag, New York, 1990, 123.

6. **Hearn, C. J. and Lukatelich, R. J.,** Dynamics of Peel-Harvey Estuary, southwest Australia, in *Coastal and Estuarine Studies 38: Residual Currents and Long Term Transport,* Cheng, R., Ed., Springer-Verlag, New York, 1990, 431.

7. **Hearn, C. J., McComb, A. J., and Lukatelich, R. J.,** Coastal lagoon ecosystem modelling, in *Coastal Lagoon Processes,* Kjerfue, B., Ed., Elsevier, Amsterdam, 1994, 471.

8. **Isaji, T., Spalding, L. S., and Stace, J.,** Tidal exchange between a coastal lagoon and offshore waters, *Estuaries,* 8, 203, 1985.

9. **Meetam, P. and Hearn, C. J.,** Physical data from Lake Songkhla, Thailand during the early dry season 1988, Department of Mathematics Report, The University of Western Australia, Perth, Western Australia, 1988.

10. **Officer, C. B.,** *Physical Oceanography of Estuaries (and Associated Waters),* John Wiley & Sons, New York, 1976, chap 4.

11. **Stommel, H. and Farmer, H. G.,** Abrupt change in width in two layer open channel flow, *J. Mar. Res.,* 11, 205, 1952.

12. **Wood, I. R.,** Selective withdrawal from a stably stratified fluid, *J. Fluid Mech.,* 32, 209, 1968.

13. **Wood, I. R.,** A lock exchange flow, *J. Fluid Mech.,* 42, 671, 1970.

14. **Wood, I. R. and Lai, K. K.,** Flow of layered fluid over a broad crested weir, *J. Hyd. Div. ASCE,* 98, 87, 1972.

15. **Lai, K. K. and Wood, I. R.,** A two-layer flow through a contraction, *J. Hyd. Res. IAHR,* 13, 19, 1975.

16. **Wood, I. R. and Simpson, J. E.,** Jumps in layered miscible fluids, *J. Fluid Mech.,* 140, 329, 1984.

17. **Armi, L.,** The hydraulics of two flowing layers with different densities, *J. Fluid Mech.,* 163, 27, 1986.

18. **Farmer, D. M. and Armi, L.,** Maximal two-layer exchange through a contraction with barotropic net flow, *J. Fluid Mech.,* 164, 27, 1986.

19. **Farmer, D. M. and Armi, L.,** Maximal two-layer exchange over a sill and through the combination of a sill and contraction with barotropic flow, *J. Fluid Mech.,* 164, 53, 1986.

20. **Whitehead, J. A., Leetmaa, A., and Knox, R. A.,** Rotating hydraulics of strait and sill flows, *Geophys. Fluid Dyn.,* 6, 101, 1974.

21. **Aubrey, D. G.,** Hydrodynamic controls on sediment transport in well-mixed bays and estuaries, in *Lecture Notes on Coastal and Estuarine Studies 16: Physics of Shallow Estuaries and Bays,* van de Kreeke, J., Ed., Springer-Verlag, New York, 1986, 245.

22. **Beer, T. and Black, R. E.,** Water exchange in Peel Inlet, Western Australia, *Aust. J. Mar. Freshwater Res.,* 30, 145, 1979.

23. **Black, R. E., Lukatelich, R. J., McComb, A. J., and Rosher, J. E.,** Exchange of water, salt, nutrients and phytoplankton between Peel Inlet, Western Australia, and the Indian Ocean, *Aust. J. Mar. Freshwater Res.,* 32, 709, 1981.

24. **McComb, A. J., Atkins, R. P., Birch, P. B., Gordon, D. M., and Lukatelich, R. J.,** Eutrophication in the Peel-Harvey estuarine system, Western Australia, in *Nutrient Enrichment in Estuaries,* Neilsen, B. J. and Cronin, L. E., Eds., Humana Press, Clifton, NJ, 1981, 332.

25. **Hearn, C. J.,** The ocean flushing of estuaries through long channels, *Appl. Math. Model.,* submitted.

26. **Stigerbrandt, A.,** On the effect of barotropic current fluctuations on the two-layer transport capacity of a constriction, *J. Phys. Oceanog.,* 7, 118, 1977.

27. **Hearn, C. J., Lukatelich, R. J., and Hunter, J. R.,** Dynamical models of the Peel-Harvey estuarine system, Environmental Dynamics Report ED-86-152, The University of Western Australia, Perth, 1986.

28. **Black, K. P.,** Hydrodynamics and salt transport in an estuarine channel, Part 3, numerical model, Tech. Rep. No. 11, Victorian Institute of Marine Science, Melbourne, Australia, 1990.

29. **Black, K. P. and Hatton, D. N,** Hydrodynamics and salt transport in an estuarine channel, Part 2, field data analysis, Tech. Rep. No. 10, Victorian Institute of Marine Science, Melbourne, Australia, 1990.

30. **Meetam, P. and Hearn, C. J.,** A tidal model of Lake Songkhla, Thailand, Department of Mathematics Report, The University of Western Australia, Perth, 1988.

31. **LeBlond, P. H.,** On tidal propagation in shallow rivers, *J. Geophys. Res.,* 83, 4714, 1978.

32. **LeBlond, P. H.,** Forced fortnightly tides in shallow rivers, *Atmosphere-Ocean,* 17, 253, 1979.

33. **Friedrichs, C. T., Aubrey, D. G., and Speer, P. E.,** Impacts of relative sea level rise on evolution of shallow estuaries, in *Coastal and Estuarine Studies 38: Residual Currents and Long Term Transport,* Cheng, R., Ed., Springer-Verlag, New York, 1990, 105.

34. **Schroeder, W. W., Wiseman, W. J., Jr., and Dinnel, S. P.,** Wind and river induced fluctuations in a small shallow tributary estuary, in *Coastal and Estuarine Studies 38: Residual Currents and Long Term Transport,* Cheng, R., Ed., Springer-Verlag, New York, 1990, 481.

35. **Stommel, H. and Farmer, H. G.,** Control of salinity in an estuary by a transition, *J. Mar. Res.,* 12, 13, 1953.

# Chapter 11

# Changes in Major Plant Groups Following Nutrient Enrichment

*Marilyn M. Harlin*

## CONTENTS

I. Introduction ........................................................................................................................173
II. Freshwater Systems ...........................................................................................................173
   A. Changes in Plant Community Structure Over Time.........................................................173
   B. Replacements in Plant Communities After Withdrawal of Nutrients .............................174
   C. Effects of Grazing and Seasonality on Outcome ...........................................................175
III. Marine Systems .................................................................................................................175
   A. Response to Nutrient Inputs ...........................................................................................175
   B. Reversal, or Response to Reduction of Nutrient Load ...................................................178
   C. Effects of Grazing and Physical Parameters .................................................................179
IV. Concepts Common to Freshwater and Marine Systems.....................................................179
   A. Releasing a Resource Limitation ...................................................................................179
   B. Perennial and Ephemeral Macrophytes in Competition ...............................................180
   C. Nutrient Fluxes and C:N:P Ratios .................................................................................181
   D. The Expression of Species Diversity ............................................................................182
V. Future Directions for Research ..........................................................................................182
VI. Conclusions........................................................................................................................182
References ...............................................................................................................................183

## I. INTRODUCTION

Nutrient-rich water bodies around the world generally contain large masses of phytoplankton, macroalgae and/or vascular plants. However, are there patterns that permit predictions on which of these groups will appear with the addition of a previously limiting nutrient? This chapter addresses how major plant groups shift when nutrient loads increase and when they decrease.

The consequences of nutrient enrichment in freshwater communities have been studied more thoroughly than those in marine communities. Investigators have capitalized upon natural experiments, and they have tested hypotheses with factorial designs in the field and in the laboratory. First, I shall present some of the classic work in freshwater systems along with more recent experimental work. Then I shall address the less extensive documentation of changes in plant communities in marine systems including experiments in the laboratory and field. Finally, I shall attempt to relate concepts involving nutrient enrichment common to both systems and to suggest directions for future research.

## II. FRESHWATER SYSTEMS

### A. CHANGES IN PLANT COMMUNITY STRUCTURE OVER TIME

Changes in community structure in freshwater habitats have been extensively documented for Lake Washington, Seattle, Washington, U.S., where Edmonson[1] recorded major deterioration in the lake following secondary sewage effluent from 1941 through 1963. Secchi disc depth readings decreased from 3.7 m to 1.0 m when phytoplankton dramatically increased in quantity and blue-green algae such as *Oscillatoria rubescens* appeared. In England, Moss[2] observed an increase in epiphyte biomass and phytoplankton with a concurrent decrease in rooted vascular macrophytes in response to fertilization and the presence of fish in a region beset by sewage and agricultural runoff. Over a 7-year period when phytoplankton increased in Lake Wingra, Wisconsin, U.S., the rooted species, *Myriophyllum spicatum*, decreased by one third its original biomass,[3] but in the Norfolks Broads, England, *Myriophyllum* was the

0-8493-6839-1/95/$0.00+$.50

last macrophyte to disappear.[4] Whereas there is agreement that shading is responsible for the decline of rooted plants with eutrophication, Hough et al.[5] attribute the decline in Shoe Lake, southeastern Michigan, U.S., to increases in epiphytes and free-floating macroalgae rather than phytoplankton.

It is now generally accepted that phosphorus is the limiting nutrient in freshwater communities in North America.[6] The response to this limiting nutrient is greater in shallow than in deep water communities[7] and in nearshore areas more than offshore.[8] By knowing the phytoplankton assemblages present, Stoermer et al.[9] have been able to predict specific suites of changes in plant species. An examination of microfossils in Lake Michigan[10] revealed a sequence that appeared to reflect progressive eutrophication between 1954 and 1964 after which diatoms became silica limited. Fossil records from lake cores in England show that over time phytoplankton diatoms, have replaced epiphytic diatoms possibly because the epiphytes cause a decline in their host plants.[11]

Wetzel[12] developed a classic model depicting the generalized change in productivity of plant groups in lakes with increasing nutrient load (Figure 1). Overlying nutrient loads and grazing are seasonal parameters which play critical roles in the outcome in freshwater communities.[13] In 1989, a set of papers was published on a series of studies in experimental ponds in Norfolk Broadland, England.[14-16] These ponds were used to separate components in a closed ecosystem and demonstrate experimentally that the outcome following enrichment could be either phytoplankton or macrophytes, depending on other variables in the systems, and will be discussed later in this chapter.

Two other types of freshwater systems need to be mentioned. Flowing waters, such as rivers and streams, have their own indicators to nutrient load.[17] Rather than an appearance of different species, one sees large growths of green macroalgae such as *Cladophora glomerata* with high nutrient levels. This species is otherwise present but "in tiny fragments". In the last century, the salinity in the Great Lakes has tripled with a consequence of the invasion of 8 marine benthic macroalgae (2 red algae, 4 green algae, 2 brown algae) and 11 phytoplankton (10 diatoms, 1 coccolithophorid).[18] The results of changing ions complicate the study of nutrient pollution in these same waters.

## B. REPLACEMENTS IN PLANT COMMUNITIES AFTER WITHDRAWAL OF NUTRIENTS

Where a loss of macrophytes is associated with a high concentration of phytoplankton, then conversely, when the phytoplankton count declines, the number of macrophytes would be expected to increase. Such a shift was documented experimentally when high nitrogen and phosphorus loads led to phytoplankton and lower nutrient loads led to vascular plants.[19] In Lake Washington, municipal sewage was diverted to

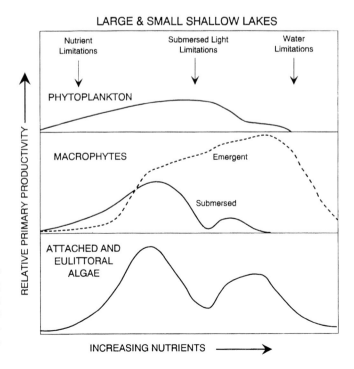

**Figure 1** Generalized relationship of primary productivity of major plant groups in shallow lakes of increasing nutrient enrichment of the whole lake ecosystem. (From Wetzel, R. G., *Arch. Hydrobiol. Beih. Ergebn. Limnol.*, 13, 145, 1979. With permission.)

**Table 1** Contribution to Production in Lawrence Lake, Michigan by Plant Components

| Component | Mean daily (mg C m$^{-2}$ d$^{-1}$) | Mean annual (kg C Lake$^{-1}$ yr$^{-1}$) | Contribution (%) |
|---|---|---|---|
| Phytoplankton | 119 | 2,154 | 13 |
| LZ[a] algae <1 m | 2,001 | 5,512 | 34 |
| LZ algae 1–5 m | 500 | 5,968 | 37 |
| Macrophytes <5 m | 241 | 2,701 | 16 |
| Total | | 16,335 | 100 |

[a] LZ = littoral zone.

From Burkholder, J. M. and Wetzel, R. G., *Arch. Hydrobiol. Suppl.*, 83, 1, 1989. With permission.

the sea in 1961, and nutrient input dropped to zero from 1963 to 1965.[20] By 1979, Secchi disk depth had increased to 7.8 m. By 1975, the lake appeared to have recovered, and the plant community had changed from a previously high phytoplankton community to a vascular plant community. The organisms in the water column no longer shaded the young plants, and a rich nutrient supply was available to rooted plants.[1] Through systematic fertilization of a series of experimental lakes in Ontario, Canada, Shindler[22] definitively demonstrated that phosphate was the nutrient critical to eutrophication and that removal of this substance would restore the original community. In Sweden sewage is chemically treated to reduce phosphate.[23]

Using autoradiography, Burkholder et al.[24] elegantly partitioned the microalgal components on surfaces of aquatic plants and their mimics. These workers showed that the physical position on the thallus determined the ability of the epiphyte to remove phosphorus. Any overlying material significantly interfered with the ability of attached microalgae to take up $^{33}$P phosphate. Based on a comparison between living and nonliving "hosts", Burkholder and Wetzel[25] suggest that macrophytes may be an important nutrient source for epiphytes in oligotrophic lakes, but not in eutrophic water where they are more likely to obtain nutrients directly from the water column. Of the total plant carbon fixed in phosphorus-limited Lawrence Lake, Michigan, U.S., epiphytic algae were calculated to contribute 71% (Table 1)[26] with blue-green algae contributing half the total algal carbon. Planktonic algae sustained their viability in attached form. *In situ* experiments in a shallow reservoir in South Carolina, U.S., established the synergistic effects of clays with phosphorus loading.[27] Epiphyte productivity increased to 90% of total productivity; most of this increase was in the blue-green algal fraction.

## C. EFFECTS OF GRAZING AND SEASONALITY ON OUTCOME

Matrices have been used to separate effects of grazing and nutrient enrichment on phytoplankton. Marks and Lowe[28] tested the effects of herbivorous snails (*Elimia linescens*) on the results of nutrient enrichment. With enrichment, diatom communities shifted to green algae (*Stigeoclonium tenue*). In the presence of snails, green algae increased from 64 to 92% compared with the controls. In neither case were original stages maintained. Vanni and Temte[13] showed that the relative strength of these parameters varied with seasonal cycle. For example, grazing effects were low in summer (when nitrogen and phosphorus inputs were low). The work in Lake Washington demonstrates the seasonal flux in sediments and its impact,[20] where the principle vectors of phosphorus and nitrogen to the sediment were diatom cells, e.g., *Fragilaria crotonensis* and *Asterionella formosa*.

## III. MARINE SYSTEMS

## A. RESPONSE TO NUTRIENT INPUTS

In shallow estuaries and lagoons, in contrast to freshwater systems, high phytoplankton numbers are not the usual response to nutrient enrichment. However, they do occur, as for example, with duck excrement at Moriches Bay, New York, U.S.,[29,30] and due to a fertilizer plant in Cockburn Sound, Australia.[31] More common are extensive mats of green macroalgae such as *Enteromorpha* and *Ulva*, which have been documented near sewage outfalls as early as the beginning of this century.[32] In Boston Harbor, Massachusetts, these massive growths of sea lettuce smothered shellfish by reducing water flow and resulted in intolerable stenches of hydrogen sulfide released during degradation of the seaweed.[33] *Ulva* is found throughout the world in water contaminated with domestic sewage.[34] Wastes can serve as a source of nitrogen for massive growths of red algae[35] as well as green algae.

Growth of green seaweeds in response to measured inputs of nitrogen have been documented in microcosms[36-38] and in the field.[39] Waite and Mitchell[36] concluded from flow-through growth chambers that the specific concentration of nitrogen and the ratio of ammonium to nitrate will influence which alga appears. They showed that ammonia (from sewage) favored the green algae because it is easier to use than nitrate (from agriculture runoff) but that nitrate reductase could be induced in the presence of nitrate. They demonstrated that phosphate could also stimulate growth, but only when ammonium was sufficient.

In a northern temperate lagoon in Rhode Island, U.S., Harlin and Thorne-Miller[39] added low loads of ammonium, nitrate and phosphate separately to the water column in uniform beds of mixed *Zostera marina* and macroalgae. In all plots given ammonium, dense mats of free-floating *Enteromorpha plumosa* and *Ulva lactuca* developed (Table 2). Biomass of leaf and rhizomes of *Zostera* were stimulated slightly, especially where current reached 12 cm s$^{-1}$, but tissue nitrogen concentration did not change. *Ruppia maritima* fared poorly in competition with the green algae, as reported for neighboring New Hampshire.[40] In the Rhode Island lagoon[39] *Gracilaria tikvahiae*, a perennial red alga, did not grow better in ammonium, but pigment increased. Nitrate additions enhanced the growth of the green seaweeds but not the angiosperms, and *Gracilaria* showed a marginal response to this nutrient. Phosphate enhanced growth in the angiosperms but not the macroalgae. None of the nutrient supplements noticeably altered the species composition of either epiphytic or planktonic algae associated with the beds, although small increases in their numbers were detected. After Orth[41] supplied a mixed commercial fertilizer to sediments of *Zostera* beds in Chesapeake Bay, Virginia, eelgrass biomass tripled, but massive algal growths did not appear. The difference in response between these two experimental systems may reflect the shape of the water body as well as where the nutrients were applied.

This pattern of drift macroalgae is typical for lagoons in widely different areas, such as Florida,[42] Ireland,[43] Sweden,[44] Italy,[45] and Australia.[31] In shallow Massachusetts bays experiencing eutrophication, Costa[46] reported a change from a *Zostera* community to one dominated by drift algae such as *Cladophora* and *Gracilaria*. Whereas nutrient enrichment enhances macroalgal growth, seaweeds in moderate densities can coexist with *Zostera*.[47]

A series of neighboring lagoons in Rhode Island examined for their community flora show a range of patterns.[47] At one extreme is a closed lagoon (Trustom Pond) characterized by 3 genera of angiosperms (*Potamogeton*, *Ruppia* and *Zanichellia*), 5 genera of green algae (primarily *Enteromorpha*, *Cladophora* and *Chara*) and the absence of red and brown algae. At the other extreme, Ninigret Pond is continuously connected with the sea. *Zostera* is the dominant angiosperm with *Ruppia* in fringe areas. Ninigret Pond also supports the largest number of algal species: 6 genera of green algae (primarily *Enteromorpha* and *Chaetomorpha*), 11 genera of red algae (primarily *Agardhiella* and *Gracilaria*) and 4 genera of brown algae (primarily *Ectocarpus* and *Pilayella*). The most striking characteristic of the algae in all the lagoons is that they are generally unattached. In addition to the angiosperms and seaweeds are the phytoplankton which in these lagoons contribute as much as 40% of the total primary production (Table 3). In selected areas the waters are opaque with phytoplankton, while the seagrass population is low.

A sequence of changes since the opening of a lagoon (Ninigret Pond) to the ocean has been documented by comparing the present flora with earlier records.[48] Before the breachway was constructed in 1952, the dominant species were *Potamogeton* spp., *Chara* spp. and *Ruppia maritima*,[49] the same species that now dominate Trustom Pond. Within the first 10 years of Ninigret Pond's permanent opening, a dramatic transition was made from fresh (or brackish) water species to those typical of marine waters.[50] *Potomogeton* and *Chara* apparently disappeared, *Ruppia* occupied a greatly reduced area, and *Zostera marina* was present in large areas of the lagoon. Two filamentous green algae, *Chaetomorpha linum* and *Cladophora rudolphiana*, were major contributors to the plant biomass, and several species of red and brown algae were noted. Two decades after Conover's[50] observations, the distribution of *Ruppia* and *Cladophora* decreased significantly.[48] There was, as expected, no trace of *Potamogeton* or *Chara*; the biomass of *Zostera* had increased nearly an order of magnitude, and red algae accounted for 14% of the total biomass. An increase in algal biomass led to decline in rooted submerged plants in Chesapeake Bay, Virginia.[51]

In Cockburn Sound, Australia,[31] the area covered by diverse (10 species) seagrasses declined 97% from 1962 to 1978. Seagrass loss from the Sound appears to have accompanied increased epiphyte loads, after which phytoplankton blooms became marked. Both epiphytes and phytoplankton limited light available to angiosperms. This large floristic change in Cockburn Sound was attributed to the installations

**Table 2** Response of Freefloating Macroalgae to Nutrient Enrichment in Ninigret Pond, Rhode Island

| Treatment (Station I) | *Enteromorpha plumosa* | *Ulva lactuca* | *Gracilaria tikvahiae* |
|---|---|---|---|
| Control | 1.0 (1.3) | 9.0 (14.8) | 1.7 (3.0) |
| Nitrate | *20.0* (18.8) | *62.6* (50.5) | 2.1 (5.3) |
| Ammonium | *118.9* (82.9) | *36.8* (29.1) | 5.8 (8.2) |
| Phosphate | 2.9 (4.7) | 0 (0) | 8.2 (7.8) |

*Note:* From each treatment 10 quadrats were harvested on August 8, 1980. Plant tissue is expressed as g dry wt m$^{-2}$ (SD). Values that differ from controls (P <0.05) are italicized.

From Harlin, M. M. and Thorne-Miller, B., *Mar. Biol.*, 65, 221, 1981. With permission.

**Table 3** Proportion of Plankton Productivity (g $O_2$ m$^{-2}$ d$^{-1}$) Measured in a Mixed *Zostera*-Macroalgal Bed in a Coastal Lagoon in Rhode Island

| | Total minus plankton | Plankton only | Total[a] |
|---|---|---|---|
| Gross production | | | |
| June | 3.4 | 5.1 | 8.5 |
| July | 2.4 | 9.0 | 11.4 |
| August | 6.4 | 4.4 | 10.8 |
| October | 3.4 | 1.6 | 5.0 |
| Net 24 h production | | | |
| June | 2.1 | 1.3 | 3.4 |
| July | 1.1 | 0.8 | 2.0 |
| August | 1.7 | 0.3 | 2.0 |
| October | 0.3 | 0.2 | 0.5 |

*Note:* Values were determined using diel oxygen rate-of-change curves.

[a] Includes plankton, epiphytes, sediments, *Zostera*, macroalgae.

From Harlin, M. M., Thursby, G. B., and Thorne-Miller, B., *Freshwater Marine Plants of Rhode Island,* Sheath, R. S. and Harlin, M. M., Eds., Kendall/Hunt, Dubuque, IA, 1988, chap. 8.

of a sewage treatment plant in 1966, and 3 years later, a nitrogen fertilizer plant. In Peel/Harvey Estuary, Australia, extremely dense mats of *Cladophora*, green macroalgae and blue-green algae have developed in response to phosphorus loading. Their density restricts the seagrasses, *Halophila* and *Ruppia*, whereas in an estuary on the southwestern coast *Ruppia* had increased.[31] Possibly *Ruppia* in the latter case was not competing with other species. In Rhode Island, *Ruppia* appeared unexpectedly on tidal deltas where *Zostera* had been manually removed:[52] normally *Ruppia* was found only in refuge areas unsuitable for *Zostera.*

An entire coral reef system died when a large population of the benthic green macroalga *Dictyosphaera cavernosa* grew in response to sewage effluent.[53] At the outfalls phytoplankton removed the inorganic nitrogen, which subsequently became transported as organic nitrogen and entered the food chain. It emerged as inorganic nitrogen in the central portion of the reef where the water was clear enough to support macroalgae.

Borum[54] observed quantitative changes in the biomass of epiphytes, macrophytes and phytoplankton along a nutrient gradient in coastal Denmark. A mechanism to account for this replacement was offered by Phillips et al.,[4] in which they suggest that shading by epiphytes causes the decline in the host plant. Epiphytic bacteria and algae increase with nutrient loading. In the Norfolk Broads, the ability of *Myriophyllum* to be the last surviving macrophyte was based on its ability to reach the water surface and use higher light intensity. Support for light limitation from macroalgae comes from comparison of eutrophic with oligotrophic lakes in Michigan.[5] In the marine environment Backman and Barilotti[55] reduced both vegetative and flowering shoots in *Zostera* by artificially shading the plants. Bulthuis and Woelkerling[56] measured the biomass accumulation and shading effects of epiphytes on *Heterozostera.*

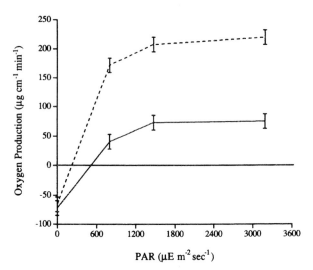

**Figure 2** Rates of photosynthesis of segments from *Posidonia australis* leaves, with (above) and without (below) epiphytes, at different light intensities. Vertical bars are standard errors; n = 5. (From Silberstein, K., Chiffings, A. W., and McComb, A. J., *Aquat. Bot.*, 24, 355, 1986. With permission.)

When extensive, this factor can lead to reduction of the host plant, as shown for *Posidonia*, where photosynthesis is drastically reduced with epiphytic cover (Figure 2).[57] Dennison[58] integrated the length of time *Zostera* needs at varying light intensities; a general guideline is that it cannot grow in light less than one Secchi disc depth.

## B. REVERSAL, OR RESPONSE TO REDUCTION OF NUTRIENT LOAD

The consequence of nutrient load reduction was documented for Cockburn Sound, Australia.[31] Once the cause of the decline in seagrasses was established, the industrial inputs of nutrients were stopped. As a result, the Sound began to return to its previous status.

In Mumford Cove, Connecticut (U.S.), *Ulva lactuca* had become a virtual monoculture since effluent from a city water treatment facility entered this shallow estuary.[59] Local residents succeeded in obtaining a court injunction to restore the water quality in the cove. In 1987, sewage discharge was rerouted to the Thames River. A model developed for this cove predicted that nutrients in flux with the sediments would sustain the *Ulva* population indefinitely, which understandably alarmed the residents. This expensive diversion and citizen concern supplied the opportunity to track the course of a nuisance macroalga and to test whether the sediments could sustain a massive biomass in the future. Contrary to predictions of the original model, but consistent with new projections, the *Ulva* population dropped precipitously. The speed of the recovery was even faster than anticipated. Within 1 year aerial cover dropped from 74% to 9%, and of this value <3% of the total plant biomass was *Ulva*. A monthly survey (May through November) in 1988 at seven stations on a distance gradient down the cove showed *Ulva* in largest biomass closest to the treatment facility but only in the earlier sampling periods (Figure 3), when it made up over half the total macrophyte biomass. The initial population of this seaweed could have been supported by nutrients from the sediment. In contrast to previous years, by summer the biomass dropped to values consistent with healthy estuaries of the same size. There did not appear to be a reservoir of nutrients left in the sediments in 1989, since the early *Ulva* bloom did not recur in 1989.[60] Of the 29 species tabulated: 2 were angiosperms, 6 chlorophytes, 9 phaeophytes, and 12 rhodophytes. *Gracilaria tikvahiaeae* was the most abundant red alga. Residents have expressed concern that the angiosperm *Zostera marina* might appear up cove and become the new "nuisance" growth. A few seedlings (<1/m$^{-2}$) have appeared on the sediment in the upper cove, but their future is not yet known.

Nowicki and Oviatt[61] used mesocosms to test for effects of nitrogen addition to estuaries. Nutrient enriched mesocosms initially retained 30% of added nitrogen and phosphorus compared to the controls. But after 6 months only 5–15% of the nutrients added were retained, mesocosms with sediment stored less than those without sediment and more phosphorus was retained than nitrogen. More nutrients were stored in winter than in summer. Denitrification was <10% nitrogen input, and there was selective loss of nitrogen through remineralization and algal uptake. These conclusions corroborate those on low sediment retention in both freshwater systems,[62] and the shallow estuary in Mumford Cove.[59]

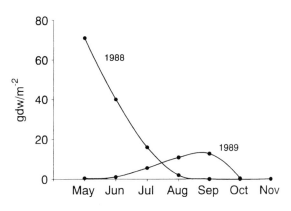

**Figure 3** *Ulva* biomass (g dry weight 0.1 m⁻²) in Mumford Cove, Connecticut in 1988 and 1989, following sewage diversion. Ninety percent of the biomass in 1989 was at Station 1, closest to the water treatment facility. Only material in the upper 4 stations is included. (From French, D. P., Harlin, M. M., Pratt, S., Rines, H., and Puckett, S., *Mumford Cove Water Quality: 1989 Monitoring Study of Macrophytes and Benthic Invertebrates*, Applied Science Associates, Narragansett, RI, 1989, 56. With permission.)

## C. EFFECTS OF GRAZING AND PHYSICAL PARAMETERS

Gastropods, isopods and amphipods control plant species community by grazing the epiphytes.[63-65] These animals enhance seagrass growth and production,[66] because epiphytes no longer shade the host plant or act as a drag force. In addition, remineralization is enhanced by their activity.

In association with nutrients and grazers, physical factors such as light and temperature control seagrass communities.[67] A model connecting these interactions was suggested by Wetzel and Neckles.[68] For green macroalgae, light and temperature limit nitrogen uptake *per se* less than they limit growth,[69] and tissue nitrogen is the critical factor determining growth.[70] In the laboratory, Parker[71] demonstrated quantitatively how water motion affects growth and nitrogen uptake. In the field, Thorne-Miller and Harlin[72] found that productivity to biomass ratios (P:B) in beds of *Zostera marina* decreased as a function of distance from the breachway, supporting the concept that water circulation within a lagoon impacts the expression of a community.

In controlled experiments in the field and laboratory, Williams and Ruckelshaus[73] separated nutrient addition in the water column from that in the sediment and both from grazing pressures. They found that the outcome depends upon the quantity and quality of nutrient and whether grazers removed the resulting epiphytic growth. Epiphyte biomass did not respond to water column enrichment and had little effect on growth of *Zostera marina* or *Z. japonica*. The former species but not the latter was stimulated by nitrogen addition to the sediments. Laboratory experiments excluding the isopod grazer *Idotea* allowed the epiphytes to grow up with nutrient addition. The presence of the isopod in the field may have restricted epiphyte growth.

Neckles[74] used microcosms to separate grazing from nutrient load during four different seasons. This design allowed her to separate biological and physical factors and ask two basic questions. Did the presence of grazers affect the consequences of eutrophication? Were the results season dependent? She showed that when grazers were present, the effect of nutrient enrichment can be significantly less than when grazers were absent (Figure 4). This difference was true for both later and early summer, but not true for fall and early spring. The effect was greatest on the oldest leaves. One of her variables was the grazing community itself. The species and numbers were selected from natural seagrass beds at the time of each experiment; for example *Bittium varium* was added each summer, *Mitrella lunata* in the fall, and *Idotea baltica* each season. The highest concentration of grazers (11,400 m⁻²) was applied in late summer, the season in which she found the clearest segregation in her data. The variables paralleled those in the field, and thus the results can be used to make predictions in the environment she was testing.

The concept of interaction was extrapolated into a simulation model based on these experimental data.[74] This model predicts that without grazers seagrass population will decrease during eutrophication because the epiphytes will have affected its host adversely. With grazers, however, eutrophication will not have a strong impact because the epiphytes will be removed.

## IV. CONCEPTS COMMON TO FRESHWATER AND MARINE SYSTEMS

## A. RELEASING A RESOURCE LIMITATION

In the preface to his book, *Resource Competition and Community Structure*, David Tilman[75] states that he "presents a simple, graphical theory of multispecies competition for resources" which he believes "may be used to predict the dependence of the species composition and species richness of communities

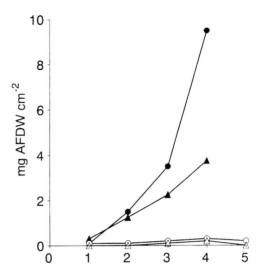

**Figure 4** Epiphyton response to microcosm treatments with *Zostera* September 4, 1987. Triangles = ambient nutrient concentrations; circles = enriched nutrient concentrations; solid symbols = grazers absent; open symbols = grazers present. (From Neckles, H. A., Ph.D. dissertation, Virginia Institute of Marine Science, The College of William and Mary, 1990. With permission.)

on the availabilities of limiting resources." He goes on to say that the majority of life-sustaining elements are essential resources for autotrophs, and these must be taken in separately (in contrast to the combined forms ingested by heterotrophs.) Growth rate is, therefore, determined by that which is most limiting. So, when two or more species compete for the same resource, the one with the lowest requirement (R*) will dominate at equilibrium. Extending this concept to an environment in which nutrient status changes, we would expect to see a concurrent adjustment in the flora. Sometimes this shift is between plant groups and sometimes between species.

One example of a replacement of dominant species with changes in nutrient application was taken from Tilman's[76] work on freshwater diatoms from Lake Michigan. *Asterionella formosa* has a lower requirement for silica and *Cyclotella meneghiana* for phosphorus. In laboratory culture with low silica concentrations the former was dominant whereas in low phosphorus concentrations the latter was dominant (Figure 5). When neither nutrient was limiting, both species were present. Conversely in high concentrations of silica, cultures became *Asterionella* and in high phosphate, *Cyclotella*.

Schelske and Stoemer with their colleagues have predicted complex assemblages of phytoplankton experimentally from a N-P factorial load.[9,77] Not only have they described shifts from diatoms to blue-green algae but also changes within these groups. Similarly, populations of nitrogen-fixing blue-green algae appeared in waters low in nitrogen, whereas species that do not fix nitrogen appear in higher N:P ratios (data from Edmonson and from Smith reviewed in Tilman).[75] The pattern differs in marine environments. Howarth[78] suggests that one of the biggest reasons that estuaries are nitrogen limited is that nitrogen fixers do not make up the deficit that they do in eutrophic lakes.

## B. PERENNIAL AND EPHEMERAL MACROPHYTES IN COMPETITION

An example of a shift in dominance of seaweed species was examined with laboratory cultures of *Ulva lactuca* and *Chondrus crispus*[79] using a de Wit factorial design. The objective of this study was to determine physical conditions in which *Chondrus* would be a more successful competitor so that *Ulva* might be reduced in aquaculture systems. When nutrients were administered equally, *Chondrus* had lower light and temperature requirements. The outcome bears on observed seasonal changes in the field.

Perennial seaweeds can store nitrogen[70,80-84] and to differing capacities. This ability allows the perennial forms a head start when other limiting resources are released. Just which macroalgae species will dominate under specific nutrient regimes is not yet predictable. Rapid growth of an annual brown seaweed *Chordaria flagelliformis* during periods of low nitrogen concentration in Nova Scotia, Canada, has been attributed to high tissue concentration of polymeric nitrogen early in the growing season.[85] This explanation was based on outdoor cultivation experiments uncoupling growth from uptake kinetics. Temporal variability can be explained by rapid short-term responses for nitrogen-starved plants.[86] This line of investigation needs considerable research.

What is established is that the opportunistic green algae such as species of *Ulva* and *Enteromorpha* show high uptake kinetics[87,88] and grow rapidly in response to nitrogen input in the laboratory[38,89] and

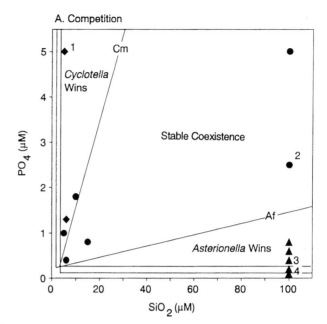

A. Competition

**Figure 5** The results of competition for silicate and phosphate. Stars show experiments in which *Asterionella* was competitively dominant, dots where the species coexisted, and diamonds where *Cyclotella* was dominant. (From Tilman, D., *Resource Competition and Community Structure,* Copyright ©1982 by Princeton University Press, Princeton, NJ. With permission from Princeton University Press.)

field.[39] *Enteromorpha* has been tied to nutrient flux from water column to sediment.[90] A macroalga such as *Gracilaria* that also grows well with supplementary nitrogen in the laboratory[38] can lose in the field in competition with *Ulva* that has been stimulated with supplementary nitrogen[39] but probably only so long as the nutrient is supplied. The highest uptake rates have been measured on short-lived opportunistic algae, filamentous algae or those with many hair-like extensions.[83] This list includes *Cladophora, Enteromorpha, Scytosiphon, Dictyosiphon* and *Ceramium.* These findings are consistent with the positive correlation between uptake of inorganic nitrogen and surface area to volume ratios in seaweeds.[91]

## C. NUTRIENT FLUXES AND C:N:P RATIOS

Short[92] has shown that eelgrass removes ammonium from an interstitial pool in the sediment. Where he removed leaves and sealed the sediment surface, ammonium was in higher concentration than where eelgrass beds had not been disturbed. Whereas *Zostera* preferentially removes ammonium over nitrate,[93] nitrate from groundwater runoff can induce nitrate reductase with the consequence of localized eutrophication.[94] Banks of decomposing seaweeds overlying sediments generate ammonium and phosphate under low oxygen conditions.[95] The induction of nitrate reductase in seaweeds has been used as an indicator of nitrogen-limitation in field populations of the kelp *Laminaria digitata.*[96] Atmospheric nitrogen loading accounts for 20–30% of biologically available nitrogen and is an effective stimulant of phytoplankton blooms.[97]

The amount of nutrients required to support benthic marine plants is theoretically less than that for phytoplankton.[98] In an analysis of 92 plants from 5 phyla and 9 locations, the median C:N:P ratio was 550:30:1, showing that macrophytes are far more N and P depleted than phytoplankton with their "Redfield ratio" of 106:16:1. That is five times the C:P and three times the C:N, resulting in low specific growth rates. Based on ash-free dry weight the composition of benthic plants is 5% protein, 80% carbohydrate and 5% lipid.[98] However, the N:P ratio can vary among seaweed species in the same location.[99] Algae near wastewater outfalls generally have higher internal nitrogen concentration than those in unpolluted areas.[70] Where nitrogen is sufficient, phosphorus can be limiting to macroalgae, the most sensitive stage being sexual fertilization.[100] This phosphorus requirement could select against certain macroalgae within shallow systems.

Nowicki[101] concludes from dynamics of nitrogen flux in mesocosm experiments that reversal of eutrophication is faster than had been previously anticipated. Indeed, the speed at which the Mumford Cove estuary returned to normal conditions was less than a year.[60] Certainly, water depth and circulation are factors in the speed of restoration. Species that do not overwinter are less dependent on nutrients in the sediment in lakes[102] or estuaries.[59,60] These conditions fit with the rapid recovery observed in Lake Washington and the switch in dominance of plant groups in England's Broads.

## D. THE EXPRESSION OF SPECIES DIVERSITY

In freshwater communities the "switch" can be induced by nutrients, but shading is not necessarily the mechanism invoked. The outcome is altered by the rest of the community. Where grazing is prevalent, phytoplankton can be cropped.[14,15] Introduction of fish that graze on zooplankton can alter the community by interfering with the zooplankton grazing pressure. The same effect is achieved by poisoning the grazers with pesticides.[16] Freshwater bodies are more homogeneous than estuaries in which water connects with the open sea, but I see no reason why shallow water and partially confined systems should not react in similar patterns.

For over 100 years, we have known that nutrient enrichment leads to decreased species diversity, with the type of enrichment influencing the species which become dominant.[103] This pattern appears to be true for aquatic as well as terrestrial systems, where highest diversity occurs in relatively resource-poor habitats. Tilman[75] points out that the most species-rich plant communities in the world occur on very nutrient-poor soils (the Fynbos of South Africa and the heath scrublands of Australia). Examples in aquatic systems include the diverse phytoplankton communities in the unproductive nutrient-poor Sargasso Sea. In contrast, productive nutrient-rich temperate waters, including those of upwelling systems, support a lower number of species (reviewed in Tilman).[75] The most extreme naturally occuring example among the phytoplankton is the near monoculture (95–99%) of *Skeletonema costatum* reported for Casco Bay, Maine, U.S., during the 2 to 3 months in which nitrogen is the only limiting factor.

So, what explains the diversity in low nutrient media? Although the least equipped species always goes to extinction, it is also true that there is no limit to the number of species on a limiting resource.[104] Coexistence, according to Tilman,[104] depends upon the tradeoffs that the plant has developed. Those traits that make the organism less fit in one circumstance make it more fit in another. Perennial plants, for example, store nitrogen, whereas ephemerals rush through their life cycle rapidly. Steady nutrient enrichment logically favors forms with the most rapid nutrient removal[105] but those species with the ability to store a limiting nutrient during period of abundance and use it later are probably favored during a fluctuating nutrient regime. In addition to the nutrient input, grazers and physical factors are parameters that regulate the outcome.[106]

A terrestrial system studied in Rothamsted, England, for 130 years (reviewed in Tilman[75]) provides the most complete long-term experimental information available on the effect of nutrient additions. The intent of these experiments is to predict species of grass that will grow competitively with specific mineral additions. When nutrient fertilization stopped, plant diversity increased. Not surprisingly, nitrogen-fixing plants (mostly legumes) competed more successfully than plants that did not fix nitrogen (grasses) on plots with soils that were poor in nitrogen but rich in other resources (e.g., phosphorus, light, water). Comparably, nitrogen-fixing blue-green algae are found in low nitrogen water in freshwater systems (Edmonson in Tilman[75]). In shallow estuaries blue-green algae can be abundant,[31,42] but Paerl[107] suggests that their presence may be associated more with low oxygen than with nutrient status.

## V. FUTURE DIRECTIONS FOR RESEARCH

Those of us examining shallow marine systems need to follow the lead of investigators on freshwater systems and partition the components to understand the mechanisms involved. We need to learn the consequences upon particular algal groups (phytoplankton and macroalgae), the vascular plants and bacteria within the same area. We must continue to separate out the role of animals to learn their impact upon the resulting community. We need multiple replicates in controlled environments — in both the field and laboratory. Remaining questions include: what are the effects of size and shape of the water body and the water flow within it on the ability to recover? Why are certain species more successful than others? What are the long-term consequences of nutrient enrichment? How can algae be used as indicators of water quality? For example, Lyngby[108] has used the nutrient tissue concentration of a red alga, *Ceramium rubrum*, as a tool to establish critical nutrients, which in Denmark changes from phosphorus in May to nitrogen in summer. Finally, how can we apply what we learn towards restoring eutrophic water bodies to healthy communities?

## VI. CONCLUSIONS

In shallow marine systems nutrient enrichment favors algae, while submerged angiosperms become light limited (Figure 6). *Zostera marina*, for example, cannot maintain its population in transparency less than

## SHALLOW MARINE SYSTEMS

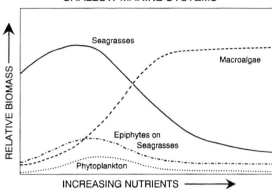

**Figure 6** Generalized shift in biomass of major plant groups with increasing nutrient input to shallow marine systems. Occasionally the phytoplankton dominate, but usually macroalgae — especially ephemeral green taxa — take off, while submersed, rooted plants decline through competition either for nitrogen or for light.

one Secchi disc depth. Sometimes the shading is from a phytoplankton bloom and sometimes from epiphytes upon seagrass blades. The most common shift is toward massive growths of macroalgae, especially green algae. In addition to shading angiosperms, large mats of these macroalgae can slow water current and compete for nutrients. Nitrogen appears to be the critical nutrient in marine systems, but phosphorus also plays a role in localized areas. Ammonium derived from sewage is most easily used by plants, but nitrogen reductase is induced in both seaweeds and seagrasses, permiting them to use nitrate from agriculture runoff. In accounting for nutrient status we must include the nitrogen in biological compounds as well as that in the water column. Water quality at the specific site regulates the outcome of nutrient enrichment, as do seasonality and the presence of animals.

## REFERENCES

1. **Edmonson, W. T.,** Phosphorus, nitrogen, and algae in Lake Washington after diversion of sewage, *Science,* 169, 373, 1970.
2. **Moss, B.,** The effects of fertilization and fish on community structure and biomass of aquatic macrophytes and epiphytic algal populations: an ecosystem experiment, *J. Ecol.,* 64, 313, 1976.
3. **Jones, R. C., Walti, K., and Adams, M. S.,** Phytoplankton as a factor in the decline of the submersed macrophyte *Myriophyllum spicatum* L. in Lake Wingra, Wisconsin, U.S., *Hydrobiologia,* 107, 213, 1983.
4. **Phillips, G. L., Eminson, D., and Moss, B.,** A mechanism to account for macrophyte decline in progressively eutrophied freshwaters, *Aquat. Bot.,* 4, 103, 1978.
5. **Hough, R. A., Fornwall, M. D., Negele, B. J., Thompson, R. L., and Putt, D. A.,** Plant community dynamics in a chain of lakes: principal factors in the decline of rooted macrophytes with eutrophication, *Hydrobiologia,* 173, 199, 1989.
6. **Schelske, C. L.,** Role of phosphorus in Great Lakes eutrophication: is there a controversy?, *J. Fish. Res. Board Can.,* 36, 286, 1979.
7. **Vollenweider, R. A.,** Scientific fundamentals of the eutrophication of lakes and flowing waters, with particular reference to nitrogen and phosphorus as factors in eutrophication, *OECD,* Paris, 1968.
8. **Stoermer, E. F., Schelske, C. L., and Feldt, L. E.,** Phytoplankton assemblage differences at inshore versus offshore stations in Lake Michigan, and their effects on nutrient enrichment experiments, *Proc. 14th Conf. Great Lakes Res.,* International Association for Great Lakes Research, Ann Arbor, MI, 1971, 114.
9. **Stoermer, E. F., Ladewski, B. G., and Schelske, C. L.,** Population responses of Lake Michigan phytoplankton to nitrogen and phosphorus enrichment, *Hydrobiologia,* 57, 249, 1978.
10. **Stoermer, E. F., Wolin, J. A., Schelske, C. L., and Conley, D. J.,** Siliceous microfossil succession in Lake Michigan, *Limnol. Oceanogr.,* 35, 959, 1990.
11. **Moss, B.,** Algal and other fossil evidence for major changes in Strumpshaw Broad, Norfolk, England in the last two centuries, *Br. Phycol. J.,* 14, 262, 1979.
12. **Wetzel, R. G.,** The role of the littoral zone and detritus in lake metabolism, *Arch. Hydrobiol. Beih. Ergebn. Limnol.,* 13, 145, 1979.
13. **Vanni, M. J. and Temte, J.,** Seasonal patterns of grazing and nutrient limitation of phytoplankton in a eutrophic lake, *Limnol. Oceanogr.,* 35, 697, 1990.
14. **Balls, H., Moss B., and Irvine, K.,** The loss of submerged plants with eutrophication. I. Experimental design, water chemistry, aquatic plant and phytoplankton biomass in experiments carried out in ponds in the Norfolk Broadland, *Freshwater Biol.,* 22, 71, 1989.

15. **Irvine, K., Moss, B., and Balls, H.,** The loss of submerged plants with eutrophication. II. Relationships between fi and zooplankton in a set of experimental ponds, and conclusions, *Freshwater Biol.*, 22, 89, 1989.

16. **Stansfield, J., Moss, B., and Irvine, K.,** The loss of submerged plants with eutrophication. III. Potential role of organochlorine pesticides: a palaeological study, *Freshwater Biol.*, 22, 109, 1989.

17. **Whitton, B. A.,** Plants as indicators of river water quality, in *Biological Indicators of Water Quality*, James A. and Evison L., Eds., John Wiley & Sons, New York, 1979, chap. 5.

18. **Sheath, R. G.,** Invasions into the Laurentian Great Lakes by marine algae, *Arch. Hydrobiol. Beih. Ergebn. Limnol.*, 25, 165, 1987.

19. **Mulligan, H. F. and Baranowski, A.,** Growth of phytoplankton and vascular aquatic plants at different nutrient levels, *Verh. Int. Ver. Limnol.*, 17, 802, 1969.

20. **Edmonson, W. T. and Lehman, J. T.,** The effect of changes in the nutrient income on the condition of Lake Washington, *Limnol. Oceanogr.*, 26, 1, 1981.

21. **Carignan, R. and Kalff, J.,** Phosphorus sources for aquatic weeds: water or sediments, *Science*, 207, 987, 1980.

22. **Shindler, D. W.,** Eutrophication and recovery in experimental lakes: implications for lake management, *Science*, 184, 897, 1974.

23. **Ahl, T.,** Natural and human effects on trophic evolution, *Arch. Hydrobiol. Beih. Ergebn. Limnol.*, 13, 259, 1979.

24. **Burkholder, J. M., Wetzel, R. G., and Klomparens, K. L.,** Direct comparison of phosphate uptake by adnate and loosely attached microalgae within an intact biofilm matrix, *Appl. Environ. Microbiol.*, 56, 2882, 1990.

25. **Burkholder, J. M. and Wetzel, R. G.,** Epiphytic alkaline phosphatase on natural and artificial plants in an oligotrophic lake: re-evaluation of the role of macrophytes as a phosphorus source for epiphytes, *Limnol. Oceanogr.*, 35, 736, 1990.

26. **Burkholder, J. M. and Wetzel, R. G.,** Epiphytic microalgae on natural substrata in a hardwater lake: seasonal dynamics of community structure, biomass and ATP content, *Arch. Hydrobiol. Suppl.*, 83, 1, 1989.

27. **Burkholder, J. M. and Wetzel, R. G.,** Response of periphyton communities to clay and phosphate loading in a shallow reservoir, *J. Phycol.*, 373, 1991.

28. **Marks, J. C. and Lowe, R. L.,** The independent and interactive effects on snail grazing and nutrient enrichment on structuring periphyton communities, *Hydrobiologia*, 185, 9, 1989.

29. **Ryther, J. H.,** The ecology of phytoplankton blooms in Moriches Bay and Great South Bay, Long Island, New York, *Biol. Bull.*, 106, 190, 1954.

30. **Ryther, J. H. and Dunstan, W. M.,** Nitrogen, phosphorus, and eutrophication in the coastal marine environment, *Science*, 171, 1008, 1971.

31. **Shepherd, S. A., McComb, A. J., Bulthuis, D. A., Neverauskas, V., Steffensen, D. A., and West, R.,** Decline of seagrasses, in *Biology of Seagrasses*, Larkum, A. W. D., McComb, A. J., and Shepherd S. A., Eds., Elsevier, New York, 1989, chap. 12.

32. **Letts, M. C. and Adeney, W. E.,** Pollution of estuaries and tidal waters Appendix VI, Fifth Rep. of the Commissions, Royal Commission on Sewage Treatment. H.M. Stationary Office, London, 1908 (reviewed in Reference 36.)

33. **Sawyer, C. M.,** The sea lettuce problem in Boston Harbor, *J. Water Poll. Cont. Fed.*, 37, 1122, 1965.

34. **Ho, Y. B.,** Mineral element content in *Ulva lactuca* L. with reference to eutrophication in Hong Kong coastal waters, *Hydrobiologia*, 77, 43, 1980.

35. **Asare, S. O.,** Animal waste as a nitrogen source for *Gracilaria tikvahiae* and *Neoagardhiella baileyi* in culture, *Aquaculture*, 21, 87, 1980.

36. **Waite, T. and Mitchell, R.,** The effect of nutrient fertilization on the benthic alga *Ulva lactuca*, *Bot. Mar.*, 15, 151, 1972.

37. **Guist, G. G. and Humm, H. J.,** Effects of sewage effluents on growth of *Ulva lactuca*, *Fl. Sci.*, 39, 267, 1976.

38. **Harlin, M. M., Thorne-Miller, B., and Thursby, G. B.,** Ammonium uptake by *Gracilaria* sp. (Florideophyceae) and *Ulva lactuca* (Chlorophyceae) in closed system fish culture, in *Proc. Int. Seaweed Symp.*, Jensen, A. and Stein J. R., Eds., Science Press, Clifton, NJ, 1978, 285.

39. **Harlin, M. M. and Thorne-Miller, B.,** Nutrient enrichment of seagrass beds in a Rhode Island coastal lagoon, *Mar. Biol.*, 65, 221, 1981.

40. **Richardson, F. D.,** The ecology of *Ruppia maritima* L. in New Hampshire (U.S.) tidal marshes, *Rhodora*, 82, 403, 1980.

41. **Orth, R. J.,** Effect of nutrient enrichment on the growth of the eelgrass *Zostera marina* in the Chesapeake Bay, Virginia, *Mar. Biol.*, 44, 187, 1977.

42. **Virnstein, R. W. and Carbonara, P. A.,** Seasonal abundance and distribution of drift algae and seagrasses in the Mid-Indian River Lagoon, Florida, *Aquat. Bot.*, 23, 67, 1985.

43. **Whelan, P. M. and Cullinane, J. P.,** The algal flora of a subtidal *Zostera* bed in Ventry Bay, Southwest Ireland, *Aquat. Bot.*, 23, 41, 1985.

44. **Wallentinus, I.,** Environmental influences on benthic macrovegetation in the Trosa-Asko area, northern Baltic Proper. II. The ecology of macroalgae and submersed phanerogams, *Contrib. Asko Lab. Univ. Stockholm*, 25, 1979, 210p.

45. **Sfrisco, A., Marcomini, A., and Pavoni, B.,** Relationships between macroalgal biomass and nutrient concentrations in a hypertrophic area in the Venice Lagoon, *Mar. Environ. Res.*, 22, 297, 1987.
46. **Costa, J. E.,** Eelgrass (*Zostera marina* L.) in Buzzards Bay: Distribution, Production, and Historical Changes in Abundance, Ph.D. dissertation, Boston University, Boston, 1988.
47. **Harlin, M. M., Thursby, G. B., and Thorne-Miller, B.,** Submerged macrophytes in coastal lagoons, in *Freshwater and Marine Plants of Rhode Island*, Sheath, R. S. and Harlin, M. M., Eds., Kendall/Hunt, Dubuque, IA, 1988, chap. 13.
48. **Thorne-Miller, B., Harlin, M. M., Thursby, G. B., Brady-Campbell, M. M., and Dworetsky, B. A.,** Variations in the distribution and biomass of submerged macrophytes in five coastal lagoons in Rhode Island, U.S.A., *Bot. Mar.*, 26, 231, 1983.
49. **Wright, T. J., Cheadle, V. I., and Palmatier, E. A.,** *A Survey of Rhode Island's Salt and Brackish Water Ponds and Marshes*, Pitman Robertson Pamphlet No. 2, R.I. Department of Agriculture and Conservation Division of Fish and Game, 1949.
50. **Conover, J. T.,** *Environmental Relationships of Benthos in Salt Ponds (Plant Relationships)*. Vol. 1 and 2., Tech. Rep. No. 3., Graduate School of Oceanography, University of Rhode Island, Kingston, RI, 1964.
51. **Twilley, R. R., Kemp, W. M., Staver, K. W., Stevenson, J. C., and Boyton, W. R.,** Nutrient enrichment of estuarine submersed vascular plant communities. 1. Algal growth and effects on production of plants and associated communities, *Mar. Ecol. Prog. Ser.*, 23, 179, 1985.
52. **Harlin, M. M., Thorne-Miller, B., and Boothroyd, J. C.,** Seagrass-sediment dynamics of a flood-tidal delta (U.S.A.), *Aquat. Bot.*, 14, 127, 1982.
53. **Laws, E. A.,** Man's impact on the marine nitrogen cycle, in *Nitrogen in the Marine Environment*, Carpenter, J. E. and Capone, D. G., Eds., Academic Press, New York, 1983, chap. 13.
54. **Borum, J.,** The quantitative role of macrophytes, epiphytes, and phytoplankton under different nutrient conditions in Roskilde Fjord, Denmark, *Proc. Int. Symp. Aquat. Macrophytes*, The Netherlands: Faculty of Science, Nijmegen, 35, 1983.
55. **Backman, T. W. and Barilotti, D. C.,** Irradiance reduction: effects on standing crops of the eelgrass *Zostera marina* in a coastal lagoon, *Mar. Biol.*, 34, 33, 1976.
56. **Bulthuis, D. A. and Woelkerling, W. J.,** Biomass accumulation and shading effects of epiphytes on leaves of the seagrass, *Heterozostera tasmanica*, in Victoria, Australia, *Aquat. Bot.*, 16, 137, 1983.
57. **Silberstein, K., Chiffings, A. W., and McComb, A. J.,** The loss of seagrass in Cockburn Sound, Western Australia. III. The effect of epiphytes on productivity of *Posidonia australis* Hook. f., *Aquat. Bot.*, 24, 355, 1986.
58. **Dennison, W. C.,** Effects of light on seagrass photosynthesis, growth and depth distribution, *Aquat. Bot.*, 27, 15, 1987.
59. **French, D. P., Harlin, M. M., Gundlach, E., Pratt, S., Rines, H., Jayko, K., Turner, C., and Puckett, S.,** *Mumford Cover Water Quality: 1988 Monitoring Study and Assessment of Historical Trends*, Applied Science Associates, Narragansett, RI, 1989, 126.
60. **French, D. P., Harlin, M. M., Pratt, S., Rines, H., and Puckett, S.,** *Mumford Cove Water Quality: 1989 Monitoring Study of Macrophytes and Benthic Invertebrates*, Applied Science Associates, Narragansett, RI, 1989, 56.
61. **Nowicki, B. L. and Oviatt, C. A.,** Are estuaries traps for anthropogenic nutrients? Evidence from estuarine mesocosm, *Mar. Ecol. Prog. Ser.*, 66, 131, 1990.
62. **Duarte, C. M. and Kalff, J.,** Influence of lake morphometry on the response of submerged macrophytes to sediment fertilization, *Can. J. Fish. Aquat. Sci.*, 45, 216, 1988.
63. **Howard, R. K.,** Impact of feeding activities of epibenthic amphipods on surface-fouling of eelgrass leaves, *Aquat. Bot.*, 14, 91, 1982.
64. **Cattaneo, A.,** Grazing on epiphytes, *Limnol. Oceanogr.*, 28, 124, 1983.
65. **Orth, R. J. and Van Montfrans, J.,** Epiphyte-seagrass relationships with an emphasis on the role of micrograzing: a review, *Aquat. Bot.*, 18, 43, 1984.
66. **Van Montfrans, J., Wetzel, R. L., and Orth, R. J.,** Epiphyte-grazer relationships in seagrass meadows: consequences for seagrass growth and production, *Estuaries*, 7, 289, 1984.
67. **Hillman, K., Walker, D. I., Larkum, A. W. D., and McComb, A. J.,** Productivity and nutrient limitation, in *Biology of Seagrasses*, Larkum, A. W. D., McComb, A. J., and Shepherd, S. A., Eds., Elsevier, New York, 1989, chap. 19.
68. **Wetzel, R. L. and Neckles, H. A.,** A model of *Zostera marina* L. photosynthesis and growth: simulated effects of selected physical-chemical variables and biological interactions, *Aquat. Bot.*, 26, 307, 1986.
69. **Duke, C. S., Litaker, W., and Ramus, J.,** Effects of temperature, nitrogen supply, and tissue nitrogen on ammonium uptake rates of the chlorophyte seaweeds *Ulva curvata* and *Codium decorticatum*, *J. Phycol.*, 25, 113, 1989.
70. **Hanisak, M. D.,** The nitrogen relationship of marine macroalgae, in *Nitrogen in the Marine Environment*, Carpenter, E.J. and Capone, D.G., Eds., Academic Press, New York, 1983, chap. 19.
71. **Parker, H. S.,** Influence of relative water motion on the growth, ammonium uptake and carbon and nitrogen composition of *Ulva lactuca* (Chlorophyta), *Mar. Biol.*, 63, 309, 1981.
72. **Thorne-Miller, B. and Harlin, M. M.,** The production of *Zostera marina* in a coastal lagoon in Rhode Island, U.S.A., *Bot. Mar.*, 27, 539, 1984.

73. **Williams, S. L. and Ruckelshaus, M. H.**, Effects of nitrogen and grazing on the interaction between eelgrass (*Zostera marina* and *Z. japonica*) and epiphytes, unpublished data, 1989.

74. **Neckles, H. A.**, Relative Effects of Nutrient Enrichment and Grazing on Epiphyton-Macrophyte (*Zostera marina* L.) *Dynamics*. Ph.D. dissertation, Virginia Institute of Marine Science, The College of William and Mary, Gloucester Point, VA, 1990.

75. **Tilman, D.**, *Resource Competition and Community Structure*, Princeton University Press, Princeton, NJ, 1982.

76. **Tilman, D.**, Resource competition between planktonic algae: an experimental and theoretical approach, *Ecology*, 58, 338, 1977.

77. **Schelske, C. L., Rothman, E. D., Stoermer, E. F., and Santiago, M. A.**, Responses of phosphorus limited Lake Michigan phytoplankton to factorial enrichments with nitrogen and phosphorus, *Limnol. Oceanogr.*, 19, 409, 1974.

78. **Howarth, R. W.**, Nutrient limitation of net primary production in marine ecosystems, *Annu. Rev. Ecol.*, 19, 89, 1988.

79. **Enright, C. T.**, Competitive interaction between *Chondrus crispus* (Florideophyceae) and *Ulva lactuca* (Chlorophyceae) in *Chondrus* aquaculture, in *Proc. Int. Seaweed Symp.*, Jenson, A. and Stein, J. R., Eds., Science Press, Clifton, NJ, 1978.

80. **Chapman, A. R. O. and Craigie, J. S.**, Seasonal growth in *Laminaria longicruris*: relations with dissolved inorganic nutrients and internal reserves of nitrogen, *Mar. Biol.*, 40, 197, 1977.

81. **Bird, K. T., Habig, C., and DeBusk, T.**, Nitrogen allocation and storage patterns in *Gracilaria tikvahiae* (Rhodophyta), *J. Phycol.*, 18, 344, 1982.

82. **Asare, S. O. and Harlin, M. M.**, Seasonal fluctuations in tissue nitrogen for five species of perennial macroalgae in Rhode Island Sound, *J. Phycol.*, 19, 254, 1983.

83. **Wallentinus, I.**, Comparisons of nutrient uptake rates for Baltic macroalgae with different thallus morphologies, *Mar. Biol.*, 80, 215, 1984.

84. **Lapointe, B. E.**, Strategies for pulsed nutrient supply to *Gracilaria* cultures in the Florida keys: interactions between concentration and frequency of nutrient pulses, *J. Exp. Mar. Biol. Ecol.*, 93, 211, 1985.

85. **Probyn, T. A. and Chapman, A. R. O.**, Summer growth of *Chordaria flagelliformis* (O. F. Muell.) C. Ag. Physiological strategies in a nutrient stressed environment, *J. Exp. Mar. Biol. Ecol.*, 73, 243, 1983.

86. **Williams, S. L. and Herbert, S. K.**, Transient photosynthetic responses of nitrogen-deprived *Petalonia fascia* and *Laminaria saccharina* (Phaeophyta) to ammonium resupply, *J. Phycol.* 25, 515, 1989.

87. **O'Brien, M. C. and Wheeler, P. A.**, Short term uptake of nutrients by *Enteromorpha prolifera* (Chlorophyceae), *J. Phycol.*, 23, 547, 1987.

88. **Bjornsater, B. R. and Wheeler, P. A.**, Effect of nitrogen and phosphorus supply on growth and tissue composition of *Ulva fenestrata* and *Enteromorpha intestinalis* (Ulvales, Chlorophyta), *J. Phycol.*, 26, 603, 1990.

89. **Harlin, M. M.**, Nitrate uptake by *Enteromorpha* spp. (Chlorophyceae): applications to aquaculture systems, *Aquaculture*, 15, 373, 1978.

90. **Owens, N. J. P. and Stewart, W. D. P.**, *Enteromorpha* and the cycling of nitrogen in a small estuary, *Estuarine Coastal and Shelf Sci.* 17, 287, 1983.

91. **Rosenberg, G. and Ramus, J.**, Uptake of inorganic nitrogen and seaweed surface area:volume ratios, *Aquat. Bot.*, 19, 65, 1984.

92. **Short, F. T.**, The response of interstitial ammonium in eelgrass (*Zostera marina* L.) beds to environmental perturbations, *J. Exp. Mar. Biol. Ecol.*, 68, 195, 1983.

93. **Thursby, G. B. and Harlin, M. M.**, Leaf-root interaction in the uptake of ammonia by *Zostera marina*, *Mar. Biol.*, 72, 109, 1982.

94. **Maier, C. M. and Pregnall, A. M.**, Increased macrophyte nitrate reductase activity as a consequence of groundwater input of nitrate through sandy beaches, *Mar. Biol.*, 107, 263, 1990.

95. **Lavery, P. S. and McComb, A. J.**, Macroalgal-sediment nutrient interactions and their importance to macroalgae nutrition in a eutrophic estuary, *Estuarine Coastal Shelf Sci.*, 32, 281, 1991.

96. **Davidson, I. R. and Stewart, W. D. P.**, Studies on nitrate reductase activity in *Laminaria digitata* (Huds.) Lamour. II. The role of nitrate availability in the regulation of enzyme activity, *J. Exp. Mar. Biol. Ecol.*, 79, 65, 1984.

97. **Paerl, H. W., Rudek, J., and Mallin, M. A.**, Stimulation of phytoplankton production in coastal waters by natural rainfall inputs: nutritional and trophic implications, *Mar. Biol.*, 107, 247, 1990.

98. **Atkinson, M. J. and Smith, S. V.**, C:N:P ratios of benthic marine plants, *Limnol. Oceanogr.*, 23, 568, 1983.

99. **Kornfeldt, R. A.**, Relation between introgen and phosphorus content of macroalgae and the waters of Northern Oresund, *Bot. Mar.*, 25, 197, 1982.

100. **Hoffmann, A. J., Avila, M., and Santelices, B.**, Interactions of nitrate and phosphate on the development of microscopic stages of *Lessonia nigrescens* Bory (Phaeophyta), *J. Exp. Mar. Biol. Ecol.*, 74, 177, 1984.

101. **Nowicki, B. L.**, The Fate of Nutrient Inputs: Evidence from Estuarine Mesocosms, Ph.D. dissertation, University of Rhode Island, Kingston, 1991.

102. **Canfield, D. E., Jr., Langeland, K. A., Maceina, M. J., Haller, W. T., Shireman, J. V., and Jones, J. R.**, Trophic state classification of lakes with aquatic macrophytes, *Can. J. Fish. Aquat. Sci.*, 40, 1713, 1983.

103. **Lawes, J. and Gilbert, J.**, Agricultural, botanical and chemical results of experiments on mixed herbage of permanent grassland, conducted for many years in succession on the same land, *Phil. Trans. R. Soc. London*, 171, 189, 1880.

104. **Tilman, D.,** Constraints and trade-offs: toward a predictive theory of competition and succession, *Oikos*, 58, 3, 1990.

105. **Fujita, R. M.,** The role of nitrogen status in regulating transient ammonium uptake and nitrogen storage by macroalgae, *J. Exp. Mar. Biol. Ecol.*, 92, 283, 1985.

106. **Olson, A. M. and Lubchenco, J.,** Competition in seaweeds: linking plant traits to competitive outcomes, *J. Phycol.*, 26, 1, 1990.

107. **Paerl, H. W., Crocken, K. M., and Prufert, L. E.,** Limitation of $N_2$ fixation in coastal marine waters: Relative importance of molybdenum, iron, phosphorus, and organic matter availability, *Limnol. Oceanogr.*, 32, 525, 1987.

108. **Lyngby, J. E.,** Monitoring of nutrient availability and limitation using the marine macroalga *Ceramium rubrum* (Huds.) C. Ag., *Aquat. Bot.*, 38, 153, 1990.

# Chapter 12

# The Commercial Fisheries in Three Southwestern Australian Estuaries Exposed to Different Degrees of Eutrophication

*R. A. Steckis, I. C. Potter, and R. C. J. Lenanton*

## CONTENTS

I. Introduction ................................................................................................................189
II. Descriptions of the Three Estuaries ........................................................................190
    A. Peel-Harvey Estuary ..........................................................................................190
    B. Swan Estuary ......................................................................................................192
    C. Leschenault Inlet ................................................................................................192
III. Estuarine Fisheries in Southwestern Australia ......................................................193
IV. Influence of Eutrophication on the Commercial Fisheries ....................................193
    A. Influence of Macroalgae ....................................................................................193
    B. Influence of the Blue-Green Alga *Nodularia spumigerna* ..............................197
V. Conclusions ..............................................................................................................198
References ......................................................................................................................201

## I. INTRODUCTION

Estuaries in many parts of the world receive nutrient runoff from urban settlement, agricultural land and forest clearing and thus have become eutrophic.[1-5] This form of eutrophication is referred to as "cultural" eutrophication. The ways in which estuaries respond to nutrient enrichment are determined by their physical, chemical and geological characteristics.[6] These include the depth of the water column, the flushing rate of the system, the amount of chemical conversion (oxidation or reduction) of the nutrients and the sedimentary characteristics of the benthos. Thus, for example, estuaries which experience considerable nutrient enrichment, but have a deep entrance and large tidal exchange, e.g., San Francisco Bay, do not exhibit the extreme manifestations of eutrophication, such as the production of large algal blooms and macroalgal growths. In other words, the high flushing rates in those systems prevent the nutrients from accumulating, and thus increasing primary productivity.[2,7] On the other hand, those estuaries that are shallower and have a much smaller tidal interchange with the open sea, such as Chesapeake Bay on the east coast of North America, and even more particularly, the Peel-Harvey Estuary in southwestern Australia, which has a very narrow entrance channel (a large artifical entrance was opened in 1994), exhibit a far greater response to nutrient enrichment. These are reflected by the development of large blooms of phytoplankton and changes in the abundance and species composition of the macrophytes.[1,4,8]

Large increases in macrophyte growth can lead to corresponding increases in fish production, either by providing more food, and thus an increase in growth rate, and/or more shelter and thereby higher survival rates.[1,9-11] Fish abundance and biomass, and sometimes species richness, are often correlated with macrophyte biomass and therefore tend to be higher in vegetated than unvegetated areas.[10,12]

An increase in phytoplankton, such as nanoplankton and diatoms, usually leads to increases in zooplankton biomass and thus provides more food near the base of the food chain.[13] Artificial enrichment of certain Canadian streams with nitrogen and phosphorus has been shown to result in an increased standing crop of both periphyton and benthic invertebrates and, as a consequence, a faster growth rate among the resident salmonid fish.[14] Keller et al.[15] have demonstrated that an increased production of diatoms following nutrient enrichment leads to increases in the zooplankton that graze on those diatoms and thus more food for the planktivorous juvenile menhaden (*Brevoortia tyrannus*) which therefore grow faster. The fish communities that tend to be advantaged by increases in phytoplankton production brought

0-8493-6839-1/95/$0.00+$.50
© 1995 by CRC Press, Inc.

about by nutrient input are those that are represented by "smaller more rapid reproducers" which occupy lower trophic levels, i.e., planktivores.[16] Examples of fisheries exploiting such species are those for herring and sprat in the Baltic Sea and for Atlantic menhaden (*Brevoortia tyrannus*) in Chesapeake Bay.[17,18]

Excessive blooms of phytoplankton and growth of epiphytic diatoms in eutrophic systems reduce the amount of light available for photosynthesis and can thus lead to declines in seagrass meadows and other macrophyte communities and thereby to declines in the production of some species of fish through the destruction of spawning habitats.[1,19] This type of situation, along with other effects of pollution such as anoxia, have contributed to a decline in spawning and nursery areas and thus, the numbers of larval and juvenile striped bass (*Morone saxatilis*) in Chesapeake Bay.[20]

The food web can be dramatically altered by eutrophication. For example, nutrient enrichment of Cockburn Sound in Western Australia (Figure 1) has resulted in the food web changing from one based on the detritus produced by macrophytes to one based on phytoplankton. This has led to a significant change in fish community structure.[21] The diets of many commercially important species rely on the animals and plants associated with a detritus-based food web. These species include King George whiting (*Sillaginodes punctata*), cobbler (*Cnidoglanis macrocephalus*) and yelloweye mullet (*Aldrichetta forsteri*). However, the current success of the Cockburn Sound bait fishery is related to the abundance of the planktivorous Western Australian pilchard (*Sardinops sagax neopilchardus*) and scaly mackerel (*Sardinella lemuru*). These species belong to the category which Nichols et al.[16] regard as the type most likely to dominate in a eutrophic system.

Marked increases in nutrient inputs to aquatic systems sometimes lead to the production of large blooms of blue-green algae such as *Nodularia spumigena*.[4,22] These can have detrimental effects on fish stocks through depleting the oxygen levels in the benthos and/or by producing toxins.[23-25] An example of a commercial species that tends to be disadvantaged by eutrophication is the Norwegian lobster (*Nephrops norvegicus*) in the Baltic Sea, which can suffer severe declines as a result of the anoxic conditions produced by blooms of *Nodularia spumigena*.[26]

From the above account of the way in which macroalgal growths and phytoplankton blooms can influence fish communities, it is evident that eutrophication can have both detrimental and beneficial effects on commercial fisheries, depending on the circumstances.[9,27]

This chapter describes the commercial fishery in the now highly eutrophic Peel-Harvey Estuary (Figure 1). This fishery operates predominantly in the two main basins of the estuary. Particular emphasis has been placed on determining the extent to which the catch, effort and catch per unit effort (CPUE), changed in this system following both the development of massive macroalgal growths in the early 1970s and the production of large seasonal blooms of diatoms and the blue-green alga *Nodularia spumigena* since the late 1970s. Comparisons are made between the CPUEs for the fishery in the Peel-Harvey with those of the Swan Estuary, 50 km to the north, and with those of Leschenault Inlet, an estuary 85 km to the south. Although the Swan Estuary is nutrient enriched,[28] the effects of eutrophication are not evident in the large basins where the fishery is mainly based. In other words, the macroalgal growths and the phytoplankton blooms are not nearly as pronounced in this region of the Swan Estuary as in the basins of the Peel-Harvey. The Leschenault Inlet is eutrophic[29] and, like the Peel-Harvey Estuary, now contains large growths of macrophytes but, unlike that system, does not support conspicuous blooms of blue-green algae. Comparisons within and between the fisheries in these three systems help clarify some of the ways in which macroalgal growths and blue-green algal blooms can affect estuarine fisheries.

## II. DESCRIPTIONS OF THE THREE ESTUARIES

### A. PEEL-HARVEY ESTUARY

The Peel-Harvey Estuary, which was formed by the coalescing of three estuarine systems as a result of the sea level rises that started occurring about 6000 years ago, assumed its current morphology about 2000 years ago.[30,31] This estuary, with an area of 133 km², is one of the largest estuarine systems in temperate Australia.[25] It comprises two large basins, the Peel Inlet and the Harvey Estuary, which are generally less than 2 m deep, and the saline reaches of three main tributary rivers (Figure 1). The Murray and Serpentine Rivers discharge into the Peel Inlet, while the Harvey River flows into the Harvey Estuary (Figure 1). These river systems collect water from a greater catchment with an area of 11,378 km². Water supply dams have been constructed on the Harvey and Serpentine Rivers, but the potential reduction in total riverine discharge into the Peel-Harvey[8] has been offset by a considerable increase in runoff due to clearing in the catchment.

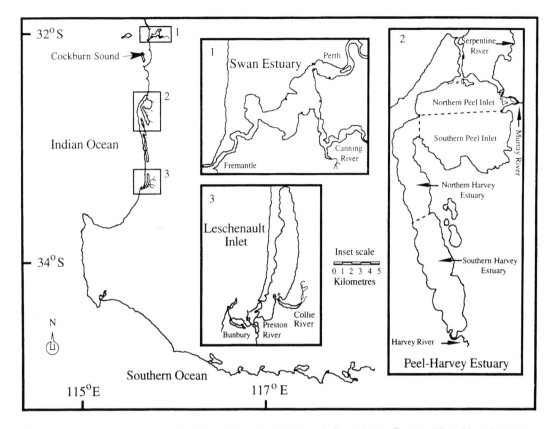

**Figure 1** The location and morphology of the Swan Estuary (1), Peel-Harvey Estuary (2) and Leschenault Inlet (3).

The climate in the region of the Peel-Harvey Estuary, and therefore also of the Swan Estuary and Leschenault Inlet, is Mediterranean, with hot dry summers and mild wet winters.[28,29,31] Average annual rainfall in the area of the estuary is 825 mm, most of which falls between May and October.[31] Direct rainfall accounts for 23–30% of the freshwater input into the estuary, ground water discharge less than 1%, with riverine inputs accounting for the rest. Evaporation is high (approximately 1980 mm $y^{-1}$),[31] producing hypersaline conditions in some shallow regions during the dry summer and autumn.[32]

Water circulation of the Peel-Harvey Estuary is driven by tidal currents, wind, density-induced circulation and river flow.[33] Wind exerts an important influence on the hydrodynamics of the estuary and causes resuspension of sediments and thus nutrient exchange between the sediments and the water column.[34] The estimated residence time is 75 days, which means that between 2.5 and 4.9 times of the volume of the estuary is flushed each year, depending on the extent to which mixing occurs.[33]

Prior to the 1970s, the dominant macrophytes in the Peel-Harvey Estuary were aquatic angiosperms, such as *Ruppia megacarpa* and *Halophila ovalis*.[30] The extreme manifestations of eutrophication exhibited by the Peel-Harvey Estuary since the late 1960s, as a result of nutrient enrichment through runoff from surrounding agricultural land, include the production of massive growths of green macroalgae and very large blooms of phytoplankton such as diatoms and blue-green algae.[8,35,36] However, the massive growths of macroalgae occur mainly in the Peel Inlet, while the blooms of phytoplankton, such as diatoms, and the extensive blooms of blue-green algae are found predominantly in the Harvey Estuary.[5,8] Macroalgal production in the Peel-Harvey system rose sharply in the 1970s, mainly as a result of growth of the green alga *Cladophora montagneana*.[4] This alga was subsequently replaced by other green algae, such as *Chaetomorpha linum, Ulva rigida* and *Enteromorpha intestinalis* which, during the 1980s, accounted for over 85% of the total macrophyte biomass in the Peel-Harvey.[37] In contrast to the situation with macroalgae, conspicuous blooms of the blue-green alga *Nodularia spumigena* did not start appearing until 1978 and these are highly seasonal, being largely restricted to the late spring and early summer.[4,8] (Refer to Chapter 2 in this book for a detailed account of the eutrophication in this estuary.[8])

The Peel-Harvey Estuary supports the most important commercial and recreational estuarine fishery in Western Australia.[9,38] For example, the commercial catch in the Peel-Harvey Estuary in 1990/91 was 317 t, which represented 37% of the total catch taken by the commercial estuarine fisheries of the state in that year.[39] The main commercial teleost species are yelloweye mullet (*Aldrichetta forsteri*), sea mullet (*Mugil cephalus*) and cobbler (*Cnidoglanis macrocephalus*).[9] Seasonal catches of two crustaceans, i.e., blue manna crab (*Portunus pelagicus*) and western king prawn (*Penaeus latisulcatus*), are also important.[40,41] The main recreational species are cobbler, blue manna crab, western king prawn, mulloway (*Argyrosomus hololepidotus*), tailor (*Pomatomus saltatrix*) and King George whiting (*Sillaginodes punctata*).[40-42]

## B. SWAN ESTUARY

The Swan Estuary was formed as a drowned river valley about 6000 years ago[43] and is consequently deeper than the Peel-Harvey Estuary.[44] This estuary, which flows through the city of Perth, the capital of Western Australia (Figure 1), receives water from the Avon River and Swan coastal catchments, which together cover an area of approximately 129,000 km². The zone of estuarine influence extends approximately 60 km upstream in the Swan River.[28] The upper estuary, which comprises the tidal reaches of the Swan and Canning Rivers, expands into the large basins of the middle estuary, which in turn enter a long narrow inlet channel before discharging into the sea at Fremantle (Figure 1). The basin of the Swan Estuary comprises marginal shallows of 2 to 3 m depth and channels that are up to 21 m in depth.[45] Average annual rainfall for Perth is 872 mm, with the majority of the rain falling between the months of May and August.[28,46] This estuary is exposed to a highly seasonal pattern of river flow, resulting in the upper parts of the estuary becoming fresh in winter and markedly saline in summer.[47] The salinity in the waters of this estuary rarely exceed full strength sea water (35‰) and, in those areas where this does occur, it is never by more than 3–4‰.[47]

Since nutrients tend to settle in the sediments of the deep channels of the broad expanses of the basins of the Swan, they are below the photic zone and are therefore not available to either macrophytes or phytoplankton. Furthermore, the Swan Estuary is far better flushed than the Peel-Harvey. Thus, although the Swan Estuary is nutrient-enriched, and its upper reaches can be regarded as eutrophic, the basins and entrance channel are essentially mesotrophic.[48] The basins of the estuary, in which the commercial fishery is largely based, do not experience the same large phytoplankton blooms and massive growths of macrophytes that occur in the basins of the Peel-Harvey Estuary.

The Swan Estuary contains an important commercial and recreational fishery. The commercial fishery is similar to that of the Peel-Harvey and is likewise managed as a limited entry fishery. The total commercial catch in the financial year 1990/91 was 112 t.[39] The major commercial species have typically been the sea mullet, cobbler, yelloweye mullet, Perth herring (*Nematalosa vlaminghi*) and to a lesser extent the blue manna crab.[49-52] The most important recreational species are the cobbler, black bream (*Acanthopagrus butcheri*), tailor, mulloway, western school prawn (*Metapenaeus dalli*), blue manna crab and western king prawn.[52,53]

## C. LESCHENAULT INLET

Leschenault Inlet is a long, coastal "lagoon-like" estuary covering an area of 27 km² and rarely exceeds 2 m in depth.[37] This estuary, which is located in an interdunal depression, comprises the drowned lower reaches of the Preston and Collie River systems. The overall catchment area of the estuary is 2900 km². The mouth of this estuary, which is represented by an artificial cut constructed in 1951, is located just north of the town of Bunbury, approximately 150 km south of Perth and 85 km south of Mandurah (Figure 1). Leschenault Inlet, which is supplied by the Collie and Preston Rivers, remains essentially marine for much of the year, with the northern (upper) end becoming hypersaline in summer.[37]

Average annual rainfall in the region is 875 mm, most of which falls between May and August. Evaporation is lower than in the Peel-Harvey Estuary, averaging 1200 mm annually.

Glover[29] describes Leschenault Inlet as mesotrophic to mildly eutrophic. Although the areal biomass of macrophytes in Leschenault Inlet is similar to that in the Peel-Harvey (123 and 160 g dry wt m⁻², respectively), the species compositions of the macrophytes in the two systems differ markedly.[37] Thus, in contrast to the Peel-Harvey, the macrophytes of Leschenault Inlet contain a larger contribution by seagrasses (>30% of biomass) and brown algae (>20% of biomass) and a far lower contribution by green algae.[37]

The commercial fishery of Leschenault Inlet in the financial year 1990/91 landed 123 t, i.e., almost the same as that of the Swan Estuary and nearly half that of the Peel-Harvey Estuary.[39] The major commercially exploited species are yelloweye mullet, sea mullet, Perth herring, King George whiting, western sand whiting (*Sillago schomburgkii*), cobbler and the blue manna crab. The two whiting species, together with cobbler, western school prawn, western king prawn and the blue manna crab, are major components of the recreational fishery of Leschenault Inlet.

## III. ESTUARINE FISHERIES IN SOUTHWESTERN AUSTRALIA

The fisheries based in the estuaries of Western Australia are among the oldest commercial fisheries in the State.[44] The fishery in the Swan Estuary was established as early as 1829, i.e., not long after settlement and was followed in the years up to 1930 by the development of fisheries in the Peel-Harvey, Leschenault Inlet and a number of estuaries along the south coast of the State. Prior to 1899, no distinction was made between commercial and recreational fishermen.[54] Since 1899, those people who were catching fish for sale have been required to be licensed and thus became distinguished from recreational fishermen.[54] The commercial estuarine fisheries were mainly established to provide fish for the populations that settled in the vicinity of those estuaries.[44] Initially, these fisheries were small and prevented from expansion by the lack of infrastructure required to allow transport of the catch to the major population center of Perth.

The development of a commercial fishery for the western rock lobster (*Panulirus cygnus*) in marine waters in the 1950s provided the estuarine net fishermen with a market for the plentiful two mullet species and Perth herring as bait for this crustacean. The subsequent rapid expansion of the rock lobster fishery resulted in the commercial estuarine fisheries in the Peel-Harvey Estuary, Swan Estuary and Leschenault Inlet developing into large bait fisheries. These fisheries, together with those for Australian salmon (*Arripis truttaceus*) and Australian herring (*Arripis georgianus*) in inshore marine waters, provided the bulk of the rock lobster bait through to the late 1980s, when the importation of cheaper frozen bait led to a decline in the demand for locally caught bait. The decline in bait prices has resulted in the commercial estuarine fishermen turning to more valuable table species, such as crabs, prawns and cobbler, to help replace their loss of income from bait species. This change has the potential to bring commercial fishermen into conflict with the very large recreational fishing sector.

A study commissioned in 1969 suggested that there were more commercial fishermen operating in southwestern Australian estuaries than could be economically supported.[44] The Western Australian Department of Fisheries recommended legislation to reduce the number of fishermen in some estuaries to a level at which individual licensed fishermen could be economically viable, and to prevent the movement of commercial fishermen between estuaries.[44] This legislation was enacted in 1969 and has resulted in a steady reduction in the number of licensed fishermen in those estuaries.

Nineteen of the 42 estuaries in temperate Western Australia are currently fished commercially. The most important are the Peel-Harvey Estuary, Swan Estuary and Leschenault Inlet, respectively.[44,55] In 1991, these three estuaries contributed about 54% of the total commercial estuarine catch in Western Australia.[39] The commercial estuarine fisheries of Western Australia accounted for 12.1% by weight of the total finfish catch in the state in the years between 1976 and 1984.[55] During those years, the contribution made by species that are either resident in estuaries or use estuaries at some stage of their life cycles, represented 20.3% by weight of the total commercial finfish catch of the State.[55]

The commercial estuarine fisheries of Western Australia are typically multi-species and employ more than one type of fishing gear to catch fish.[44,55] Gill and haul netting are used to catch teleosts and the blue manna crab, while beam tide nets are used to obtain the western king prawn. Commercial fishermen generally operate from outboard powered dinghies.[44] The only changes in the way in which estuarine commercial fishermen have operated in Western Australia during the 1970s and 1980s have been the introduction of monofilament nets and the use of trailerized dinghies. The latter makes the fishermen more mobile and thus better able to follow the movements of fish.

## IV. INFLUENCE OF EUTROPHICATION ON THE COMMERCIAL FISHERIES

### A. INFLUENCE OF MACROALGAE

The total annual catch in the Peel-Harvey Estuary between 1952 and 1970 lay between 322 and 454 t (Figure 2). It rose sharply to reach a maximum of 995 t in 1976. Large catches were obtained in each

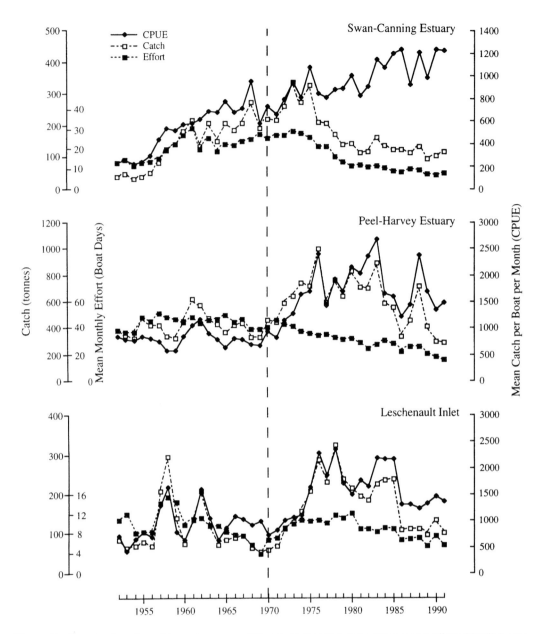

**Figure 2** Annual values for catch, and mean monthly values for effort and catch per unit effort (CPUE) for the commercial fisheries in Swan Estuary, Peel-Harvey Estuary and Leschenault Inlet between 1952 and 1991. The vertical dashed line represents the approximate time when the abundance of macroalgae in the Peel-Harvey Estuary started to increase markedly.

subsequent year and were as high as 891 t in 1983. Although a substantial catch was obtained in 1988, the catches have essentially declined since 1983. The decline in 1989–1991 to the levels of the 1960s largely reflects the influence of the fall in the value of bait fish.

Effort remained relatively stable between 1953 and the early 1970s and then, as a result of legislation, declined progressively (Figure 2). The decline in the last 3 years of the study period, i.e., 1989–1991, also reflects a reduction in effort as a consequence of the reduced value of bait fish. CPUE followed similar trends as catch.

The annual values for catch, effort and CPUE between 1952 and 1970 tended to be more variable in Leschenault Inlet than in the Peel-Harvey. However, the subsequent trends shown by these three variables in Leschenault Inlet were similar to those exhibited in the Peel-Harvey.

In contrast to the situation in both the Peel-Harvey and Leschenault Inlet, the catches and CPUE in the Swan Estuary rose less sharply in the 1970s and the catch declined markedly after 1975.

Analysis of variance showed that the CPUEs for the total fin fishery and for the fishery for *Aldrichetta forsteri* differed significantly among the 1960s, 1970s and 1980s in the Peel-Harvey Estuary, Swan Estuary and Leschenault Inlet (Table 1). Tukey's *a posteriori* test showed that, in all but the case of *A. forsteri* in Leschenault Inlet, these CPUEs were significantly higher in both the 1970s and 1980s than in the 1960s. Furthermore, the CPUE for *A. forsteri* in Leschenault Inlet in the 1980s was significantly higher than in the 1960s, and the values for the 1970s, were markedly higher than the 1960s, but not sufficiently so as to be significant at $p < 0.05$.

There was a significant difference between the CPUEs for *Mugil cephalus* in the three decades in the case of the Peel-Harvey and Swan estuaries, but not with Leschenault Inlet (Table 1). The CPUEs for this species were significantly higher in both the 1970s and 1980s than in the 1960s in the Peel-Harvey. The CPUEs in the Swan Estuary for the 1980s were significantly higher than those in the 1970s, but not than those in the 1960s. Although the differences between decades were not significant in Leschenault Inlet, it is worth noting that the CPUEs in this system in the 1970s and 1980s were still 34.5 and 75% greater than in the 1960s, respectively.

The CPUEs for *C. macrocephalus* did not show any significant interdecade differences in either of the Peel-Harvey and Swan estuaries (Table 1). This is almost certainly related to the large 95% confidence limits, which in turn reflect the "boom and bust" cycles that are undergone by the fishery for this valuable species.[56] These occur as a result of intense fishing pressure and a relatively slow recovery rate, which is due to a combination of the fishing of individuals before they reach maturity, the minimum time taken to reach maturity (3 years) and the very low fecundity.[56,57] Extreme fishing pressure almost certainly accounts for the marked decline in the CPUE for *C. macrocephalus* in Leschenault Inlet between the 1960s and 1980s (Table 1).

It is relevant that the increase in CPUEs for the total fishery and of the fishery for the three main species in the 1970s was much greater in the Peel-Harvey than in the Swan Estuary and that it was only in the former system that massive macroalgal growths developed during the 1970s.

ANOVAs showed that the CPUEs for the total fin fishery and for the fishery for *A. forsteri*, *M. cephalus* and *C. macrocephalus* differed significantly among the three systems in the 1970s and 1980s,

**Table 1** Arithmetic Mean CPUE (Catch in kg per Boat per Month per Year) ±95% Confidence Limits, for the Total Fin Fishery and the Fishery for *Aldrichetta forsteri, Mugil cephalus* and *Cnidoglanis macrocephalus*, in the 1960s, 1970s and 1980s in the Peel-Harvey, Swan and Leschenault Estuaries.[a] Significance Levels for the Results of ANOVAs (log[n+1] Transformed Data) are also shown[a]

| Species | 1960s | 1970s | 1980s | Significance levels |
|---|---|---|---|---|
| Total fin fishery | | | | |
| Peel-Harvey | 826 ± 120 | 1490 ± 342 | 1915 ± 336 | <0.001 |
| Swan | 678 ± 82 | 841 ± 83 | 1048 ± 101* | <0.001 |
| Leschenault Inlet | 976 ± 196 | 1455 ± 425 | 1661 ± 278 | <0.01 |
| *A. forsteri* | | | | |
| Peel-Harvey | 259 ± 56 | 555 ± 224 | 829 ± 227 | <0.001 |
| Swan | 45 ± 14 | 78 ± 17 | 110 ± 45 | <0.001 |
| Leschenault Inlet | 379 ± 87 | 444 ± 121 | 823 ± 163* | <0.001 |
| *M. cephalus* | | | | |
| Peel-Harvey | 196 ± 36 | 426 ± 114 | 656 ± 70 | <0.001 |
| Swan | 198 ± 46 | 189 ± 24 | 252 ± 39* | <0.05 |
| Leschenault Inlet | 145 ± 42 | 195 ± 64 | 254 ± 71 | n.s. |
| *C. macrocephalus* | | | | |
| Peel-Harvey | 169 ± 132 | 284 ± 118 | 212 ± 127 | n.s. |
| Swan | 113 ± 43 | 103 ± 30 | 67 ± 15 | n.s. |
| Leschenault Inlet | 163 ± 154 | 91 ± 39 | 27 ± 9** | <0.05 |

[a] The values for those decades that are underlined are significantly higher than the 1960s. * = The value for the 1980s is significantly higher than the value for the 1970s. ** = The value for the 1980s is significantly lower than those for the 1960s and 1970s.

**Table 2**  Significance Levels for the Results of ANOVAs (log[n+1] Transformed Data) for Mean CPUE (Catch in kg per Boat per Month per Year) for the Total Fin Fishery and for the Fishery for *Aldrichetta forsteri, Mugil cephalus* and *Cnidoglanis macrocephalus* in the Peel-Harvey, Swan and Leschenault Estuaries in the 1960s, 1970s and 1980s[a]

| Species | Estuaries | Significance levels |
|---|---|---|
| All species: | | |
| 1960s | Swan Peel-Harvey <u>Leschenault</u> | <0.01 |
| 1970s | Swan <u>Peel-Harvey Leschenault</u> | <0.01 |
| 1980s | Swan <u>Peel-Harvey Leschenault</u> | <0.001 |
| *A. forsteri* | | |
| 1960s | Swan <u>Peel-Harvey Leschenault</u> | <0.001 |
| 1970s | Swan <u>Peel-Harvey Leschenault</u> | <0.001 |
| 1980s | Swan <u>Peel-Harvey Leschenault</u> | <0.001 |
| *M. cephalus* | | |
| 1960s | Swan Peel-Harvey Leschenault | n.s. |
| 1970s | Swan <u>Peel-Harvey</u> Leschenault | <0.001 |
| 1980s | Swan <u>Peel-Harvey</u> Leschenault | <0.001 |
| *C. macrocephalus* | | |
| 1960s | Swan Peel-Harvey Leschenault | n.s. |
| 1970s | Swan <u>Peel-Harvey</u> Leschenault | <0.01 |
| 1980s | Swan <u>Peel-Harvey</u> Leschenault* | <0.001 |

[a]  Those estuaries that are underlined indicate that their values are significantly higher than those for the Swan. *The value for Leschenault Inlet is significantly lower than the value for both the Swan and Peel-Harvey.

and the same was true for the total fishery and the fishery for *A. forsteri* in the 1960s (Table 2). While CPUEs for the total fin fishery and the fisheries for *M. cephalus* and *C. macrocephalus* in the Peel-Harvey did not differ significantly from those of the corresponding fisheries in the Swan Estuary in the 1960s, they were significantly greater in the Peel-Harvey in both the 1970s and 1980s (Table 2). These differences should be considered in the context of the fact that the 1970s was the period when massive macroalgal growths appeared in the Peel-Harvey, and that these continued into the 1980s, if not quite so profusely. No such macroalgal growths developed in the Swan Estuary during the corresponding period. The CPUEs for the fishery for *A. forsteri* were significantly higher in the Peel-Harvey than in the Swan in all 3 decades.

The CPUEs for the total fin fishery and for the fishery for *A. forsteri* were significantly greater in Leschenault Inlet than in the Swan Estuary in all 3 decades. It would thus appear relevant that the development of extensive growths of seagrasses and brown algae was initiated earlier in Leschenault Inlet than in the Peel-Harvey, i.e., in the late 1950s or in the 1960s rather than the 1970s. The onset of these growths was apparently related to the construction of a new entrance to the Leschenault Inlet in the 1950s which increased tidal exchange with the ocean, thereby making the environment in the inlet more marine and thus probably providing improved conditions for the growth of marine macrophytes. Although the CPUEs for *M. cephalus* in Leschenault Inlet did not differ significantly from those in the Swan Estuary in any of the 3 decades, it is worth reiterating that the increase in CPUE in the Leschenault during the 30 years of this study was greater than in the Swan Estuary (Table 1), in which there was no comparable development of macrophytes.

The annual catches of western king prawns in the Peel-Harvey Estuary rose sharply from 1.2 to 11 t between 1952 and 1962 to 24 t in 1964 (Figure 3). The annual catches subsequently declined markedly from about 10 t in 1966 to 1968 to less than 5 t between 1969 and 1978 (Figure 3). These latter very low catches correspond to a period when *Cladophora montagneana* was the dominant macroalga in Peel Inlet, i.e., in the region where king prawns are usually abundant.[41] It is almost certainly relevant that this macroalga, which coated the sandy substrate surface with algal balls and a thick ooze of decomposing algae at that time, inhibited the king prawns from burrowing into the substrate.[41,58] The conclusion that *Cladophora* had a detrimental affect on the western king prawn is consistent with the fact that in 1979 and the 1980s the subsequent change to more free-floating types of macroalgae that occupy shallower

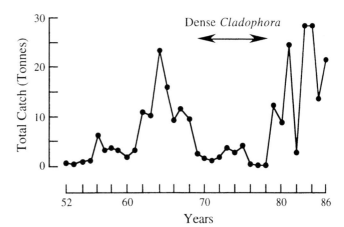

**Figure 3** Annual catches of the western king prawn (*Penaeus latisulcatus*) in the Peel-Harvey Estuary between 1952 and 1986 (Modified from Potter, I. C., Manning, R. J. G., and Longeragan, N. R., *J. Zool. Soc. London*, 223, 419, 1991. With permission.)

waters of the estuary, such as *Chaetomorpha linum* and *Enteromorpha intestinalis*,[58] was accompanied by a marked rise in the catches of king prawns.[41]

## B. INFLUENCE OF THE BLUE-GREEN ALGA *Nodularia spumigena*

The major annual peak in chlorophyll *a* in the water column of the Peel-Harvey Estuary has been shown to be due to blooms of *Nodularia spumigena*.[4] The trends shown by chlorophyll *a* between January 1984 and December 1989 thus demonstrate that the density of this blue-green alga produced a sharp peak between mid-October and mid-December in all years except 1987 (Figure 4). The trends also demonstrate that, during those years, the density of *Nodularia spumigena* was greater in the Harvey Estuary than in the southern part of Peel Inlet, which in turn was far greater than in the northern part of Peel Inlet. This parallels the situation in earlier years and reflects the presence in the Harvey Estuary of an optimal combination of salinity, nutrients and temperature for the growth of this blue-green alga.[5]

In an attempt to elucidate the behavior of fish in the presence of *Nodularia*, comparisons were made between catches of fish taken by seine net throughout the Peel-Harvey Estuary between the middle of spring and the end of summer of a year in which there was no *Nodularia* bloom (1979/80) with those obtained in the corresponding period in 2 years in which the blooms were large and extended into February (1980/81 and 1981/82).[25] Catches at the *Nodularia*-affected sites were lower in the 2 *Nodularia* years than in the year when *Nodularia* blooms did not occur (Figure 5). Conversely, catches at sites in the entrance channel and the rivers, where *Nodularia* does not proliferate, were higher in the 2 years when *Nodularia* blooms occurred than in the year when no such blooms were recorded (Figure 5). In other words, there is strong circumstantial evidence that, when fish in the Harvey Estuary and lower southern Peel Inlet are exposed to blue-green algal blooms, they move out into other areas where such blooms are not present.

Analysis of the data on catches in the *Nodularia*-affected areas showed that the densities of fish caught at sites with chlorophyll *a* levels exceeding 100 $\mu$g L$^{-1}$ were significantly lower than the numbers caught at sites with chlorophyll *a* levels less than 100 $\mu$g L$^{-1}$.[25]

Commercial fishermen in the Peel-Harvey Estuary employ gill nets when water clarity is low and use haul nets when clarity is high and schools of fish can be located visually.[38] Thus, in the northern Peel Inlet, the fishermen begin switching from gill netting to haul netting in October as the water starts to clear as a result of declining freshwater discharge (Figure 6). A marked change from gill to haul netting occurred approximately 1 month later in the southern Peel Inlet and 2 to 3 months later in the northern and southern Harvey Estuary (Figure 6). The delays in the switch in fishing method reflect the fact that in October to December, the development of *Nodularia* blooms in the Harvey Estuary, and to a lesser extent in the southern Peel Inlet, result in an extension of the period when water clarity is poor and thus in an inability to locate fish visually.

Netting in the presence of *Nodularia* leads to the fouling of the nets, which makes them difficult to clean and, under extreme circumstances, causes them to sink. The fishermen who work in the Harvey Estuary thus tend to move up into the Peel Inlet during dense *Nodularia* blooms, a feature that is reflected by a decline in both gill and haul net effort in the Harvey Estuary at such times, that is, November and December (Figure 6). This movement of fishermen to the Peel Inlet contributes to the marked increase in haul net effort in that part of the estuary at that time (Figure 6).

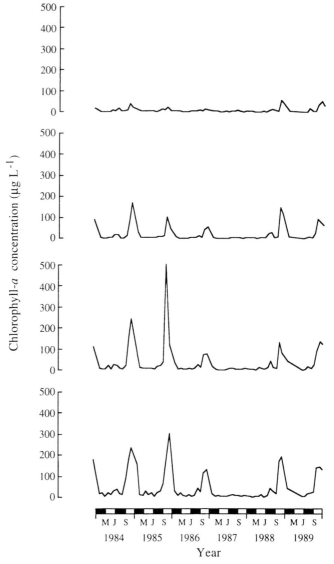

**Figure 4** Chlorophyll *a* concentrations recorded in the northern and southern regions of both the Peel Inlet and Harvey Estuary in each month between January 1984 and December 1989. The black rectangles on the x axis represent winter and summer and the white rectangles represent spring and autumn.

The CPUE in Peel Inlet gradually increases during the late spring and mid-summer, i.e., in the period when pronounced *Nodularia* blooms typically occur in the Harvey Estuary and are followed by a decline in oxygen levels as the *Nodularia* decomposes. Although the mean CPUE for haul netting in the northern Harvey Estuary in December was quite high, there was no conspicuous rise in the preceding months. Furthermore, the high mean CPUE for haul netting in December was due to exceptionally large catches that were taken by haul netting in clear patches of water in 1984 and 1985. These catches were taken at a time when *Nodularia* was in rapid decline and chlorophyll *a* levels had fallen below 100 µg L$^{-1}$ (Figure 4). It should also be recognized that, since fish tend to avoid dense *Nodularia* blooms and fishermen make every effort to avoid *Nodularia*, the CPUE for the limited amount of fishing that is carried out in the Harvey Estuary during *Nodularia* blooms do not represent the relative abundance of fish throughout the whole of that part of the estuary during those periods.

## V. CONCLUSIONS

It has been emphasized that the large basins of the Swan Estuary, where the commercial fishery in that system is based, have not exhibited the massive growth of macrophytes that have occurred in the last 20 years in the Peel-Harvey Estuary and 25–35 years in Leschenault Inlet. The values obtained for the CPUEs for the fishery in the Swan Estuary can thus be used as a "control" to elucidate whether any

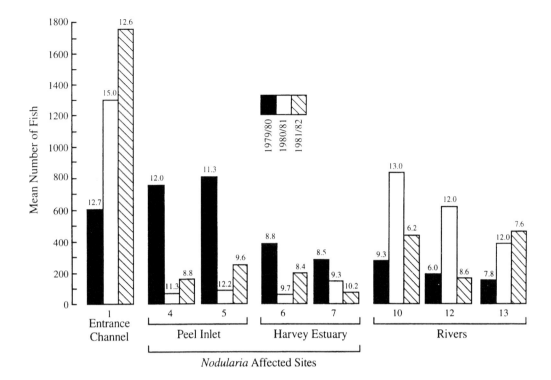

**Figure 5** Geometric means for the number of fish caught at various sites in the Peel-Harvey Estuary between mid-spring and the early autumn of 1979–80, 1980–81 and 1981–82. The mean number of species is given above each histogram. (From Potter, I. C. et al., *Mar. Pollut. Bull.*, 14, 228, 1983. With permission.)

changes in the CPUEs in the other two estuaries are likely to have been related to the more extreme manifestations of eutrophication displayed by those two systems. However, it should be recognized that, during the 1970s, commercial fishermen in all southwestern Australian estuaries have increasingly replaced multifilament nets with monofilament nets and turned to using trailerized dinghies. These changes increased fishing efficiency to some extent and thus help account for rises in CPUE for the total fishery in the Swan Estuary during the 1970s and 1980s.

A comparison between the results of the statistical tests given in Table 2 show that, while the CPUE for the total fishery in the Peel-Harvey and Swan Estuaries in the 1960s did not differ significantly, this was not the case in the 1970s or 1980s when, in both of those latter decades, the CPUEs for the former estuary were significantly higher. It would therefore appear highly relevant that massive growths of macroalgae first appeared in the late 1960s and have extended through the 1970s and 1980s.[8] The CPUEs for the total fishery in Leschenault Inlet were greater than those for the Swan Estuary, not only in the 1970s and 1980s, but also in the 1960s. It is therefore noteworthy that the macrophyte growth started to become conspicuous earlier in Leschenault Inlet than in the Peel-Harvey, i.e., in the 1960s rather than the 1970s. Since the macrophyte growths in Leschenault Inlet are due to seagrasses and brown algae, rather than to green macroalgae as in the Peel-Harvey, a range of different plant growths can be accompanied by high CPUEs in southwestern Australian estuaries.

The presence of higher CPUEs in macrophyte-dominated waters could reflect an increased catchability of fish, faster growth rates (and thus a greater biomass of fish) or enhanced protection from piscivorous birds (and thus reduced mortality) or a combination of these effects. Fishermen in the Peel-Harvey felt that the fish became more abundant with the advent of macroalgae in the 1970s, and that the CPUE therefore did not mainly reflect increased catchability (B. Toussaint, personal communication).[59] The growth rates of the main fish species in the Peel-Harvey were not greater than in the Swan Estuary.[9] Thus, increases in the biomass of populations of those species in the Peel-Harvey, resulting from the effects of nutrient enrichment, would not appear to have been due mainly to increases in the growth rate of fish, as is usually the cause when the biomass of fish rises in response to such nutrient enrichment. Although this could imply that the growth rates for the main species in the Peel-Harvey may have been approaching

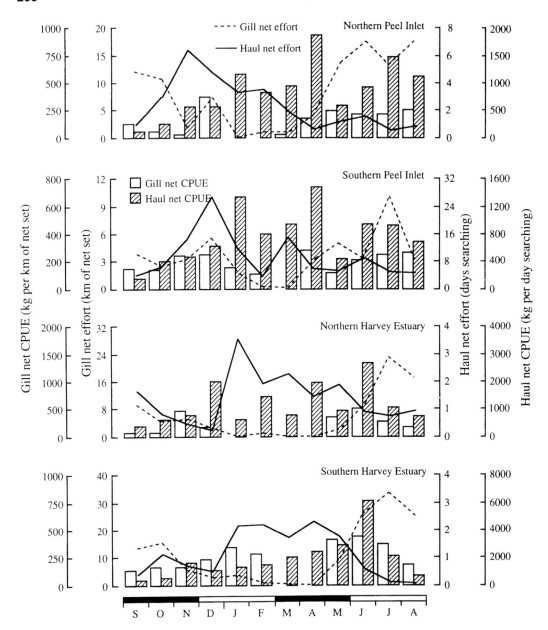

**Figure 6** Mean monthly effort and CPUE for combined catches of yelloweye and sea mullet taken with gill and haul netting in the northern and southern regions of both the Peel Inlet and Harvey Estuary. Data are means for values between October 1982 and December 1990 and do not include values for 1987 when there was no *Nodularia* bloom.

their optima, it seems at least equally likely that the rise in CPUE following eutrophication reflects an increase in the abundance of fish.[9] Since there is a very large piscivorous bird community in the region of the Peel-Harvey, it therefore seems possible that the greatly increased cover provided by macroalgae from avian predators led to a reduction in predation and thus an increase in abundance.[9]

The far higher CPUEs recorded for the fisheries in the Peel-Harvey and Leschenault Inlet than in the Swan Estuary reflect in part the greater CPUEs recorded for the yelloweye mullet in the first two systems. The yelloweye mullet is an omnivore,[60] grows quickly, is relatively low in the food chain and has a high reproductive rate and is the type of species that would be expected to do well in a eutrophic environment dominated by macrophytes.[16] However, it is also highly relevant that, while the CPUEs for the sea mullet and cobbler in the Peel-Harvey did not differ significantly from those in the Swan Estuary in the 1960s, they were both significantly greater in the former system in both the 1970s and 1980s. The apparent increases in abundance of sea mullet and of cobbler (at least *vis à vis* the Swan) in the Peel-Harvey during the 1970s and 1980s would be consistent with the fact that both of these species would be likely to benefit from the decomposition of large amounts of plant material. The former species is a detritivore, while the latter feeds largely on benthic invertebrates.[52,60]

Although there is strong circumstantial evidence that macrophytes can be beneficial for estuarine fin fisheries in southwestern Australia, there is also good evidence to suggest that species which tend to coat the substrate with both living and decaying material, e.g., *Cladophora montagneana*, can have a detrimental effect on the fishery for the western king prawn, presumably through inhibiting burrowing.[41]

In contrast to the situation with macroalgae, the effects of *Nodularia spumigena* are highly seasonal. However, the dense blooms of this blue-green alga, which appeared in the late spring and early summer in all but three of the years between 1977 and 1993, apparently had a considerable impact on the fish fauna and fishery during those blooms. There is for example circumstantial evidence that fish tend to move away from *Nodularia*-affected sites, especially when the chlorophyll *a* levels that reflect the presence of this blue-green alga rise above 100 $\mu$g L$^{-1}$.[25] It is also conspicuous that the advent of *Nodularia* leads to fishermen transferring their activities from the Harvey Estuary into those parts of the Peel Inlet where the blooms are far less dense.[38] This movement is triggered to a large extent by their desire to avoid having their nets "clogged" by *Nodularia* and to be able to use their preferred method of fishing, i.e., haul netting, in the clearer waters of Peel Inlet in spring.

## REFERENCES

1. **Orth, R. J. and Moore, K. A.,** Chesapeake Bay: an unprecedented decline in submerged aquatic vegetation, *Science,* 222, 51, 1983.
2. **Conomos, T. J., Smith, R. E., Peterson, D. H., Hager, S. W., and Schemel, L. E.,** Processes affecting seasonal distributions of water properties in the San Francisco Bay estuarine system, in *San Francisco Bay: The Urbanized Estuary,* Conomos, T. J., Ed., Pacific Division, American Association for the Advancement of Science, San Francisco, 1979, 115.
3. **Jaworski, N. A.,** Sources of nutrients and the scale of eutrophication problems in estuaries, in *Nutrient Enrichment in Estuaries,* Nielsen, B., and Cronin, A., Eds., Humana Press, Clifton, NJ, 1981, 83.
4. **McComb, A. J., Atkins, R. P., Birch, P. B., Gordon, D. M., and Lukatelich, R. J.,** Eutrophication in the Peel-Harvey estuarine system, Western Australia, in *Nutrient Enrichment in Estuaries,* Nielsen, B., and Cronin, A., Eds., Humana Press, Clifton, NJ, 1981, 323.
5. **Hillman, K., Lukatelich, R. J., and McComb, A. J.,** The impact of nutrient enrichment of nearshore and estuarine ecosystems in Western Australia, *Proc. Ecol. Soc. Aust.,* 16, 39, 1990.
6. **Pritchard, D. W., and Schubel, J. R.,** Physical and geological processes controlling nutrient levels in estuaries, in *Nutrient Enrichment in Estuaries,* Nielsen, B., and Cronin, A., Eds., Humana Press, Clifton, NJ, 1981, 47.
7. **Conomos, T. J.,** Properties and circulation of San Francisco Bay waters, in *San Francisco Bay: The Urbanized Estuary,* Conomos, T. J., Ed., Pacific Division, American Association for the Advancement of Science, San Francisco, 1979, 47.
8. **McComb, A. J.,** The Peel-Harvey estuarine system, Western Australia, in *Eutrophic. Shallow Estuaries and Lagoons,* McComb, A.J., Ed., CRC Press, Boca Raton, FL, 1995, chap. 2.
9. **Lenanton, R. C. J., Potter, I. C., Loneragan, N. R., and Chrystal, P. J.,** Age structure and changes in abundance of three important species of teleost in a eutrophic estuary (Pisces: Teleostei), *J. Zool. London,* 203, 311, 1984.
10. **Lubbers, L., Boynton, W. R., and Kemp, W. M.,** Variations in structure of estuarine fish communities in relation to abundance of submersed vascular plants. *Mar. Ecol. Prog. Ser.,* 65, 1, 1990.
11. **Wilson, K. A., Able, K. W., and Heck, K. L.,** Predation rates on juvenile blue crabs in estuarine nursery habitats: evidence for the importance of macroalgae (*Ulva lactuca*), *Mar. Ecol. Prog. Ser.,* 58, 243, 1990.
12. **Humphries, P., Potter, I. C., and Loneragan, N. R.,** The fish community in the shallows of a temperate estuary: relationships with the aquatic macrophyte *Ruppia megacarpa* and environmental variables, *Estuarine Coastal Shelf Sci.,* 34, 325, 1992.
13. **Hernroth, L.,** Zooplankton in the Baltic Sea, *Mar. Pollut. Bull.,* 12, 206, 1981.
14. **Johnston, N. T., Peffin, C. J., Slaney, P. A., and Ward, B. R.,** Increased juvenile salmonid growth by whole river fertilization, *Can. J. Fish. Aquat. Sci.,* 47, 862, 1990.
15. **Keller, A. A., Doering, P. H., Kelly, S. P., and Sullivan, B. K.,** Growth of juvenile Atlantic menhaden, *Brevoortia tyrannus* (Pisces: Clupeidae) in MERL mesocosms: effects of eutrophication, *Limnol. Oceanogr.,* 35, 109, 1990.
16. **Nichols, F. H., Cloern, J. E., Luoma, S. N., and Peterson, D. H.,** The modification of an estuary, *Science,* 231, 567, 1986.
17. **Hansson, S. and Rudstam, L. G.,** Eutrophication and Baltic fish communities, *Ambio,* 19, 123, 1990.

18. **Wright, D. A. and Phillips, D. J. H.,** Chesapeake and San Francisco Bays. A study in contrasts and parallels, *Mar. Pollut. Bull.,* 19, 405, 1988.

19. **Cambridge, M. L., Chiffings, A. W., Brittan, C., Moore, L., and McComb, A. J.,** The loss of seagrass in Cockburn Sound, Western Australia. II. Possible causes of seagrass decline, *Aquat. Bot.,* 24, 269, 1986.

20. **Setzler-Hamilton, E. M., Whipple, J. A., and Macfarlane, R. B.,** Striped bass populations in Chesapeake and San Francisco Bays: two environmentally impacted estuaries, *Mar. Pollut. Bull.,* 19, 466, 1988.

21. **Dybdahl, R. E.,** Technical report on fish productivity. An assessment of the marine faunal resources of Cockburn Sound, *Department of Conservation and Environment of Western Australia Report,* 4, 1979, 96 pp.

22. **Arnold, D. E.,** The ecological decline of Lake Erie, *N.Y. Fish Game J.,* 16, 27, 1969.

23. **Larsson, U., Elmgren, R., and Wulf, F.,** Eutrophication and the Baltic Sea: causes and consequences, *Ambio,* 14, 9, 1985.

24. **Larkin, P. A. and Northcote, T. G.,** Fish as indices of eutrophication, in *Eutrophication: Causes, Consequences, Correctives,* Rolich, G. A., Ed., National Academy of Sciences, Washington, DC, 1969, 256.

25. **Potter, I. C., Loneragan, N. R., Lenanton, R. C. J., and Chrystal, P. J.,** Blue-green algae and fish population changes in a eutrophic estuary, *Mar. Pollut. Bull.,* 14, 228, 1983.

26. **Baden, S. P., Loo, L.-O., Pihl, L., and Rosenberg, R.,** Effects of eutrophication on benthic communities including fish: Swedish west coast, *Ambio,* 19, 113, 1990.

27. **Lee, G. F., Jones, P. E., and Jones, R. A.,** Effects of eutrophication on fisheries, *Rev. Aquat. Sci.,* 5, 287, 1991.

28. **Thurlow, B. H., Chambers, J., and Klemm, V. V.,** Swan estuarine system. Environment use and the future, *W. Aust. Waterways Comm. Rep.,* 9, 1986, 463 pp.

29. **Glover, R. P.,** Preliminary Nitrogen and Phosphorous Flux Study, Leschenault Inlet, Western Australia, unpublished honors thesis, Murdoch University, Western Australia, 1980.

30. **Hodgkin, E. P., Birch, P. B., Black, R. E., and Humphries, R. B.,** The Peel-Harvey estuarine system study (1976-1980), *Department of Conservation and Environment of Western Australia Rep.,* 9, 1980.

31. **Semeniuk, C. A. and Semeniuk, V.,** The coastal landforms and peripheral wetlands of the Peel-Harvey estuarine system, *J. Roy. Soc. W. Aust.,* 73, 9, 1990.

32. **Loneragan, N. R., Potter, I. C., Lenanton, R. C. J., and Caputi, N.,** Spatial and seasonal differences in the fish fauna in the shallows of a large Australian estuary, *Mar. Biol.,* 92, 575, 1986.

33. **Anon.,** Peel Inlet and Harvey Estuary management strategy: Environmental review and management program — Stage 2, Kinhill Engineers Pty. Ltd., State Printing Division, Perth, Western, Australia, 1988, 182 pp.

34. **Gabrielson, J. O., and Lukatelich, R. J.,** Wind-related resuspension of sediments in the Peel-Harvey estuarine system, *Estuarine Coastal Shelf Sci.,* 20, 135, 1985.

35. **Birch, P. B.,** Phosphorus export for coastal plain drainage into the Peel-Harvey estuarine system of Western Australia, *Aust. J. Mar. Freshwater Res.,* 33, 23, 1982.

36. **Knox, G. A.,** *Estuarine Ecosystems: A Systems Approach,* Vol II, CRC Press, Boca Raton, FL, 1986, 180.

37. **Lukatelich, R. J.,** Leschenault Inlet macrophyte abundance and distribution, *W. Aust. Waterways Comm. Rep.,* 15, 27, 1989.

38. **Lenanton, R. C. J., Loneragan, N. R., and Potter, I. C.,** Bluegreen algal blooms and the commercial fishery of a large Australian estuary, *Mar. Pollut. Bull.,* 16, 477, 1985.

39. **Anon.,** State of the Fisheries December 1991, Report to the Parliament of Western Australia, Fisheries Department of Western Australia Annual Report, Perth, 1992.

40. **Potter, I. C., Chrystal, P. J., and Loneragan, N. R.,** The biology of the blue manna crab *Portunus pelagicus* in an Australian estuary, *Mar. Biol.,* 78, 75, 1983.

41. **Potter, I. C., Manning, R. J. G., and Loneragan, N. R.,** Size, movements, distribution and gonadal stage of the western king prawn (*Penaeus latisulcatus*) in a temperate estuary and local marine waters, *J. Zool. Soc. London,* 223, 419, 1991.

42. **Potter, I. C., Loneragan, N. R., Lenanton, R. C. J., Chrystal, P. J., and Grant, C. J.,** Abundance, distribution and age structure of fish populations in a Western Australian estuary, *J. Zool. London,* 200, 21, 1983.

43. **Collins, L. B.,** Geological evolution of the Swan Estuarine system, in *The Swan River Estuary: Ecology and Management,* John, J., Ed., Curtin University Environmental Studies Group Rep. No. 1, Perth, Western Australia, 1987, 9.

44. **Lenanton, R. C. J.,** The commercial fisheries of temperate Western Australian estuaries: Early settlement to 1975, *Rep. Fish. Dep. West. Aust.,* 62, 1984.

45. **Hodgkin, E. P.,** The hydrology of the Swan River Estuary: salinity the ecological master factor, in *The Swan River Estuary: Ecology and Management,* John, J., Ed., Curtin University Environmental Studies Group Rep. No. 1, Perth, Western Australia, 1987, 34.

46. **Riggert, T. L.,** *The Swan River Estuary Development, Management and Preservation,* Swan River Conservation Board, Perth, Western Australia, 1978, chap. 1.

47. **Loneragan, N. R. and Potter, I. C.,** Factors influencing community structure and distribution of different life-cycle categories of fishes in shallow waters of a large Australian estuary, *Mar. Biol.,* 106, 25, 1990.

48. **Hodgkin, E. P. and Vicker, E.,** A history of algal pollution in the estuary of the Swan River, in *The Swan River Estuary: Ecology and Management*, John, J., Ed., Curtin University Environmental Studies Group Rep. No. 1, Perth, Western Australia, 1987, 65.

49. **Chubb, C. F., Potter, I. C., Grant, C. J., Lenanton, R. C. J., and Wallace, J.,** Age structure, growth rates and movements of sea mullet, *Mugil cephalus* L., and yelloweye mullet, *Aldrichetta forsteri* (Valenciennes), in the Swan-Avon River System, Western Australia, *Aust. J. Mar. Freshwater Res.*, 32, 605, 1981.

50. **Chubb, C. F., Hall, N. G., Lenanton, R. C. J., and Potter, I. C.,** The fishery for Perth Herring *Nematalosa vlaminghi* (Munro), *Dept. Fish. Wildl. West. Aust. Rep.*, 66, 1984, 17 pp.

51. **Loneragan, N. R., Potter, I. C., and Lenanton, R. C. J.,** The fish and fishery of the Swan Estuary, in *The Swan River Estuary: Ecology and Management*, John, J., Ed., Curtin University Environmental Studies Group Rep. No. 1, Perth, Western Australia, 1987, 178.

52. **Nel, S. A., Potter, I. C., and Loneragan, N. R.,** The biology of the catfish *Cnidoglanis macrocephalus* (Plotosidae) in an Australian estuary, *Estuarine Coastal Shelf Sci.*, 21, 895, 1985.

53. **Potter, I. C., Penn, J. W., and Brooker, K. S.,** Life cycle of the western school prawn, *Metapenaeus dalli* Racek, in a Western Australian estuary, *Aust. J. Mar. Freshwater Res.*, 37, 95, 1986.

54. **Lenanton, R. C. J.,** The inshore-marine and estuarine licensed amateur fishery of Western Australia, *Fish. Res. Bull. West. Aust.*, 23, 33pp, 1979.

55. **Lenanton, R. C. J. and Potter, I. C.,** Contribution of estuaries to commercial fisheries in temperate Western Australia and the concept of estuarine dependence, *Estuaries*, 10, 28, 1987.

56. **Laurenson, L., Potter, I., Lenanton, R., and Hall, N.,** The significance of length at sexual maturity, mesh size and closed fishing waters to the commercial fishery for the catfish *Cnidoglanis macrocephalus* in Australian estuaries, *J. Appl. Ichthyol.*, 9, 20, 1993.

57. **Laurenson, L. J. B., Neira, F. J., and Potter, I. C.,** Reproductive biology and larval morphology of the marine plotosid *Cnidoglanis macrocephalus* (Teleostei), in a seasonally closed Australian estuary, *Hydrobiologia*, 268, 179, 1993.

58. **Lavery, P. S., Lukatelich, R. J., and McComb, A. J.,** Changes in the biomass and species composition of macroalgae in a eutrophic estuary, *Estuarine Coastal Shelf Sci.*, 33, 1, 1991.

59. **Toussaint, B.,** Personal communication, 1983.

60. **Thompson, J. M.,** The organs of feeding and the food of some Australian mullet, *Aust. J. Mar. Freshwater Res.*, 6, 328, 1954.

# Chapter 13

# The Role of the Sediments

## J. A. Thornton, A. J. McComb, and S.-O. Ryding

## CONTENTS

I. Introduction .................................................................................................................206
II. Characterization of Estuarine Sediments ............................................................206
   A. The Dynamics of Sedimentation ......................................................................206
      1. Relationship Between Particle Size Distribution and Geology ...............206
      2. Relationship Between Particle Size Distribution and Flow ....................206
   B. Particle Size Distribution in Estuaries .............................................................207
      1. Longitudinal Distributions ...........................................................................207
      2. Transverse and Vertical Distributions .......................................................208
      3. Temporal Distributions .................................................................................208
   C. Resuspension of Estuarine Sediments ...............................................................208
III. Sediment Nutrient Pools .......................................................................................209
   A. Accumulation of Nutrients .................................................................................209
      1. Mechanisms for Trapping Nutrients ...........................................................209
      2. Forms of Nutrients in Estuarine Sediments ..............................................211
   B. Release and Bioavailability of Nutrients ..........................................................212
      1. Resuspension .................................................................................................212
      2. Remobilization ...............................................................................................213
      3. Regeneration ..................................................................................................214
IV. Uptake, Release and Bioavailability of Trace Elements ....................................215
   A. Relationship Between Nutrients and Metals ....................................................215
   B. Uptake of Metals .................................................................................................216
   C. Metals in Estuarine Sediments ..........................................................................216
   D. Release and Bioavailability of Metals ...............................................................217
V. Discussion and Management Implications ...........................................................217
VI. Summary ..................................................................................................................219
   A. A Synthesis ..........................................................................................................219
   B. What We Know ....................................................................................................219
      1. Nutrient Fluxes Occur ..................................................................................219
      2. Nutrient Fluxes Are Driven by Physical, Chemical and
         Biological Mechanisms ...............................................................................219
      3. Nutrient Fluxes Are Complex Combinations of Processes ......................220
      4. Nutrient Fluxes Can Be Interrupted ...........................................................220
   C. What We Think We Know ..................................................................................220
      1. Nutrient Flux Processes Can Be Defined ..................................................220
      2. Nutrient Flux Processes Can be Modeled ..................................................220
      3. Nutrient Fluxes Can Be Controlled ............................................................220
   D. What We Would Like to Know ..........................................................................220
      1. How Nutrient Fluxes Interact with Biological Processes .........................220
      2. How the Concept of Scale Applies to Estuarine Nutrient Dynamics ......221
      3. How Nutrient Fluxes Can Be Managed .....................................................221
   E. Conclusion ............................................................................................................221
Acknowledgments ...........................................................................................................221
References ........................................................................................................................221

# I. INTRODUCTION

One of the basic premises of estuarine science is that the sediments of these systems play a major role as both nutrient sinks, storing vast quantities of phosphorus and nitrogen, and sources, releasing these stored nutrients to fuel the high level of productivity for which estuaries are known. Should this be true generally, then it follows that the sediments should play an even more important role in eutrophic estuaries and coastal lagoons (collectively referred to here as estuaries). In this chapter, we briefly examine some of the processes associated with this premise and review the extent to which estuarine sediments act as sinks and/or sources of nutrients and other elements.

# II. CHARACTERIZATION OF ESTUARINE SEDIMENTS

## A. THE DYNAMICS OF SEDIMENTATION

Krumbein and Sloss,[1] in their classical text, summarize much of the conceptual basis for the behavior of suspended solids in water, including the physical laws governing the movement and settling of particulates. These laws generally relate to the size, density and shape of the solids being transported, which, in turn, are a function of regional geology, weathering and in-stream processing. Recently, Walling and Moorehead[2] have reexamined these concepts in light of field evidence that appears, in some instances, to be at variance with the conceptual models commonly applied to sediment transport in streams (see below). While much of the field work to date has emphasized the overall magnitude of the suspended sediment loads, Walling and Moorehead[2] make a convincing case for the need to understand the particle size composition and its variation in space and time in order to understand the role of the sediments in such processes as nutrient and heavy metal exchange following deposition. For instance, small particles ($<2$ $\mu$m) will be dominated by secondary silicate materials while larger particles will be made up of quartz, which due to its smaller surface area per unit volume has a cation exchange capacity many times less than that of the smaller particles.

## 1. Relationship Between Particle Size Distribution and Geology

Typically, the particle size distribution in the suspended sediment load carried by any stream is a function, in part, of catchment geology. For example, the situation of the Huangfu (China) and Limpopo (Zimbabwe/Mocambique) Rivers in drainage basins dominated by coarse loess and sand results in a sediment load that is dominated by particles in excess of 60 $\mu$m, while the Barwon River (Australia), situated in a basin comprised of deeply weathered clays, has a sediment load with a mean particle size of $<2$ $\mu$m. In addition, in larger watersheds such as those of the Nile and Yellow Rivers, larger-sized particles are typically deposited in-channel before the sediments reach the estuaries of the rivers; in the Nile, for example, the mean particle size decreased from 5–10 $\mu$m to $<2$ $\mu$m over the 1,000-km reach upstream of Cairo. What this means in practice is that, while the suspended load will reflect the particle size distribution of its parent material, the internal riverine sorting process will result in the suspended sediment being selectively enriched in finer materials and depleted in coarser sediments when compared to the source materials.

Nevertheless, there is a body of evidence that further suggests that anthropogenic land use activities can selectively enhance or diminish the natural soil matrix particle size distribution by modifying the erodibility of the soil surface. Generally, this modification takes the form of a disturbance of the soil surface (for example, by tilling) that exposes a greater area of soil surface (and hence a greater number of smaller-sized particles) to the erosive effects of precipitation and wind. This latter process, known as selective erosion, as well as the former process, known as selective deposition, modify the particle size distribution of the sediments reaching the estuary. These processes favor the transport of silt and clay-sized particles into the estuarine environment, but are subject to further modification due to the potential for clay particles (especially) to aggregate, forming hydrodynamically larger particles which are then subject to in-stream deposition; for example silts and clays form small aggregates (2–63 $\mu$m in diameter) which can combine with sand grains to form larger aggregates ($>63$ $\mu$m) with a higher settling velocity.[2]

## 2. Relationship Between Particle Size Distribution and Flow

While the foregoing provides a persuasive argument relative to the origin of the particular size distribution observed in the sediments of various estuaries, it fails to address the hydrodynamic issue that has long been held to be the dominant determinant of particle size distributions in streams.[1] The process of selective deposition suggests that the stream bed load, which can be moved by channel erosion or scour, will be coarser-grained relative to the eroded terrestrial soils (generated by slope erosion) from which it was derived. Generally, unless the slope processes remain essentially constant, the degree of slope erosion will incorporate a degree of seasonality based on the amount of precipitation and runoff (i.e., the flow-related component of Manning's equation).[3] To wit, in semiarid environments, such as in the Niobrara River (U.S.), seasonal increases in runoff can result in a dominance of slope-derived suspended materials (i.e., an increase in finer particle sizes with increased flow) over the coarser, channel-derived suspended materials that dominate during drier periods — an inversion of the commonly accepted theorem that relates increased particle sizes to increased flow velocities.[2] This suggests:

1. That the nature of the suspended load is dependent on either hydraulic limitations ("transport limitation", where there is a relative lack of water to suspend and move the soil particles) or geological limitations ("supply limitation", where the soil particles being moved are in short supply relative to the amount of water available) — the latter being more common, according to Walling and Moorehead,[2] despite the traditional acceptance of the former by hydrologists; and/or
2. That the absolute particle size of the individual sediment grains being transported is less important than their effective size (e.g., as aggregates rather than discrete particles) — this being, to some degree, a function of salinity, as in the Euphrates[2] and Vaal[4] Rivers.

## B. PARTICLE SIZE DISTRIBUTION IN ESTUARIES
### 1. Longitudinal Distributions
Estuaries are hydrodynamically complex — a simple enough statement that belies the numerous complications instilled into the sediment transport model developed above for rivers and streams. In contrast to the predominantly unidirectional flow in rivers (downhill, as it were), estuaries are subject to wave action, currents and tidal flows that generate flow velocities in excess of 75 cm s$^{-1}$, reaching upwards of 650 cm s$^{-1}$ in the Bay of Fundy, for example.[5] In these highly dynamic environments, certain additional characteristics or influences become apparent in the sediment particle size distributions. These have been enumerated by Sly et al.[5] as:

1. A well-defined change in mean particle size-water depth relationships generally coincident with the effective depth of wave influence on the sediments.
2. A clear relationship between the rate of deposition of fine particles in the offshore zone that is related to the rate of decrease in intensity of the nearshore energy field (i.e., the rate of decrease in wave action and current velocity).
3. A clear relationship between the rate of very fine particle deposition and the occurrence of still-water conditions.
4. A well-defined tidal influence on the coarse particle-water depth relationship which results in coarser sediments being transported to greater depths than can be explained in terms of wave action (see 1 above) alone.

These characteristics apply to both lacustrine (e.g., the Laurentian Great Lakes) and marine (e.g., Mersey Estuary, Mackenzie Estuary) depositions;[5] the net effect being a distinctive sorting of the sediment load entering the estuary from landward with coarser particles being deposited in the nearshore area and finer particles being deposited at depth in the offshore zone.[5-7]

In addition to the primarily geochemical-physical processes discussed above, Sly[7] introduces a further modifying effect in the form of estuarine biota, noting that both salinity effects (causing fine particle aggregation and settling) and biofiltration processes, mediated by filter-feeding benthic fauna, "fix" fine sediments into the bottom deposits in estuaries. In the Mersey Estuary (U.K.), he noted a fining of the particle size distribution both to landward and seaward of the estuary; the former being a function of the high suspended load carried by the Mersey River (>2,000 mg L$^{-1}$) and the rapid flocculation/deposition of the material at slack water, a portion of this material being left behind after the tide turned. In other words, the complex dynamic processes at work in the estuary can permit seemingly anomalous accumulations of fine particulates in extremely active, high flow areas of the estuary.[7] Further seaward, the influence of wave action and generally deeper waters reestablished the more normal patterns of coarser particle deposition to landward and finer particle deposition to seaward. However, Sly[7] notes that this deposition pattern only applies along the centerline (or thalweg) of the estuary which is the most active zone hydraulically.

## 2. Transverse and Vertical Distributions

In transverse section, the boundary effects imposed across tidal flats also permit or encourage deposition of fine particulates.[7] The occurrence of muds in these areas is strongly influenced by biogenic processes as well as the lower flow velocities and the oscillating movements of the suspended load carried into and out of the estuary on a continuing basis. The filtering effects of both estuarine plants and benthic fauna, the biogenic processes, are discussed further below.

Finally, Nakata,[8] in the partially mixed Yoshii Estuary (Japan), notes a strong vertical component to the sedimentation process related to the differing vertical current velocities in the less saline river water (surface) and the more saline ocean water (bottom) that occur throughout the tidal cycle. During the ebb tide, these vertical currents are primarily downward (resulting in downwelling of the suspended sediments), while upwelling occurs during the tidal flood. At mid-depth, a similar cycle of downwelling during the ebb and upwelling during the flood was evident, although the net flux tended to be upwards at this depth. In the bottom waters (the most saline), little change was observed during either ebb or flood (hence, there was a net deposition at this depth). The overall effect was a decrease in the concentration of fine particulates (<15 μm) due to downwelling and settling during ebb flows, and an increase during flood flows, creating the frequently observed estuarine turbidity maximum. Nakata[8] found a net deposition of coarser particulates (>15 μm), primarily phytoplankton, consistent with Stokes' Law, throughout the tidal cycle at the seaward end of the estuary.

## 3. Temporal Distributions

In addition to the tidal effects noted by Nakata,[8] the seasonality factor noted by Walling and Moorehead[2] also operates in estuaries, both as the result of variations in the suspended load carried by the inflowing rivers and as the result of variations in the nearshore currents caused by seasonal changes in geostrophic variables.[9,13] Ferretti et al.[9] have observed that sedimentation in the Garigliano and Volturno Estuaries (Italy) is enhanced during the boreal winter by increased terrestrial inorganic sediment loadings and by the greater degree of particle aggregation which they observed at that season. During summer, riverine sediment loads were dominated by an organic component at a time when phytoplanktonic demand for inorganic nutrients in the coastal zone was also high. These seasonal changes in loading patterns led to seasonal changes in patterns of deposition which, in turn, created habitat pressures that affected nearshore distributions of the estuaries' benthic communities.[9,13]

Longer-term variations in the nature and composition of materials deposited in estuaries have also been observed (on an inter-decadal scale), primarily due to anthropogenic disturbances in the watershed[10,11] which affect the nature of the slope materials[2,10] (see also above), as well as historical changes due to geologic scale variations, such as sea-level movements.[7,11,14]

## C. RESUSPENSION OF ESTUARINE SEDIMENTS

Although sediment resuspension has been touched on above,[7,8] the importance of this process is such that it should be explored further before leaving the topic of estuarine sediments and sedimentation. It has been noted that estuaries are highly dynamic environments in which complex spatial and temporal interactions of their biogeochemical and physical attributes occur. Sediment resuspension plays a major role in the determination of the abundance and species composition of the biota (both benthic and planktonic flora and fauna) in the estuary, and these, in turn, influence the degree of resuspension.[7,8] Resuspension, while primarily a physical process, can be heavily influenced by biotic and anthropogenic processes.[8,9,13,14] While sediment-nutrient relationships will be discussed below, suffice it to say here that the influx of particulate and dissolved materials, including the plant nutrients N and P, from terrestrial sources stimulates primary production which in turn stimulates benthic faunal (secondary) production. Not only do these benthic filter-feeders aggregate and "fix" fine sediments into the estuary bottom,[7] but they also perturb the bottom, which enhances erosion of the surface sediments. Such bioturbation can enhance channel erosion or scour.[14] Part of this resuspended material is in the form of organic carbon which preferentially strips $Mn^{2+}$ and $Fe^{2+}$ from the sediments.[14,20] These elements, on contact with the overlying water, rapidly reoxidise and reprecipitate, scavenging nutrients (and other elements) from the water column.[14,20] This relationship, including the biotically induced seasonality, has been clearly shown in data on Fe and Mn cycling in the Gulf of Gaeta (Garigliano-Volturno Estuaries),[9,13] both elements showing a progressive dilution and movement to seaward as the result, primarily, of physical transport factors.

Generally, however, resuspension is a physical process related to the effect of the transfer of wind energy to the water surface; thus, while resuspension is usually related to wave action,[15,16] the wind is the ultimate driving force behind the process. (While this is not always true — tides,[7,8] river flows,[9,13] and boating activities[17,18] as well as bioturbation[14] being dominant in some estuaries — wind-mediated resuspension is rarely absent in any but the most wind-sheltered systems.) The role of benthic macrophytes in resuspension processes in estuaries appears to be poorly understood. Presumably, these organisms would suppress sediment suspension and enhance sediment retention as they do in riverine and nearshore regions, while the elimination of benthic angiosperms through eutrophication could lead to enhanced resuspension rates. In the Peel-Harvey Estuary (Australia; see Chapter 2), wind-induced resuspension is responsible for circulating roughly 70–90% of the estuarine sediment.[15] The net result of this resuspension is a resorting and redistribution of the sediments in such a way that the fines are generally transported shoreward and the coarser particulates seaward.[15,17,18] In both the Harvey Estuary and the Great Bay Estuary (U.S.), the larger part of the "fines" appeared to be aggregates (biotically produced in the Harvey Estuary;[15] either biotically or abiotically produced in the Great Bay Estuary[17,18]) of relatively large size but low density (compared to sandier sediments).

Finally, Santschi[14] notes that, over geologic time, many of these estuarine sediments accumulate to such an extent that they will "slump" into deeper offshore waters as the result of sea-level changes and/or other (gravitational) forces. While perhaps this latter process is not resuspension in the strict sense, it does help to account for some of the net seaward transport of terrestrial materials not otherwise accounted for in terms of shorter-term processes. Other evidence provided by Jouanneau and Latouche[19] suggests that, at least in some estuaries, extreme runoff events (>2,000 $m^3 s^{-1}$, in the Gironde Estuary) can literally blast significant quantities of fine sediments out of an estuary and into deeper, offshore waters on a much more frequent basis ($10^0–10^1$ y vs. $10^3–10^4$ y for slumping-related events).

## III. SEDIMENT NUTRIENT POOLS

### A. ACCUMULATION OF NUTRIENTS

Thus far, we have primarily examined the physical processes involved in the accumulation of sediments in estuaries, the phenomena that influence their particle size distributions and, by implication, at least part of their role in the estuarine nutrient cycle. In this section, we will further examine the processes by which nutrients are accumulated in estuarine sediments as a precursor to addressing the related issues of nutrient form and release.

### 1. Mechanisms for Trapping Nutrients

Figure 1, adapted from Hakanson,[12] provides a comprehensive summary of the numerous physical, chemical and biological processes operating in the estuarine environment. Accretion and erosion of the estuarine nutrient pool would appear to be a function of the predominantly physical and geochemical processes associated with nutrient loading in the estuary and with the deposition and processing (both biotic and abiotic) of particulate matter. Young and Comstock,[21] in reviewing the specific processes involved in the accumulation of sediment P, note that, generally:

1. Particulate P, made up of a whole spectrum of size classes, is usually more abundant than the various species of dissolved P (an exception occurs in the case of waters dominated by wastewater treatment plants).[22]
2. Particulate P, except under bloom conditions, is usually inorganic.
3. Particulate P, except under reducing and alkaline conditions in the sediments, is usually associated with amorphous oxyhydroxides of Fe (depending on the concentrations of other ligands present, such as sulfate, carbonate, fluoride and humic substances).[21,24]

Generally, P is present in low concentrations relative to the major cations, but can be present in concentrations approaching those of the metals, especially Fe, Al and the like, with which it forms colloids of varying solubilities — as the pH decreases, metallic phosphates such as $Al^{3+}$ phosphates, $Fe^{3+}$ phosphates and hydroxylapatite (calcium/magnesium hydroxyphosphate) precipitate from solution.[24] At these lower pH values (5–6), phosphate adsorption onto clays, in substitution for silicates, is favored. Young and Comstock[21] further state that this process so strongly favors the formation of $Fe^{3+}$ phosphates that $Fe^{3+}$-humic complexes are almost totally excluded, even given the presence of strongly negative humic macromolecules of $10^2–10^5$ dalton. Only when the $Fe^{3+}$ to humic acid ratio (Fe:HA) approached between 20:1–2:1 did some exclusion of P occur, while at the more common Fe:HA ratios of 200:1 found in most surface waters, orthophosphate ($HPO_4^{2-}$) remained the preferred ligand.

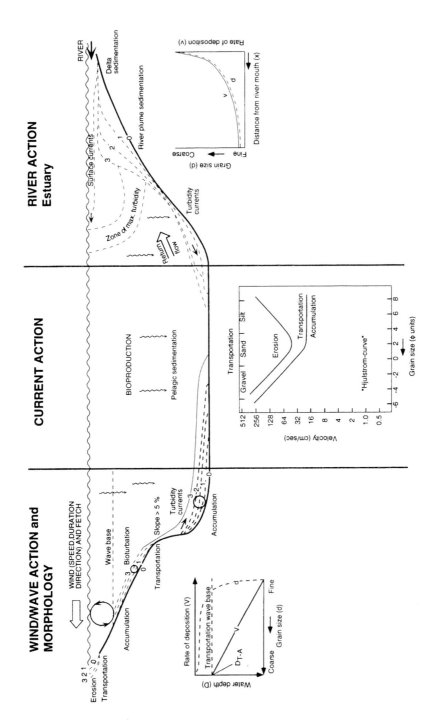

**Figure 1**  Diagrammatic representation of the major hydrodynamic and sedimentological processes operating in an estuarine and coastal area with a small tidal range. (From Hakanson, L., in *Sediments and Water Interactions*, Sly, P. G., Ed., Springer-Verlag, New York, 1986, 40. With permission.)

In short, the accumulation of phosphorus in estuarine sediments, as it is presently understood, usually involves an adsorption reaction which occurs in the water column, followed by the physical process of sedimentation. While the adsorption reaction is largely chemical in nature,[21,24] the situation does arise where the nutrient adsorbs to, or is contained in, organic materials which are also subject to the same sedimentation process (e.g., the sedimentation of algal cells noted by Nakata[8] or the transport and sedimentation of larger plant fragments of either/both terrestrial and aquatic origin[10,13]). In eutrophic systems, the rate of accumulation of organic materials in the estuarine sediments is increased relative to that observed under less enriched conditions.

Nitrogen appears to be somewhat less reactive than phosphorus in its inorganic form, forming complexes less frequently, although it is more common, as ammonium, in organic complexes in estuarine sediments. Organic nitrogen compounds, primarily associated with algal excretory products, form metallic-peptide complexes (primarily) with Fe, Cu and phosphates, which precipitate in much the same manner as the metallic-phosphate complexes discussed above. More commonly, nitrogen is transported to the sediments in the form of organic matter such as algal cells and detritus. Much of the estuarine nitrogen input is rapidly recycled in the water column, up to 50 times per day in many southeastern U.S. estuaries, with ammonium and ammonified reduced nitrogen compounds being processed in a matter of hours.[23] This rapid cycling retains dissolved inorganic nitrogen in the estuary (the ammonium residence times in these estuaries exceeding the water residence times by several-fold).

In addition to these mechanisms, allochthonous and autochthonous particulates can be trapped in estuaries by eddies and wetlands that exercise localized influences over the flow regimes (e.g., boundary effects) which encourage settling of particulates and aggregates. While the former affect the hydrodynamics of the estuaries and can be created by topographic and/or anthropogenic features, the latter are more multifaceted. Not only do wetland plants exert a hydrodynamic influence on the particulates that encourages settling, but they also can actively incorporate the nutrients into new growth. They may obtain these nutrients from either the water phase, the sediment (interstitial water) phase, or some combination of these; they may even draw upon regenerated nutrients released as excretory products by their epiphytic flora and fauna. During growth, these nutrients can be translocated to various portions of the plants, and, upon senescence and death, the nutrients may be either released in a dissolved (probably organic) form or conveyed to the sediments in particulate organic form.[9,13,19,45]

## 2. Forms of Nutrients in Estuarine Sediments

As mentioned above, the accumulation of nitrogen and phosphorus is primarily the net result of a (bio)chemical complexation process which removes these nutrients by adsorption, in some cases (especially that of nitrogen) by means of biological processes. In the Garigliano-Volturno Estuaries (the Pontine Island Basin), the highest concentrations of organic nitrogen and phosphorus were associated with the finest-grained sediments and were linked to the decomposition products of the macrophyte, *Posidonia oceanica*; concentrations of these nutrients ranged from 10–200 mg N $g^{-1}$ and from 7–363 µg P $g^{-1}$.[12] Inorganic phosphorus concentrations at the same site were an order of magnitude higher, ranging from 40–814 µg P $g^{-1}$, but this was probably a function of the surrounding volcanic bedrock geology which was rich in apatite.[9,13] In contrast, in the Harvey Estuary, the organic forms of the nutrients were more abundant, approaching 500–1100 µg N $l^{-1}$ and 50 µg P $l^{-1}$, as opposed to the inorganic forms which averaged between 5–350 µg $NO_2 + NO_3$-N $l^{-1}$, 25–400 µg $NH_4$-N $l^{-1}$ and 1–15 µg P $l^{-1}$ — the higher values typically reflecting winter conditions and the lower, summer conditions.[28] A similar seasonality was present in the Italian data set.[13]

Once in the sediments, bacterial decomposition of detrital materials reduces the elements from organic to inorganic form,[13] which undergoes diagenesis resulting in the incorporation of the nutrients into the sediments or their slow release back into the water column by diffusion.[14] This latter process, described more fully below, has recently been modeled by Lofgren[25] under both aerobic and anoxic conditions as a function of diffusion moderated by the $Fe^{3+}$-$Fe^{2+}$ iron oxidation state shift at the sediment-water boundary (i.e., primarily a physical-chemical process) while the former has been modeled as a biological-geochemical process involving microbial uptake and incorporation into alumino-silicates, calcites or clays. The release of nutrients back to the water column does, however, have a biological component to the process, as it is biochemical oxygen demand that initially generates the high electron acceptor demand that frees the Fe and nutrients from the particulate materials to which they were adsorbed. In terms of nitrogen, this demand, which strips the oxygen from the nitrate-nitrite molecules, accounts for the dominant role of ammonium in this process; the microbial action can also result in denitrification and loss of gaseous nitrogen from the system. Both functions, because of their biological components, are thus temperature-related.

## B. RELEASE AND BIOAVAILABILITY OF NUTRIENTS

While nutrient release is primarily a physicochemical process, several other processes can influence the magnitude of the release. These include resuspension (primarily associated with wind events), remobilization (primarily associated with geochemical processes), and regeneration (primarily associated with biological processes such as microbial activity, aquatic plants or bioturbation). Each of these processes, while mentioned above, is discussed in more detail below, with particular reference to their role in sediment-water nutrient cycling.

### 1. Resuspension

Resuspension of sediments in estuaries is a physical process involving both organic (detritus) and inorganic particulates being mixed, post-deposition, back into the water column by wind waves, currents or similar phenomena. Previously, the role of tidal surges in creating still water conditions (at slack water) conducive to deposition has been noted.[2] This phenomenon is peculiar to estuaries and is of importance in the release of nutrients because of the sorting action that it has on the suspended load entering the estuaries from terrestrial sources. It is especially important in the nitrogen cycle.[26] In some cases, the nutrients bound to the fine particulates can be directly extracted by algae, via the alkaline phosphatase enzyme, during bloom conditions, when dissolved forms have been depleted.[28] In other cases, especially involving nitrogen, the lack of a quantified hydrodynamic (physical) — chemical pathway linkage has led to the underestimation of denitrification losses.[27] Primarily, resuspension controls the eutrophication process through its effects on nutrient sinks — burial, nutrient export, and denitrification.[26]

Burial is probably the least important of the nutrient sinks in estuaries as it is typically the larger particles that are left behind in the deposition/resuspension process, although aggregation of fines accounts for a higher percentage of fines in these buried sediments than would typically be predicted solely on the basis of physical processes governed by Stokes' Law.[2,26] Burial affects phosphorus more than nitrogen as this former element does not have a gaseous phase as part of its biogeochemical cycle. One of the more important factors in the burial process is the macrobiota which pelletize fine particulates and package them in a form so that resuspension is minimized. In the Swedish Laholmsbukten, the most important pelletizers have been identified as mussels (*Mya arenaria* and *Cardium edula*),[26] while in the southern African Zandvlei Estuary a similar role is filled by the epizoic polychaete, *Ficopomatus enigmaticus* (see Chapter 8).[29] Floderus and Hakanson[26] suggest that the source of the pelletized materials is autochthonous, in contrast to the concept expressed by Walling and Moorehead[2] who identified this material as being of allochthonous origin. This seeming conflict can be resolved easily as the autochthonous material referred to by Floderus and Hakanson[26] is largely reprocessed allochthonous materials that have been converted to detrital form within the estuary by algal uptake — a situation noted by researchers around the world.[2,8,13,23,28] In reality, burial is likely to be the net result of a biological-physical process which removes particulate nutrients derived from both allochthonous and autochthonous source materials from the water column. Thus, in the context of resuspension processes, burial results in a net loss of materials from the water column; nutrient recycling from buried materials (remobilization) is discussed below.

Nutrient export through resuspension is connected to the net seaward transport of particulates that are not removed from the water column by deposition/burial processes. Typically, the movement can be described as "rachet-like", controlled by the ebb and flow of the tide and by wind-induced circulation. Materials are deposited and resuspended many times before being finally deposited in the nearshore zone beneath the influence of the waves and/or currents. The rate at which this net outflow of particulate nutrients to the deep ocean occurs depends on the hydrodynamic balance between river flow,[19] tidal movements, and nearshore current patterns, as well as biological processes associated with plant growth. These processes are characterized by temporary deposition sites or "ephemeral mud blankets" that are in a constant state of flux, the periodicity of which can range from daily (if predominantly tidal-related) to a few weeks (if predominantly wind circulation-related). Walling and Moorehead[2] have described the daily net seaward flux of sediment-bound nutrients in the upper reaches of the Mersey Estuary (controlled by river flow, moderated by tidal ebb and flow), while Floderus and Hakanson[26] have described the bimonthly seaward movement of particulate nutrients in the Kattegat (controlled by wind-wave orbital motions and bottom shear stresses). Superimposed on these movements is the seasonality related to both intra-annual variations in runoff and variations in wind speed and direction. There is also a biological component to this process which is discussed below in terms of nutrient regeneration.

Under calm conditions, the nutrient-rich particulates that make up the ephemeral mud blanket may remain in place for some extended period of time — long enough for diffusive exchange to take place, enriching the interstitial (pore) water with released inorganic nutrients. In the Kattegat, this interstitial water nutrient exchange (remobilization; see below) typically occurs during spring and autumn.[26] Via benthic regenerative processes, organic nitrogen is transformed into inorganic nitrogen[26] (see also below) and, together with remobilized phosphorus, is released from particulate complexes[25] into the interstitial waters, creating a steep concentration gradient within the upper few centimeters of the mud blanket. Through diffusive processes, the newly liberated phosphate and ammonium radicals move upward into the overlying water. There they can be recomplexed with (especially) iron,[25] taken up by algae and bacteria,[28,30-32] or mixed into the water column where they are available for plant growth. The amount of nutrient entering the water column in the dissolved state would seem to be a function of the mixing rate; the more rapidly the nutrients are mixed into the overlying water, the greater the probability is that they will be in the dissolved phase. If the mixing process is slow, it appears that most of the nutrients will be recomplexed and reprecipitated and/or used by plants for growth — the precise rates at which these processes occur and the partitioning of the nutrients between them is unclear from the available evidence. The rate of algal uptake apparently depends on the nutritional status of the plants, the temperature, the depth of water, the depth of light penetration, the degree of shelter/turbulence, and the degree of oxygenation.[29,31] In the Peel-Harvey Estuary, it would appear that a residence time of about 7 days is required for conditions favoring sediment-based algal nutrition to develop — a rare situation in winter but relatively common in summer (which is the period of maximum algal growth).[30] This suggests that biological uptake of the released nutrients would be an intermediate rate process compared to the more rapid processes of mixing and reprecipitation.[26,30,31]

Denitrification, while commonly associated with nutrient regeneration (see below), is the process whereby microbial decomposition processes reduce nitrate and nitrite to gaseous nitrogen, $N_2O$ and $N_2$, which can be lost to the atmosphere.[32] As previously mentioned in connection with P release,[25] this process is enhanced under anoxic conditions and the presence of an anoxic microlayer at the sediment surface. The denitrification/nitrification process in estuaries usually manifests itself as detectable concentrations of nitrite (an intermediate denitrification product) in the overlying water,[32] a nitrogen form characteristically absent in freshwaters.[33] Up to 50% of the nitrite present in the Odawa and Tama Estuaries (Japan) was produced by denitrifiers and introduced to the water column through physical mixing processes.[32] As this nitrogen form is available for ammonification and phytoplanktonic uptake, denitrification can be considered both a source and a loss factor relative to the water column and sediment-nutrient pool.

## 2. Remobilization

The role of denitrifiers in the sediment-water nutrient exchange process bridges the gap between the largely physical processes associated with resuspension and the predominantly chemical/biological processes associated with nutrient remobilization and regeneration. As might be expected, these processes can be highly seasonal as a result of their biological components. Remobilization may be distinguished from nutrient regeneration (see below) in that the latter involves a change of state (i.e., from organic to inorganic), is heterotrophic and is typically associated with sedimentary decomposition processes.[33,35] Regeneration is seasonal due to its dependence on microbiota, which, in turn, are temperature dependent. Nutrient remobilization, in contrast, is primarily driven by abiotic (chemical) processes associated with diagenesis.[14,34] It is seasonal due to its dependence, in part, on temperature and hydrodynamic processes. It is often difficult to distinguish between these processes *in situ* as they are often associated with resuspension and physical mixing processes as the foregoing discussion may have indicated.

Pagnotta et al.[34] used dilution graphs to identify phosphorus and nitrogen remobilization in the Tiber Estuary (Italy), which showed an uncharacteristic conservative (straight-line) dilution response (relative to increasing salinity) during winter. During summer, soluble reactive phosphorus and total dissolved phosphorus even showed a slight convex (source) response, as did ammonium and total inorganic nitrogen. While they interpreted the winter data as showing a shift in the equilibrium between dissolved and particulate phosphate (due to physicochemical effects), they related the summer response to the presence in the estuary of a halocline as the result of diminished river flows. The presence of the saline waters modified the sediment-water concentration gradients, generally causing them to become steeper, encouraging diffusive nutrient fluxes from the sediments into the overlying waters. Further, the degree of positive or convex deviation appeared to be related to the "age" of the saltwater wedge, with the greatest deviations being associated with periods of anoxia in the bottom waters of the estuary (i.e., in relatively "old" water). This suggested a regenerative component to the process which was confirmed through stoichiometric analyses. When compared to Kester's ratio of oxygen consumption (dO) to nitrate ($dNO_3$, as the $HNO_3$ radical) and phosphate ($dPO_4$, as the $H_3PO_4$ radical) production as the result of remobilization ($dO:dNO_3:dPO_4 = -138:16:1$), their data ($dO:dNO_3:dPO_4 = -276:-2:5$) suggested a net loss of nitrate (instead of a net gain) and a net gain in phosphate beyond that expected.[34] While unusual, this was totally consistent with what could be expected if the remobilized nutrients were subjected to further regenerative processes which converted the nitrate to ammonium under anoxic conditions, which would also free additional P from ferric phosphate complexes.

Remobilization of nutrients also includes the renewed input of nutrients due to resuspension.[14] While resuspension lifts the particulate nutrients off the bottom of the estuary, remobilization takes place in the water column as a result of the dissolution of the particulates, releasing the nutrients. To contrast this process with the regenerative processes described below, the medium of nutrient release is purely geochemical, being related to the dissolution constants of the nutrients involved. Thornton[36] used these constants to theoretically estimate the abiotic release of silica from diatom frustules in the Great Bay Estuary and concluded that this was an extremely minor input compared with the allochthonous inputs from the estuary's granitic watershed, the frustules normally sinking out of the water column before dissolution could occur (hence, the lack of dissolution in this instance led to a net loss of silicate from the water column). While these processes occur more commonly in the deep ocean, the nutrients released in this manner can be introduced into the estuarine environment through coastal upwelling.[37,42]

## 3. Regeneration

The foregoing again demonstrates the high degree of interconnectedness that exists between sediment-nutrient processes. Regenerative processes, acting on remobilized nutrients in the Tiber Estuary, radically changed the character of the nutrients released from the sediments.[34] Similarly, Callender[35] has shown that sediment phosphorus regeneration in the Potomac Estuary (U.S.) contributes up to 10% of the total phosphorus load to the estuary, a mass equivalent to 25% of that contributed by wastewater treatment plants discharging to the system. Boyer et al.[23] report similar results for nitrogen regeneration in the nearby Neuse Estuary, with up to 15% of the total nitrogen budget of the estuary being supplied by regenerative processes.

In these various systems, nutrient regeneration has been associated with bioturbation and benthic fauna, macrophytes and macroalgae, and phytoplankton and bacteria. In the latter case, that of the Neuse Estuary, nutrient regeneration was mediated by planktonic algae, dominated by cyanophytes in the fresher portions and by dinoflagellates in the more saline areas of the estuary. Inorganic nitrogen was ammonified and rapidly recycled within the water column. Turnover times ranged from 0.02–11 days, the more rapid turnover times reflecting summer conditions in the more saline waters, while the longer turnover times reflected winter demands in the fresher waters. Turnover time was inversely correlated with ammonium loading — even though ammonium concentrations remained at detectable levels (2–8 $\mu M$) throughout the year — suggesting some degree of ammonium limitation in the estuary. (However, Boyer et al.[23] also note that the period of maximum ammonium loading coincided with both the peak water-loading period and the period of minimum algal ammonium demand, which might also suggest that phytoplankton washout and/or seasonality was contributing to this inverse relationship.) Nevertheless, total planktonic ammonium demand in the system exceeded supply by 115 times, underlining the importance of regenerative processes in the nutrient budget of this estuary.

In the Australian Peel-Harvey Estuary, McComb and co-workers[28,30,31] have identified enhanced phosphate and ammonium release rates, equating to between 4–22 mg P m$^{-2}$ d$^{-1}$ and 100–200 mg N m$^{-2}$ d$^{-1}$, within beds of the macroalga, Chaetomorpha linum, in the Harvey Estuary portion of the system. The resultant higher nutrient concentrations within the beds of the macroalga were coincident with reduced oxygen conditions which were also consistent with enhanced sediment-water nutrient exchange conditions — the oxygen depletion resulting from algal respiration, decay and microbial oxygen demand (both epiphytic and at the sediment surface). Nutrient release from senescent plant material may have also contributed to the higher nutrient concentrations observed in the macrophyte beds.[30] Both microbially mediated sediment-water nutrient releases and the release of nutrients into the water column as the result of decay processes provide examples of nutrient regeneration in this system, while geochemical nutrient release from anaerobic sediments illustrates the role of this process in facilitating additional nutrient release (although in this system, the latter process was relatively minor when compared to the allochthonous and regenerative nutrient inputs). In contrast, the cyanophyte, *Nodularia spumigena*, which also exists in the Peel-Harvey system, relies heavily on sediment phosphorus during bloom conditions,[28] as does the macroalga, *Cladophora montagneana*, which until recently dominated the algal flora of the Peel Inlet (the Murray-Serpentine Estuary).[28,31]

In the Potomac Estuary,[35] the regenerative agent was the benthic fauna, particularly oligochaete and polychaete worms burrowing into the sediments and ventilating their burrows with currents produced by bodily movements or ciliate action. This ventilation process pushes nutrient-rich interstitial waters out into the overlying water column and increases the expected nutrient flux relative to the rate anticipated on the basis of diffusive processes alone. Remobilization rates and diffusive fluxes in the tidal portion of the Potomac Estuary ranged from 0.03–0.20 m$M$ P m$^{-2}$ d$^{-1}$, with remobilization accounting for about 0.06 m$M$ P m$^{-2}$ d$^{-1}$. In the lower estuary, phosphate fluxes were higher, ranging from 0.3–2.3 m$M$ m$^{-2}$ d$^{-1}$, due, primarily, to an anoxic surface layer at the sediment-water interface; during the summer months, the bottom waters of the estuary were also anoxic, which further facilitated the release of adsorbed nutrients.

The interstitial water phosphorus concentrations in the Potomac Estuary also increased from landward to seaward within the estuary, increasing from 20 μ$M$ l$^{-1}$ cm$^{-1}$ upstream to over 100 μ$M$ l$^{-1}$ cm$^{-1}$ downstream. These patterns were consistent with the observation that the tidal estuary was primarily a deposition zone, while the lower estuary was primarily a zone of net nutrient loss. This sediment-water flux in the tidal zone was sufficient to satisfy up to 30% of the algal P demand in this estuary — a similar percentage to those measured elsewhere in neighboring estuaries along the eastern coast of the United States (sediment-water nutrient exchange supplied about 50%, ranging up to 300%, of the algal nutrient demands in the Patuxent Estuary, Chesapeake Bay, for example).[35] While about 50% of this phosphorus appears to have been generated through remobilization processes, the balance is likely to have been produced as the result of biogenic regeneration of phosphorus in the sediments and water column.[35,59] Callender[35] also notes the importance of increasing salinity in lysing algal cells entering the estuary from upstream and releasing their cell nutrient contents to the water column.

## IV. UPTAKE, RELEASE AND BIOAVAILABILITY OF TRACE ELEMENTS

### A. RELATIONSHIP BETWEEN NUTRIENTS AND METALS

The same processes that govern the cycling of nutrients in aquatic systems (set out above) govern the uptake and release of metals and trace elements (collectively referred to as metals for the purposes of this portion of the text). Given the environmental significance of many of these elements, their susceptibility to bioaccumulation/biomagnification, and their often toxic natures, it would be remiss of us not to briefly address this subject in our discussion of the role of sediments in estuarine eutrophication.

Having established the productive nature of estuaries and the potential for both the rapid uptake and release of nutrients in the sediment-water system, this section will focus on relationships between the nutrient cycles discussed above and the metals and trace elements. Briefly, the nutrient cycles relating to sedimentary processes involve an aqueous phase governed by primarily biotic uptake, excretion and release processes; the aqueous phase also involves the adsorption of nutrients by suspended particulates (biotic and abiotic), particle aggregation and sedimentation. The physical processes involved form the principal transport mechanism to and from the sediment surface; e.g., currents, wind-waves, and tidal effects that transport and mix nutrients in both direction. Once deposition has occurred, there comes into play a variety of chemical processes (dissolution, diffusion and diagenetic processes) which interact with biological processes including decomposition processes to modify the sediment nutrient pool and either promote nutrient release or "fix" nutrients into the sediment sink. All of these processes are affected by seasonality (whether it be seasonal inflow, changes in current patterns or tidal regimes) and, to some degree, by temperature, oxygen regime and degree of light penetration. Hart,[44] in a recent review of trace metal uptake in aquatic systems, uses this basic mass-balance-type approach as the basis for his discussion, while numerous other authors have examined aspects of this framework in more detail (diagenetic partitioning,[45,46] metal speciation and behavior in the sediments,[47-49] and metal mobility and release).[50,51] For consistency, we adopt the same approach.

## B. UPTAKE OF METALS

Hart[44] and others[53] have noted a remarkable consistency in sediment-water interactions across various aquatic environments, due, in part, to similarities in the nature and composition of the particulates; all particulates have a slightly negatively charged (organic film) surface, and there is surprisingly little variation in the mobility of these particulates despite a wide variety of source types.[44] Hart[44] noted evidence from various areas that points to the preferential uptake of organics (typically carboxylic acid and phenolic groups) even in the presence of such competing anions as phosphate, sulfate and silicate — although Comstock and Young[21] suggest that this is only true for humic acids when the humates are present in abundance relative to the amount of dissolved iron (see above). Some of this organic film has been identified as microbial.[44,59]

Metal uptake by sediments follows the same mechanisms as identified earlier: physicochemical adsorption, biological uptake and deposition of metal-containing particulates. As discussed previously, iron (FeOOH), aluminum ($Al_2O_3$), manganese oxide ($MnO_2$) and clays are the most active in this process — and again the role of competing ligands becomes critical to the determination of the magnitudes of the adsorption reaction in any given estuary. Typically, mercury, copper, lead, zinc and cadmium, in order of decreasing dominance, are taken up by both high and low molecular weight organic complexes.[44] The order of dominance seems to vary slightly when inorganic metal oxide substrates are present, with cadmium, copper, lead and zinc being rapidly taken up by iron oxides, for example. These kinetics are especially sensitive to pH, and to calcium and magnesium concentrations — the latter primarily because of their relative superabundance in aquatic systems. Citing evidence from the Yarra Estuary (Australia), Hart[44] shows a shift in the uptake mechanisms for manganese from biotic/abiotic adsorption-based processes to geochemical-based processes as the flow proceeds seaward.

Biological uptake of metals has generally been poorly studied,[44] although bacteria have been implicated in the methylation of mercury[49] and lead[48] (the precise role of bacteria in the latter process remains controversial[48]). While bacterial uptake of metals is well established, the role of macrophytes and algae appears to be more species-specific, some species (such as *Myriophyllum* sp. and *Cladophora* sp.) being metal accumulators while others (such as *Cyclotella meneghiniana*) are metal sensitive. Other evidence would suggest that this metal sensitivity is primarily a function of where in the cell structure the metal is stored. In the metal resistant algae, *Phaeodactylum tricornutum* (zinc-tolerant) and *Chlamydomonas reinhardtii* (copper-tolerant), the accumulated metals are stored in the cell wall and enter the cell protoplasm, if at all, in extremely small quantities which are rapidly (<2 hours) expelled when the external metal concentrations are reduced.[44] Chlorophytes, diatoms and chrysomonads are reported to be the more metal tolerant algal families. Given that the metals appear to be contained in the more resistant parts of the cell structure, biotic metal uptake can form an important removal mechanism in many aquatic systems.

Finally, much of the metal content of the sediments is derived from allochthonous sources such as the catchment bedrock and soils.[7,44] This material is accumulated in the sediments through the deposition mechanisms described at the beginning of this chapter.

## C. METALS IN ESTUARINE SEDIMENTS

The shift in metal distribution from metals adsorbed to particulates and organic matter in the upper reaches of the estuary to metals associated with bedrock-derived particulates in the lower reaches, observed in the Yarra Estuary,[44] (which appears as a gradation of metal enrichment from landward, where the highest metal concentrations occur, to seaward) seems to be consistent in many other estuaries (including the Gironde,[45] Rhone[46] and Mersey[7]) although the fate of the organically bound and adsorbed metals in the seaward portions of the estuaries seems to be unclear. Sly[7] suggests that simple dilution of the predominantly terrestrial (and some autochthonous) materials is unlikely to play a major role in changing the relative composition of these sediments. El Ghobary and Latouche[45] imply that early diagenetic changes associated with the formation of (especially) iron and copper sulfides (decreasing their solubility and altering their state from organic-metal complexes to more stable chemical compounds) and/or the release of the metals to the overlying water column in a dissolved state could potentially supply an answer to this seeming mystery. This solution is less satisfactory for manganese, zinc, lead and nickel (although nickel can form both organic complexes and more stable sulfur compounds).[45,47] Most of these latter elements were associated with anthropogenic pollution episodes in the Gironde Estuary,[45] which may have some bearing on their "unusual" response pattern. Fernex et al.,[46] in their survey of the Rhone Estuary (France), applied Fick's equation to their results and concluded that sediment-water exchange of dissolved metals could indeed modify the speciation of cadmium, copper, iron and manganese in the surfacial sediments (<3 cm depth) and that a large percentage of the remaining metals were in the form of early diagenetic products — either as $CdCO_3$ or CdS, CuS or FeS. Fifty percent or more of the deposited metals were released during transport and post-deposition as dissolved metals in the nearshore environment during their study.[46] Similarly, Sadiq[47] suggests that >50% of the nickel in marine sediments will be in the form of the dissolved metal, while most of the remainder will be complexed with chloride or sulfate ions. Chakraborti et al.[48] report similar findings for alkyl lead compounds in the Scheldt Estuary (Holland), with the larger part of the complexed lead compounds occurring in a dissolved form (in contrast to freshwaters where these compounds were usually present in the sediment as organic complexes). In Tokuyama Bay (Japan), 50–90% of the mercury present in the sediments was in the form of HgS with most of the remainder occurring as elemental mercury.[49]

## D. RELEASE AND BIOAVAILABILITY OF METALS

The foregoing discussion suggests that estuaries are active zones of deposition and transformation of terrestrially derived metals, and that estuaries act as a source of dissolved metals.[45-50] While this is generally true, Forstner et al.[50] report somewhat different results from the grossly polluted[11] Elbe Estuary (Germany). In this estuary, zinc exhibited the expected dilution response, but cadmium showed an unexpected convex (source) response which suggested an active remobilization of this element within the estuary. They attributed this to the effects of a lowering of the pH (the result of both poor buffering capacity/low carbonate content and the bacterial action of the sulfur and iron bacteria, *Thiobacillus thiooxidans* and *T. ferrooxidans*) and fluctuating Eh conditions (the result of seasonal changes in the oxygen regime above the sediments). Under these conditions, geochemical mobility will favor the release of mercury, cadmium, lead and copper over other metals.[24,50]

As noted earlier, the bioavailability of these metals depends not only on their availability but also on the nature and characteristics of the biota exposed to the metals. Both plants and animals can alter the state of metals (as well as nutrients as discussed above) prior to bioaccumulation, and the elements can be positioned in different portions of the organisms after uptake.[44,51,54] Thus, major differences in the levels of contamination can be distinguished even within functional groups; for example, in the Severn Estuary (U.K.), lead concentrations were of the same order across a spectrum of deposit feeders (*Macoma balthica*, *Nereis diversicolor*), grazers (*Littorina littorea*, *Patella vulgata*) and primary producers (*Fucus vasculosus*), but differed markedly among and within these groups in the West Looe Estuary (U.K.) (ranging over nearly an order of magnitude between the deposit feeder, *Scrobicularia plana*, and its fellows *M. balthica* and *N. diversicolor*, and between the filter feeder, *Mytilus edulis*, and its fellow, *Cerastroderma edule*; varying over a smaller range for the grazers, *P. vulgata* and *L. littorea*, and for the macroalga, *F. vasculosus* — the organism with the higher level of lead being listed first in each case).[54] In most cases, the net accumulations of lead in the benthic fauna resulted from exposure to the metal in food, rather than from direct uptake from the sediments[54] — in contrast to the benthic fauna, the macroalga did accumulate the element through environmental exposure. However, Luoma[54] points out that bioaccumulation through ingestion depends on the geochemical characteristics of the substrate, a factor that probably accounts for the variations reported among and between species in the Severn and West Looe Estuaries. These findings closely parallel the results obtained by Lyngby and Brix[51] in experiments with eel grass (*Zostera marina*) in the Limfjord (Denmark). Lyngby and Brix[51] further showed that senescence and decomposition of the plant resulted in increased levels of chromium, lead and zinc (possibly due to both adsorption onto the resultant detritus and/or microbial activity occurring on the plant substrate). A concomitant decrease in cadmium concentrations suggested that this metal was associated with the cell contents of the plant which were quickly lost through lysis on the death of the plant.[51] This dual response (both an increase in some metals and a decrease in others) reinforces the need to examine the complete biogeochemical cycle of these elements in estuaries.

## V. DISCUSSION AND MANAGEMENT IMPLICATIONS

In the management of sediment-water interactions, there are usually two basic approaches that can be adopted: namely, watershed-based interventions designed to reduce the source materials of terrestrial origin, and *in situ* approaches designed to remove or negate the consequences of sediment nutrients (or metals) within an estuary. The latter are usually not viewed as viable options in most estuaries due to their hydrodynamics although there are exceptions to this generalization. However, there are also regulatory or institutional measures that can be taken to eliminate or mitigate some sources and consequences of estuarine sediment-water interactions. This section, therefore, presents an overview of those practices and procedures that have been employed in estuaries to control these sediment-water nutrient exchanges which have contributed to incidences of eutrophication. In every case, the manager/decision-maker should not overlook the "do-nothing" option, which is rarely a real possibility in most eutrophic estuaries, but which, nevertheless, remains an alternative.

The typical *in situ* approaches to managing sediment-water interactions involve nutrient inactivation or bottom-coverings.[55,56] The former uses the addition of a layer of a multivalent cationic salt to precipitate phosphorus from the water column. Iron or aluminum salts are commonly used, and can be applied either to the water surface or injected under the sediment surface using an underwater harrow-type implement. The resultant floc forms a blanket which effectively seals the sediments against any subsequent nutrient release. The latter typically involves covering the sediments with a geo-fabric, synthetic fabric (such as polyethylene, PVC or nylon), natural fabric (such as burlap or hessian) or grid (either metal or plastic) to reduce plant growth and/or interfere with the sediment-water exchange process. Both methods suffer from the same drawbacks when employed in estuaries; namely, the fact that the sediments in estuaries are in a constant state of flux, undergoing daily (tidal) and seasonal cycles of scour and deposition. In most cases, these movements would disrupt and negate the effectiveness of most sediment covers or flocculant blankets as control mechanisms. In addition, such methods may have adverse or at least unpredictable effects on the biota of estuaries.

In contrast, sediment nutrient control methods involving the removal of either the sediments or the resultant biomass can be more successfully employed in estuaries. In the case of the former, dredging is a common, if expensive, practice in those estuaries serving a recreational or commercial function (i.e., marinas and harbors). Generally, dredging is a short-term response, given the movement of sediments into, within and out of an estuary, but it can be effective when combined with upstream, terrestrial erosion control measures in those estuaries dominated by terrestrially derived sediment loads. It is often less successful when the sediments are derived from oceanic or nearshore sources, which are more difficult to intercept even with shoreline control structures.[6,58] While possible heavy metal contamination may make dredge spoil disposal difficult at times,[50] sediment removal continues to be a widely used technique.

A variant of sediment removal is the removal of aquatic plant biomass using an aquatic plant harvester. This technique has been used successfully at Zandvlei, where it is estimated that some 0.20–0.25 tonnes of phosphorus (about 0.1% of the annual net deposition of the nutrient in the estuary) is removed annually in the form of 224 tonnes of plant biomass.[57]

Other *in situ* control measures include artificial enhancement of circulation, such as flow augmentation and flushing, hypolimnetic aeration, and circulation or destratification. Some of these options require emplacement of aerators or other structures on the estuary floor, which are susceptible to surge damage or other damage related to the dynamic nature of estuaries and their sediments (although these options may be viable in some protected embayments or lagoons). The former options, which affect the horizontal circulation, may be more easily accomplished in some estuaries, especially if the obstructions to circulation are man-made. In these cases, the cost of the infrastructure and its associated use-benefits will have to be weighed against the expected benefits of more complete circulation (which may be largely aesthetic in many cases). In some cases, however, re-siting structures may have other benefits (such as increased carrying capacities in the case of aquacultural operations — although siting of aquaculture cages in or near navigation channels may have other repercussions) in addition to the water quality benefits to be gained from increased flushing.

Finally, experimental techniques such as subsediment surface injection of nitrate to enhance sediment oxidation, and biomanipulation may have specific applications in some systems. The latter techniques are designed to control the biological consequences of enrichment.

Probably the most effective approaches to controlling estuarine sediment-water nutrient dynamics are structural approaches employed in the watershed or along the coastline to minimize the supply of nutrients and sediments to an estuary.[55,56,58] The range of options is large, including zoning and planning options that integrate development and estuarine environment in a sensitive way; nonpoint source pollution control options that employ stormwater management techniques (both urban and rural; agricultural measures, for example, include no-till cultivation, strip-cropping, and contour-plowing techniques) to trap potential pollutants and nutrients before they enter an estuary; point source pollution controls applied in the watershed, along the estuary and to water craft using the estuary to interdict nutrients before they reach the estuary; and in-stream management techniques (such as stream-bank revegetation and stabilization, installation of sediment traps, and wetland-floodplain preservation policies). Some of these techniques can also be employed to seaward in the form of sand by-passes and sand traps built into harbor and estuary outlet structures, while dune revegetation and other "soft" options can be used to control the sources of sand input from the coastline and sea.[58]

While it is not possible to discuss all of the available options in detail here, suffice to repeat that a combination of options will probably be most effective in controlling the undesirable effects of sediment-water nutrient exchange. While many coastal developments have taken place over the centuries, new developments still continue in most coastal states. The opportunity therefore exists to employ good planning and land use controls to these developments and, while such controls may be expensive to retrofit, their application to new developments can be beneficial even if to no greater extent than preventing a bad situation from becoming worse.

## VI. SUMMARY

### A. A SYNTHESIS

In this chapter we have examined a number of processes affecting the release and bioavailability of nutrients and pollutants in estuarine sediments. While striving to be comprehensive insofar as the basic mechanisms are concerned, we have, of necessity, omitted some detail regarding the many pathways by which these mechanisms operate. For example, we have identified regenerative pathways associated with macroalgae, microalgae and macrobenthos, but have not considered in detail similar pathways associated with microbial flora, although their role has been touched on.[32,67] Likewise, we have not dealt with the much discussed theory of macrophytic "nutrient pumps",[50] although this is discussed in at least one of the case studies included in this volume. Other, seemingly more esoteric mechanisms, such as methane gas ebullition,[35] have not been discussed at all even though these processes may be important in some estuaries.

The physical, chemical and biological processes which have been presented summarize the major sediment-water exchange pathways by which nitrogen and phosphorus are gained or lost by the water column in estuaries and coastal lagoons. These are the processes that appear to be well understood, having been observed and measured for over half a century.[38,39] Yet even these processes are complex. For example, it is only within the last 15 years that aerobic nutrient exchange processes have been identified[40] and more recently still that their geochemical mechanisms have been modeled.[25] The interconnectedness of many of these processes is only now beginning to be understood as the limitations of laboratory-scale, "undisturbed" sediment core experiments are becoming known[32,41] and experiments are being carried out in larger "belljars" and "mesocosms".[42] Even these have limitations when compared to the ecosystem scale case studies presented in this volume, though the latter inevitably suffer from a discontinuity when it comes to linking environmental responses to specific organisms or events. Thus, all these approaches seem essential and interdependent — controlled laboratory experiments where it is possible to disentangle the influence of specific environmental variables, through to studies on natural systems where observed responses represent the integration of diverse and imperfectly defined processes. Hopefully, modeling will prove a valuable tool in integrating information over these different scales.

We would like to conclude by suggesting something of what we know, something of what we think we know and something of what we would like to know concerning the role of sediments in estuarine eutrophication.

### B. WHAT WE KNOW

### 1. Nutrient Fluxes Occur

Measurable nutrient fluxes have long been known in anaerobic waters, but have only recently been observed and quantified under aerobic conditions.

### 2. Nutrient Fluxes Are Driven by Physical, Chemical and Biological Mechanisms

Resuspension, remobilization and regeneration processes appear to be the major mechanisms by which nutrients move between the sediments and the overlying waters:

1. Physical mechanisms include sediment transport, deposition, export, burial and resuspension.
2. Physical mechanisms are a function of wind-waves, currents and tidal influences acting on the transport medium (the water), seasonality being introduced due to intra-annual variations in wind velocity and direction, tidal regimes and river hydrology.

3. Chemical mechanisms include dissolution, remineralization and diffusive exchanges driven by concentration gradients between the sediments and overlying waters, and sediment-water chemistry.

4. Chemical mechanisms are a function of metal concentrations (especially iron) in the sediments and overlying waters, salinity (in respect of aggregates), and oxygen regime — temperature also affects reaction times and rates but only to a relatively minor extent over the temperature regimes experienced in most estuaries.

5. Biological mechanisms include decomposition, lysing of living and senescent cells and transformation processes such as denitrification/ammonification and methane/hydrogen sulfide production.

6. Biological mechanisms are a function of species composition (microbiota, microalgae, macroalgae and macrobenthos have been identified as agents in this process), temperature, salinity, and oxygen status in the sediments, at the interface and in the overlying waters.

## 3. Nutrient Fluxes Are Complex Combinations of Processes

Rarely does one mechanism occur in isolation in estuaries; at the very least most sediment-water nutrient exchanges involve both biological and chemical processes, and, although not always recognized, all involve the physical processes associated with the existence and sorting of the sediment substrate at any given point in an estuary.

## 4. Nutrient Fluxes Can Be Interrupted

It is this theorem that forms the basis of most of our management actions. Since some of these actions are effective, this statement must be true (see also below).

## C. WHAT WE THINK WE KNOW
## 1. Nutrient Flux Processes Can Be Defined

Empirical relationships such as Kester's ratios,[34] the Redfield ratios,[37] and the partition models of Carignan[43] have been used with some success in identifying discrete processes in estuaries, at least insofar as quantifying the relative magnitudes of remobilization and regeneration processes. There is renewed debate over the use and design of sediment traps and related techniques for the measurement of deposition/resuspension, and on-going debate on the "best" way to measure sediment-water exchange rates (e.g., cores vs. mesocosms).

## 2. Nutrient Flux Processes Can Be Modeled

At the "micro" level, physicochemical laws governing deposition and resuspension (e.g., Stokes' Law) and diffusion (e.g., Fick's Law), among other processes, can be used to describe discrete, individual processes occurring within the sediments and estuarine water column. At the estuarine system ("macro") level, stochastic input-output models can be used to reproduce the gross functioning of the sediment-water system.[8,19,26,35] However, few "user-friendly", widely applicable models linking these levels of process appear to be available.

## 3. Nutrient Fluxes Can Be Controlled

The design and implementation of sediment-water nutrient exchange intervention mechanisms presupposes knowledge about the structure and functioning of the components of the exchange process. Unfortunately, an inability to define a desired outcome in every case highlights the small degree of understanding that we actually have about these processes.

## D. WHAT WE WOULD LIKE TO KNOW
## 1. How Nutrient Fluxes Interact with Biological Processes

While it may appear that the dominant form of nutrients in estuaries (e.g., particulate vs. dissolved, organic vs. inorganic) is related to the biotic response (e.g., macrophyte vs. phytoplankton), the precise relationship remains unclear — is the relationship between rooted plants and a high particulate organic nutrient load based on light limitation, oxygen depletion, turbulence, or some other, more complex set of interactions? Can the relationships be simplified? Generally, management interdictions address a single issue (such as nutrient loading from point sources in the Potomac Estuary[35] or sediment loading from agricultural nonpoint sources upstream of the Peel-Harvey Estuary[31]), and the consequences of such interventions will be better predicted if the relations between sediment and biological processes are clearly understood.

## 2. How the Concept of Scale Applies to Estuarine Nutrient Dynamics

Many of the estuaries discussed above are relatively small systems. Can knowledge gained in these systems be applied to estuaries such as those of the Mississippi (U.S.), the Amazon (Brazil), the Mekong (Asia) or the Nile (Africa)? The simple response is definitely yes — the laws of physics and chemistry will certainly apply — but what other considerations are introduced, and how transferrable is our knowledge and our models, between estuaries of different scales, climatic zones, and continents?

## 3. How Nutrient Fluxes Can Be Managed

Given the highly dynamic nature of these systems (which tends to negate traditional "engineering" approaches to controlling sediment-water interactions, such as alum dosing), and the fact that many estuaries are heavily used by humans, what suite of controls can be applied to the extensive nutrient loads generated within estuaries in order to achieve human use objectives? Are these objectives at odds with the ecological structure and functioning of these systems? Is there a single paradigm that can be applied to such complex areas, as there appears to be[53,60] for lakes, rivers and the deep ocean?

## E. CONCLUSION

What the foregoing suggests is that, while we have a reasonable understanding of the individual processes that relate to sedimentation and sediment-water nutrient dynamics, and a rudimentary knowledge of some basic control strategies to manage these dynamics, we have relatively little understanding of how these processes interact and how human interventions will affect them in any given situation. Our management of sediment nutrient dynamics is therefore still, largely, a "hit or miss" affair. While this may always be the case, given the vagaries of biologically driven systems, it is our hope that the case studies and information presented in this volume will help to clarify some potential responses of these systems to management actions, and help the manager and decision-maker to consider more completely the full range of options and potential consequences of action/inaction.

## ACKNOWLEDGMENTS

The opinions expressed in this chapter are those of the authors and do not necessarily reflect the opinions of the institutions which they represent.

## REFERENCES

1. **Krumbein, W. C. and Sloss, L. L.,** *Stratigraphy and Sedimentation*, Freeman and Co., San Francisco, 1963, 660 pp.
2. **Walling, D. E. and Moorehead, P. W.,** The particle size characteristics of fluvial suspended sediments: an overview, *Hydrobiologia*, 176/177, 125, 1989.
3. **Gregory, K. J. and Walling, D. E.,** *Drainage Basin Form and Process: A Geomorphological Approach*, Edward Arnold, London, 1985, 458 pp.
4. **Grobler, D. C., Toerien, D. F., and Roussouw, J. N.,** A review of sediment/water-quality interactions with particular reference to the Vaal River system, *Water SA*, 13, 15, 1987.
5. **Sly, P. G., Thomas, R. L., and Pelletier, B. R.,** Comparison of sediment energy-texture relationships in marine and lacustrine environments, *Hydrobiologia*, 91, 71, 1982.
6. **CERC [U.S. Army Coastal Engineering Research Center],** *Shore Protection Manual*, Vol. I, Department of the Army Corps of Engineers, Washington, D.C., 1984, chap. 4.
7. **Sly, P. G.,** Sediment dispersion: Part 2, Characterization by size of sand fraction and percent mud, *Hydrobiologia*, 176/177, 111, 1989.
8. **Nakata, K.,** A simulation of the process of sedimentation of suspended solids in the Yoshii Estuary, *Hydrobiologia*, 176/177, 431, 1989.
9. **Ferretti, O., Niccolai, I., Bianchi, C. N., Tucci, S., Morri, C., and Veniale, F.,** An environmental investigation of a marine coastal area: Gulf of Gaeta (Tyrrhenian Sea), *Hydrobiologia*, 176/177, 171, 1989.
10. **Arakel, A. V., Hill, C. M., Poirevicz, J., and Connor, T. B.,** Hydro-sedimentology of the Johnstone River Estuary, *Hydrobiologia*, 176/177, 51, 1989.
11. **Petts, G. E.,** *Historical Change of Large Alluvial Rivers: Western Europe*, Wiley, Chichester, U.K., 1989.
12. **Hakanson, L.,** The Swedish Coastal Zone Project: Sediment types and morphometry, in *Sediments and Water Interactions*, Sly, P. G., Ed., Springer-Verlag, New York, 1986, chap. 4.
13. **Damiani, V., Ambrosano, E., DeRosa, S., DeSimone, R., Ferretti, O., Izzo, G., and Zurlini, G.,** Sediments as a record of the input dispersal and settling process in a coastal marine environment, in *Sediments and Water Interactions*, Sly, P. G., Ed., Springer-Verlag, New York, 1986, chap. 2.

14. **Santschi, P.,** Radionuclides as tracers for sedimentation and remobilization processes in the ocean and in lakes, in *Sediments and Water Interactions*, Sly, P. G., Ed., Springer-Verlag, New York, 1986, chap. 38.

15. **Gabrielson, J. O. and Lukatelich, R. J.,** Wind-related resuspension of sediments in the Peel-Harvey estuarine system, *Estuarine, Coastal Shelf Sci.*, 20, 135, 1985.

16. **Anderson, F. E.,** Resuspension of estuarine sediments by small amplitude waves, *J. Sediment. Petrol.*, 42, 602, 1972.

17. **Anderson, F. E.,** The Effect of Boat Waves on the Sedimentary Processes of a New England Tidal Flat, University of New Hampshire-Jackson Estuarine Laboratory Tech. Rep. No. 1, Office of Naval Research, Arlington ,VA, 1974, 38 pp.

18. **Anderson, F. E.,** The Short Term Variation in Suspended Sediment Concentration Caused by the Passage of a Boat Wave Over a Tidal Flat Environment, University of New Hampshire-Jackson Estuarine Laboratory Tech. Rep. No. 2 (Final), Office of Naval Research, Arlington,VA, 1975, 40 pp.

19. **Jouanneau, J. M. and Latouche, C.,** Estimation of fluxes to the ocean from mega-tidal estuaries under moderate climates and the problems they present, *Hydrobiologia*, 91, 23, 1982.

20. **Santschi, P. H., Nyffeler, U. P., Li, Y.-H., and O'Hara, P.,** Radionuclide cycling in natural waters: Relevance of scavenging kinetics, in *Sediments and Water Interactions*, Sly, P. G., Ed., Springer-Verlag, New York, 1986, chap. 17.

21. **Young, T. C. and Comstock, W. G.,** Direct effects and interactions involving iron and humic acid during the formation of colloidal phosphorus, in *Sediments and Water Interactions*, Sly, P. G., Ed., Springer-Verlag, New York, 1986, chap. 40.

22. **Twinch, A. J., Ashton, P. J., Thornton, J. A., and Chutter, F. M.,** A comparison of phosphorus concentrations in Hartbeespoort Dam predicted from phosphorus loads derived near the impoundment and in the upper catchment area, *Water SA*, 12, 51, 1986.

23. **Boyer, J. N., Stanley, D. W., Christian, R. R., and Rizzo, W. M.,** Modulation of nitrogen loading impacts within an estuary, in *Proc. Symp. Coastal Water Resources*, Lyke, W. L. and Hoban, T. J., Eds., American Water Resources Association Technical Publications Series No. TPS-88-1, AWRA, Bethesda, MD, 1988, 165.

24. **Stumm, W. and Morgan, J. J.,** *Aquatic Chemistry: An Introduction Emphasizing Chemical Equilibria in Natural Waters*, Wiley-Interscience, New York, 1970, 583 pp.

25. **Lofgren, S.,** Phosphorus Retention in Sediments — Implications for Aerobic Phosphorus Release in Shallow Lakes, Ph.D. thesis, Institute of Limology, University of Uppsala, Sweden, 1987.

26. **Floderus, S. and Hakanson, L.,** Resuspension, ephemeral mud blankets and nitrogen cycling in Laholmsbukten, south east Kattegat, *Hydrobiologia*, 176/177, 61, 1989.

27. **Nixon, S. W.,** Remineralization and nutrient cycling in coastal marine systems, in *Estuaries and Nutrients*, Neilson, B. J. and Cronin, L. E., Eds., Humana Press, Clifton, NJ, 1981, 3.

28. **McComb, A. J. and Lukatelich, R. J.,** Inter-relationships between biological and physicochemical factors in a database for a shallow estuarine system, *Environ. Monit. Assess.*, 14, 223, 1990.

29. **Davies, B. R., Stuart, V., and deVilliers, M.,** The filtration activity of a serpulid polychaete population (*Ficopomatus enigmaticus* (Fauvel) [sic] and its effect on water quality in a coastal marina, *Estuarine, Coastal Shelf Sci.*, 29, 613, 1989.

30. **Lavery, P. S. and McComb, A. J.,** Macroalgal-sediment nutrient interactions and their importance to macroalgal nutrition in a eutrophic Estuary, *Estuarine, Coastal Shelf Sci.*, 32, 281, 1991.

31. **Lukatelich, R. J. and McComb, A. J.,** Distribution and abundance of benthic microalgae in a shallow southwestern Australian estuarine system, *Mar. Ecol. — Prog. Ser.*, 27, 287, 1986.

32. **Millis, N. F.,** Microorganisms and the aquatic environment, *Hydrobiologia*, 176/177, 355, 1988.

33. **Wetzel, R. G.,** *Limnology*, W.B. Saunders, Philadelphia, 1975, chap. 11.

34. **Pagnotta, R., Blundo, C. M., LaNoce, T., Pettine, M., and Puddu, A.,** Nutrient remobilization processes at the Tiber River mouth (Italy), *Hydrobiologia*, 176/177, 297, 1989.

35. **Callender, E.,** Benthic phosphorus regeneration in the Potomac River Estuary, *Hydrobiologia*, 92, 431, 1982.

36. **Thornton, J. A.,** The Distribution of Reactive Silicate in the Piscataqua River Estuary of New Hampshire–Maine, M.Sc. thesis, University of New Hampshire, Durham, NH, 1976.

37. **Chapman, P.,** Nutrient cycling in marine systems, *J. Limnol. Soc. South. Afr.*, 12, 22, 1986.

38. **Mortimer, C. H.,** The exchange of dissolved substances between mud and water in lakes (Parts I and II), *J. Ecol.*, 29, 280, 1941.

39. **Mortimer, C. H.,** The exchange of dissolved substances between mud and water in lakes (Parts III, IV, Summary and References), *J. Ecol.*, 30, 147, 1942.

40. **Ryding, S.-O. and Forsberg, C.,** Sediments as a nutrient source in shallow, polluted lakes, in *Interactions Between Sediments and Fresh Waters*, Golterman, H. L., Ed., Dr. W. Junk Publishers, The Hague, 1977, 227.

41. **Twinch, A. J.,** Phosphate exchange characteristics of wet and dried sediment samples from a hypertrophic reservoir: implications for the measurement of sediment phosphorus status, *Water Res.*, 21, 1225, 1987.

42. **Nixon, S. W., Oviatt, C. A., Frithsen, J., and Sullivan, B.,** Nutrients and productivity of estuarine and coastal marine ecosystems, *J. Limnol. Soc. South. Afr.*, 12, 43, 1986.

43. **Carignan, R.,** An empirical model to estimate the relative importance of roots in phosphorus uptake by aquatic macrophytes, *Can. J. Fish. Aquat. Sci.,* 39, 243, 1981.

44. **Hart, B. T.,** Uptake of trace metals by sediments and suspended particulates: A review, *Hydrobiologia,* 91, 299, 1982.

45. **El Ghobary, H. and Latouche, C.,** A comparative study of the partitioning of certain metals in sediments from four near-shore environments of the Aquitaine coast (southwest France), in *Sediments and Water Interactions,* Sly, P. G., Ed., Springer-Verlag, New York, 1986, chap. 10.

46. **Fernex, F. E., Span, D., Flatau, G. N., and Renard, D.,** Behavior of some metals in surficial sediments of the northwest Mediterranean continental shelf, in *Sediments and Water Interactions,* Sly, P. G., Ed., Springer-Verlag, New York, 1986, chap. 30.

47. **Sadiq, M.,** Nickel sorption and speciation in a marine environment, *Hydrobiologia,* 176/177, 225, 1989.

48. **Chakraborti, D., van Cleuvenbergen, R. J. A., and Adams, F. C.,** Ionic alkyllead compounds in environmental water and sediments, *Hydrobiologia,* 176/177, 151, 1989.

49. **Nakanishi, H., Ukita, M., Sekino, M., and Murakami, S.,** Mercury pollution in Tokuyama Bay, *Hydrobiologia,* 176/177, 197, 1989.

50. **Forstner, U., Ahlf, W., Calmano, W., Kersten, M., and Salomons, W.,** Mobility of heavy metals in dredged harbor sediments, in *Sediments and Water Interactions,* Sly, P. G., Ed., Springer-Verlag, New York, 1986, chap. 31.

51. **Lyngby, J. E. and Brix, H.,** Heavy metals in eelgrass (*Zostera marina* L.) during growth and decomposition, *Hydrobiologia,* 176/177, 189, 1989.

52. **Malone, L. A.,** *Environmental Regulation of Land Use,* Clark, Boardman, Callaghan, New York, 1992, chap. 2.

53. **Chapman, P. and Thornton, J. A.,** Nutrients in aquatic ecosystems: An introduction to similarities between freshwater and marine ecosystems, *J. Limnol. Soc. South. Africa,* 12, 2, 1986.

54. **Luoma, S. N.,** Can we determine the biological availability of sediment-bound trace elements?, *Hydrobiologia,* 176/177, 379, 1989.

55. **Ryding, S.-O. and Rast, W.,** *The Control of Eutrophication of Lakes and Reservoirs,* UNESCO Man and the Biosphere Series, Vol. 1, Parthenon Publishing, Carnforth, U.K., 1989, chap. 9.

56. **Olem, H. and Flock, G.,** The Lake and Reservoir Restoration Guidance Manual, U.S. Environmental Protection Agency Rep. No. EPA-440/4-90-006, Washington, D.C., 1990.

57. **Dick, R.,** *Potamogeton pectinatus* standing crop in Zandvlei Report to the Inland Waters Management Team, City Engineer's Department, Scientific Services Branch, Cape Town, 1988.

58. **CERC [U.S. Army Coastal Engineering Research Center],** *Shore Protection Manual,* Vol. 2, Department of the Army-Corps of Engineers, Washington, D.C., 1984, chap. 6.

59. **Schleyer, M. H.,** Decomposition in estuarine ecosystems, *J. Limnol. Soc. South. Afr.,* 12, 90, 1986.

60. **Thornton, J. A. and Rast, W.,** A test of hypotheses relating to the comparative limnology and assessment of eutrophication in semi- and man-made lakes, in *Comparative Resevoir Limnology and Water Quality Management,* Straskraba, M. L., Tundisi, J., and Duncan, A., Eds., Kluwer, The Hague, 1993, chap. 1.

## Chapter 14

# The Sustainable Economics of Eutrophication Control

*Timothy O'Riordan*

## CONTENTS

I. Setting the Scene ........................................................................................................... 225
II. The Norfolk and Suffolk Broads Authority ............................................................... 226
III. Causes of Eutrophication in the Broads ..................................................................... 227
IV. Experimental Measures to Control Eutrophication ................................................... 228
V. Justifying the Investments .......................................................................................... 229
VI. Toward a Sustainable Economics of Eutrophication Control ................................... 230
References ................................................................................................................................ 232

## I. SETTING THE SCENE

The new "buzzword" in environmental circles is sustainable development. In a nutshell, this is supposed to mean improving both human and (nonhuman) environmental welfare simultaneously in such a manner as to permit the globe to meet the reasonable aspirations of future generations forever.[1] This triple maximization of requirements is both a conceptual and a physical impossibility. It simply cannot be done in an industrial society, let alone a global community where poverty and underdevelopment lead to and are caused by nonsustainable demands on environmental resources. Humankind is still a colonizing species which has yet to come to terms with any level of environmental limitation, let alone global containment.

Yet such is the scale of environmental loss, and such are the economic consequences if these losses are to multiply, that attention is now seriously focused on trying to devise management approaches that are more sustainable for ecosystems. This is a vexing issue because, in the popular mind, sustainability equates with homeostasis, and even equilibrium, whereas in the natural world such states are uncommon in short-run time frames. Humanity is in danger of trying to impose an artificial construct on what is inherently an unstable and forever evolving natural world. It is of little wonder that the concept of sustainability defies satisfactory definition, let alone implementation.

Despite the many practical problems, the concept of sustainable management will not go away.[2] No matter how much the notion is fudged, it will continue to command attention in all spheres of environmental politics. There are three good reasons for this. First, there is sufficient evidence of economic and militaristic malaise associated with continued environmental stress, irrespective of the moral aspects of deprivation and real suffering.[3] Sustainability becomes a vital ingredient in all future development schemes if for no other reason than that nonsustainable resource investments are likely to prove counterproductive both economically and politically. In that sense, therefore, there is nowadays a sustainable imperative.[4]

The second reason lies in the development of new approaches to economic valuation of natural resource management.[5] Many of the techniques currently being resurrected are by no means new to economic science.[6] What is intriguing is that the usual detractors of these techniques — business, politicians and not a few ecologists — are beginning to recognize the validity of the approaches used. In short, we are moving towards a relatively new realm of ecological economics[7] where already some exciting work has been achieved.

The third reason for the persistence of the sustainability notion lies in the drive among the international community to place it at the heart of new international protocols and conventions.[8] This began with the publication of the *World Conservation Strategy* by the International Union for the Conservation of Nature in 1980,[9] was strengthened by the publication of the Brundtland Commission Report,[10] and has been further enhanced by its centrality in the recent UN Conference on Environment and Development (Rio de Janiero, June 1992). Substantial research programs are now initiated around the notion of sustainability

0-8493-6839-1/95/$0.00+$.50
© 1995 by CRC Press, Inc.

generally,[11] and the idea is becoming established in the new jargon of international agreements. This is particularly true for the European Community, which declared at its Dublin Summit on June 20, 1990 that "action by the Community and its Member States will be developed on a coordinated basis and on the principles of sustainable development and preventative and precautionary action".[12] Definable or not, sustainability has gained the status of international political currency.

This chapter makes modest claims. The application of sustainability principles to ecosystem management is still in its infancy. The expansion of sustainability principles to politics and economics is even more nascent. Yet serious attempts are being made on an exploratory basis. In England, the Norfolk and Suffolk Broads Authority was established in 1988 with a duty to continue and enhance the natural beauty of an internationally renowned wetland, to promote the enjoyment of the Broads by the public, and to protect the interests of navigation. In English law, the term "natural beauty" covers both biological, geological and aesthetic phenomena. Details of the work and challenges facing the Authority follow in the next section. At this stage it is worthy of note that the Authority is seeking to develop an approach to the restoration of eutrophic rivers and lakes based on the principles of ecological economics. This has important implications for a public body which must work in partnership with other statutory agencies and which has always to confront an annual allocation of money, determined primarily on political grounds, while pursuing an investment strategy over many years. This, in turn, requires a lot of attention to consultation, communication and education with partner agencies. The pathway to sustainability is painstaking and time consuming, but also rewarding and radical in terms of local government approaches to environmental management.

## II. THE NORFOLK AND SUFFOLK BROADS AUTHORITY

The Norfolk and Suffolk Broads Authority is unique in British local government. It is somewhat like a national park agency, but is more autonomous than any other U.K. national park body, being established under special legislation.[13] Whereas all but two of the English and Welsh national parks are subcommittees of county councils, the Broads Authority is largely independent from local government, with powers to hire or release staff, set its budget and conduct its affairs quite independently of the two county councils and six district councils represented on it. The Authority consists of 37 members, of which two are coopted from navigation interests, as the area it services is mostly tidal and popular with holiday craft. Of the 33 members that remain, 18 are appointed by the local authorities over whose areas the Authority has jurisdiction. The Countryside Commission nominates two members, English Nature one member, the local navigation body provides two members, and the National Rivers Authority, responsible for the water environment generally and for coastal defence, nominates one member. Further information on each of these contributing public bodies is set out below. Nine members are appointed by the Environment Secretary of which two must represent farming interests and three boating interests.

The financial autonomy of the Authority is limited by a requirement that 9 of the 18 local authority members must vote to approve the annual budget. In general this constitutes no problem, as the Authority consults widely with the local authority representatives when preparing its strategy. Nevertheless, when long-term programmed investments are in prospect, it cannot always be guaranteed that local authority members will accept commitments many years in advance. This is a matter for subsequent comment.

The Authority works very much on a partnership basis with other statutory agencies. This is indeed its strength, since without coordination it could not achieve its statutory aims. The key partners are the National Rivers Authority, English Nature, the local authorities and the voluntary conservation movement. The National Rivers Authority (NRA) was established in 1989 following the privatization of the water industry.[14] The NRA is a wholly public, regulatory agency responsible for setting standards for minimum water flows in rivers and groundwater, for determining the residual quality of rivers, lakes and tidal estuaries, and for ensuring that land drainage and flood protection work smoothly and efficiently. It licenses all water abstractions and all wastewater discharges, both into fresh water and sea water. It is, in essence, the environmental regulatory arm of the water industry in England and Wales. We shall see that liaison between the two authorities is vital if the objectives of eutrophication control are to be met. A decision to privatize the water industry in 1989 has actually strengthened the union between the two regulatory aspects of water management in the area, since the NRA is much more independent and aggressively financed than its predecessor which was engulfed in the industry as a whole.[15]

Other important partnerships lie with English Nature, recently split from the old NCC GB (Nature Conservancy Council for Great Britain) as a result of the controversial Environmental Protection Act of 1990.[15] The NCCE cofunds experimental research, looks after important sites for nature conservation and advises on both scientific and management matters for the Authority. The Countryside Commission for England (CCE) also split from its Welsh partner by the 1990 legislation, and is also an important source of grant aid and advice, especially over matters of public access and environmental education. We shall see that these are likely to become increasingly important aspects of any future sustainable environmental management program.

Finally, there is a host of voluntary organizations in nature conservation, recreation, angling, boating and planning matters which form the great British tradition of nongovernmental environmental organizations. These liaise with the Authority via a series of consultative panels, geared to the interests of farming, recreation and environmental protection generally. Their strength lies in the networks of support (or antagonism) they can generate, the special qualities of their local expertise, and their capacity to manage schemes and projects for which grant aid has been provided through official funds. Such organizations are the eyes and ears of the Authority in political terms, and its hands in management. The goodwill generated by such connections and associations is a vital part of the Authority's success. Much time and consideration is given to nurturing these relationships.

The Authority essentially takes on voluntaristic linkages from its operational success. This has merit in that the Authority can only move forward by consent and through full understanding and appreciation of its aims and programs. It is somewhat ironic that voluntary relationships are often morally and politically more binding than formal legal contracts. However, this is only the case where there is a genuine community of intent. The sustainability paradigm, for all its flaws, does provide such a framework.

## III. CAUSES OF EUTROPHICATION IN THE BROADS

The Broads area is a unique mix of navigable rivers, shallow lakes (known locally as broads), undrained marsh, woodland, grazing marsh (drained marshland) and more deeply drained cultivated land. The area of the Authority's remit is 20,000 ha, essentially the river valleys of the three tributaries of the River Yare (Figure 1).

The wetlands of the Broads have all the hallmarks of international ecological value. They are designated as Ramsar sites, with nomination for special protection status by the European Community. There are 3 National Nature Reserves, 26 sites of special scientific interest, 3 Royal Society for the Protection of Birds reserves, 7 sites owned and managed by the Norfolk Naturalists Trust, 2 local nature reserves, and 5 sites of local natural history interest. Figure 1 identifies all these important sites, which collectively account for about 6% of the Broads area.

The Broads region is also important for its economic value. Local industries dependent on the marshes and rivers system include the boat building and hiring trade associated with the holiday market, the reed-craft industry, agriculture, and tourism generally. Some £30 million annually is spent on the leisure industry, the reed thatching business is worth £350,000 per year, and agriculture nets some £20 million annually, of which £2 million is spent supporting farmers to maintain grazing marshes, drainage ditches rich in flora, and continue to provide, against the trend of economic reality, a 19th century marsh landscape.

The fens and rivers, with their capacity to absorb organic waste and to scavenge nutrients by incorporating them into the aquatic biomass also serve a valuable function so long as the rate of pollution discharge does not exceed the cleansing capacity of the aquatic ecosystem. There is also a natural economy. The wetlands provide enjoyment and inspiration for people who value this unique natural world for its own sake. We shall see that it is this economy above all that needs to be sustained.

Eutrophication is endemic in the region. Even without the disturbing hand of humanity, the wetlands would eventually have been colonized by the typical hydroseric succession. Traditional maintenance of the reedswamp for horse litter and thatch, plus the cleaning out of the marsh drainage ditches, coupled to a regular flush of spring-fed water into the broads (which are medieval peat cuttings subsequently flooded) have served to stem the natural tendency toward enrichment and infill.

What accelerated the process was the introduction of sewage treatment works to service the expanding rural villages post war, the steady arable conversion of the marshes, together with the excessive addition

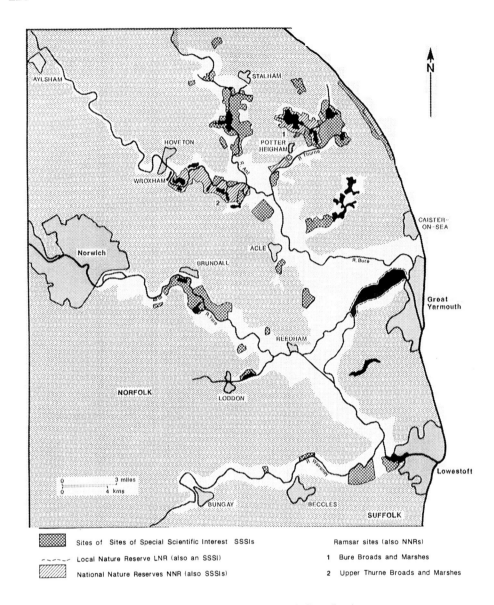

**Figure 1**  Protected sites in Broadland.

of nitrogenous fertilizer, and the inadequate control of stock wastes, notably slurry from piglots and cattle barns. Subsequent research showed that not only did this nutrient enrichment degrade the ecological diversity of the broads and rivers, it also led to rapid sedimentation of the shallow broads and to the loss of fringing reedswamp that protected much of the river banks from erosion caused by wind and boat wash.

## IV. EXPERIMENTAL MEASURES TO CONTROL EUTROPHICATION

By the late 1970s all but 4 of the 52 broads had suffered from advanced eutrophication, and about a quarter of the protective reedswamp had disappeared.[17] This triggered an exploration for remedial measures. Four stages were followed. First came *isolation* for the main rivers in the hope that spring-fed flushes would help to restore a protected broad. This was not a solution, as many of the broads are navigable and the universality of nutrient enrichment was extremely difficult to counter. Second came *phosphate stripping* by the sewage treatment works. This was tried out on an experimental basis using ferric sulfate as a removal agent. The results were remarkable in terms of phosphate removal, with an overall reduction of emissions by 90%, but inconclusive because of the release of phosphate from the rich detritus of algal remains. Hence algal concentrations hardly decreased, though species composition changed with a reduction in both centric diatoms and blue-greens.[18]

The third tactic, therefore, was *mudpumping*, the physical removal of sediment from some of the broads to stem phosphorus inflow. This proved expensive but workable, though it will only be possible on a few broads because of the cost and scale of the remedial measures involved. Finally, there is the stage of *biomanipulation*, the temporary removal of fish in order to allow zooplankton to flourish, and the planting of indigenous macrophytes in restored and protected areas. At present four broads are receiving the latter three treatments on a lengthy experimental basis. The tactic of isolation has been dropped now that phosphorus levels in one of the main rivers have been substantially reduced.

Table 1 illustrates the current budget of the Broads Authority, and Table 2 lists the costs to date of eutrophication limitation programs. It is worthy of note that until 1992 the phosphate-stripping effort has been co-financed by the private water industry, English Nature and the Authority, and that the NRA are licensing and monitoring sewage works discharges in the spirit of this important scientific experiment. From 1992 on, all phosphate stripping costs are paid for by the private water company, Anglia Water pk.

## IV. JUSTIFYING THE INVESTMENTS

The total cost to date of attempting to combat eutrophication in the Broads exceeds £5 million. On an annual basis a budget of over £300,000 is allocated for this measure, about a quarter of the total Authority budget. How can this investment be justified? Figure 2 illustrates the case.

The Authority tries to see environmental management in the round. The well-being of the rivers and broads is not just a moral issue: it is also a matter of hard-headed economics. The navigation value of the area, so vital a part of the Authority's remit and a significant element in the local economy, would suffer mightily if eutrophication gained a significant and possibly irreversible hold. The annual cost of dredging alone is £140,000, of which £60,000 can be attributed to accelerated enrichment. The reedswamp fringing the rivers and broads provides a vital protection against wind and wave action. The cost of restoring what is already lost, as part of a quasi-natural engineering exercise using plastic webbing as a stabilizing agent through which reeds can grow, would exceed £25 million, or over £125,000 per year for 40 years at present values. The river walls are vulnerable to erosion so must be patched up on a regular basis. The budget head for this, financed by the NRA with local authority assistance, is also £125,000 annually. As sea levels tend to rise following atmospheric warming, and as the North Sea becomes more stormy, so this level of protection will have to be maintained with even more commitment in future years.

In essence the cost of not doing anything about eutrophication would exceed £300,000 annually, irrespective of the aesthetic value attributed to clear water and diverse aquatic vegetation. The question therefore arises: is it worth it to save the Broads from ecological, monetary and economic hardship? The answer is that the "value" of the area should be somehow worth at least £300,000 annually. This raises the awkward matter of how can society value assets whose benefits cannot be priced in conventional markets by the technique of willingness to pay? There is a vast and lengthy economic literature on this topic, luckily conveniently summarized by Professor David Pearce and his colleagues.[6] The aim is to devise a method for finding a substitute or surrogate for willingness to pay to retain an area of wetlands and traditional agricultural scenery for future generations to enjoy. The technique used is called contingent valuation, and is essentially a sophisticated way of asking people to bid a figure representing how much a day's visit is worth. There are a number of variants to the contingent valuation approach. These include bidding games to evaluate how much an area is worth as a bequest in its present state, where the cost of not developing it (the opportunity cost) is incorporated in the calculation. Furthermore, there is an approach known as existence, value through which participants are asked to estimate how much an area means to them in terms of simply having it there to enjoy because its existence is guaranteed. In summary, contingent valuation provides an estimate of what an amenity and an ecological asset is deemed to be worth just because it is there.

The NRA and the Broads Authority are combining to produce a rigorous estimate of the valuation of the Broads area based on these surrogate worth values. It is significant that any benefit attached to the investment in the broads is now being judged in terms of its environmental significance. Accordingly, any

**Table 1**  Functional Strategies for the Broads Authority 1988/89 to 1993/94 (£ × 10³)

|  | 1988/89 | 1989/90 | 1990/91 | 1991/92 | 1992/93 | 1993/94 |
|---|---|---|---|---|---|---|
| Conservation generally | 244 | 357 | 443 | 544 | 544 | 722 |
| Information and interpretation | 164 | 225 | 247 | 278 | 413 | 371 |
| Recreation and access | 175 | 155 | 226 | 232 | 226 | 199 |
| Planning and development | 64 | 65 | 80 | 88 | 98 | 114 |
| Administration | 223 | 339 | 341 | 367 | 390 | 401 |
| Total | 871 | 1,143 | 1,340 | 1,571 | 1,673 | 1,809 |

**Table 2**  Costs of Controlling Eutrophication in the Broads Area 1994/95 (£ × 10³)

| | |
|---|---|
| Phosphate reduction | 150 |
| River bank protection | 175 |
| Broads restoration | 225 |
| Dredging | 100 |
| Experimental research | 75 |
| Total | 725 |

*Note:* These expenditures apply to the contributory programs of a number of partner authorities, not to the Broads Authority alone.

proposal to diminish these events, for example by adding new amounts of nutrient to the Broads, would be considered a negative benefit and would be costed as such. How the tables are now turning in the topsy-turvy world of cost benefit analysis! Research elsewhere[19] suggests that it could be worth around £10–20 per day per person, or over £500,000 in a year.

These economic surrogates to the value of eutrophication-reduction programs are still very crude. However, they do indicate that the net present value of the investments over a 20-year time frame would clearly justify their costs. Not too much store should be placed on the actual figures as some more work still needs to be completed. The Broads Authority experience does show that sustained investment based on carefully staged research can be justified, however, at least to the point of safeguarding a selected set of rivers and flood-protecting banks.

## VI. TOWARD A SUSTAINABLE ECONOMICS OF EUTROPHICATION CONTROL

The sustainability issue here lies in restoring a wetland to some of its natural functions so that, in the long term, the wetland ecosystem will meet social as well as natural needs. This is not to suggest that the wetlands have no other function than to serve humanity. That would be a cruel denial of the notion of sustainability. However, it is to indicate that there are real economic gains to be enjoyed from the reduction of eutrophication, not only in terms of saved expenditures that will rise in the future. There are also major benefits for a leisure society that places values on the amenities of the natural beauty, and which recognizes through the lengthy experiment such as that now taking place in the Broads, that prevention is indeed better than cure.

This leads to a renewed emphasis on the precautionary principle, namely, the application of avoidance remedies in advance of scientific certainty, both as a cost-effective measure and as an act of moral rightness. There is still a lot of debate over how far the precautionary principle should be involved when scientific uncertainties abound, and the consequences of hasty commitment could be very expensive. The concept of precaution is not new, indeed it derives from principles of "polluter pays" and prevention at source that were enshrined in the First Environmental Action Plan of the European Community in 1973.[20,21] But the idea has broadened to embrace three additional themes. The first is *minimum risk*, namely, devising strategies that seek to avoid uncertainty. The second is *progressive experimentation,* where investments in "no-regret" strategies are initiated because they are deemed cost-effective in their own right. The reduction of nitrogenous fertilizer in already over fertilized fields would be a case in point, even though, to the applicator, it would still pay to add more. Third, the *burden of proof* falls on the would

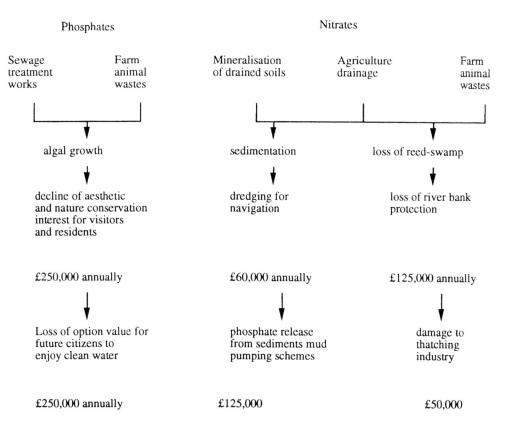

**Figure 2** The economics of eutrophication control in Broadland.

be polluter to prove no harm rather than the potential victim to demonstrate possible harm. The Broads experiments are really aimed at limiting further damage, even when that damage cannot be proved.

Clearly, what must also be done is to bring in the public, who are not only the enjoyers of the investments, but also, ultimately, the source of public revenue. This has meant an extensive program of interpretation as well as improved access facilities at some of the experimental sites (some sites are purely privately owned). In addition, it means the preparation of functional strategies for budgetary commitment, not just of the Broads Authority, but also of its partner agencies, that lie across the normal annual financing barriers. Table 1 outlines the Authority's functional strategy for the next 5 year period.

To obtain success in this long-term financial commitment is not easy. National governments frequently control what they believe ought or ought not to be spent in the cause of maintaining what they believe to be a viable economy. Local authorities have their own financial worries to overcome. This means that more effort needs to be spent on the economics of sustainable ecological restoration so that the longer term gains of preventative and remedial action can be shown to be cost-effective.

Various measures for doing this present themselves. Contingent valuation studies do have their place. However, they need to be conducted carefully because they can seriously mislead if improperly handled. Likewise, there is no substitute for careful experimentation. Currently, the Authority and the NRA are looking at the scope for iron dosing as a more cost-effective means of removing surface phosphorus in sediment. This is a device to precipitate the phosphorus in surface sediment onto subsurface particles so as to avoid the release of phosphate under low oxygen conditions in the hypolimnion. Third, there is a tremendous need for imaginative interpretation, not just at the sites undergoing treatment. There is also a possibility for creating images of cleaner waters and reed-restored banks either via the arts or by means of interactive computer videos which allow taxpayers or tomorrow's citizens (today's youngsters) to play with different "cost" inputs to create various environmental outcomes over 20 year periods up to 100 years hence. This should be a marvellous opportunity to combine the best of applied science and technology to the creative arts and to genuine negotiated participation in ecological futures. The pathways are barely identified, let alone explored.

# REFERENCES

1. **Brundtland, H. (Chmn.),** *Our Common Future.* Report of the World Commission on Environmental and Development, Oxford University Press, Oxford, 1987, chap. 1.
2. **O'Riordan, T.,** The politics of sustainability, in *Sustainable Environmental Management.* Turner, R. K., Ed., Belhaven Press, London, 1989, chap. 2.
3. **Myers, N.,** Environment and security, *Foreign Affairs,* 74, 23–41, 1989.
4. **Adams, W.,** *Green Development: Environment and Sustainability in the Third World,* Routledge, London, 1990, chaps. 1–3.
5. **Pearce, D. W. and Turner, R. K.,** *Economics of Natural Resources and the Environment,* Harvester Press, London, 1990.
6. **Pearce, D. W., Markandya, A., and Barbier, E.,** *Blueprint for a Green Economy,* Earthscan Publications, London, 1990, chap. 1.
7. **Constanza, R., Ed.,** *Ecological Economics,* University of Columbia Press, New York.
8. **McCormick, J.,** *The Global Environmental Movement,* Belhaven Press, London, 1989.
9. **International Union for the Conservation of Nature,** *World Conservation Strategy,* IUCN, Geneva, 1980.
10. **Brundtland, G. H. (Chmn.),** *Our Common Future,* Oxford University Press, Oxford, 1987.
11. **Commission of the European Communities,** Programme of socio-economic research on sustainable development, DG XII Brussels, 1991.
12. **Brinkhorst, L. V.,** Environment policy and the European Community. *Town and Country Planning,* 60, 18–21, 1991.
13. *The Norfolk and Suffolk Broads Act 1988,* HMSO, London, 1988, chap. 4.
14. **Kinnersley, D.,** *Troubled Waters,* Hilary Slipman, London, 1988.
15. **Penning-Rowsell, E. and Parker, D.,** *Water Planning in England and Wales,* Allen and Unwin, London, 1983.
16. **Ratcliffe, D.,** Messy business, *Ecos,* 11, 41, 1990.
17. **Broads Authority,** *Broads Plan,* Broads Authority, Norwich, 1987, chap. 3.
18. **Broads Authority,** *Water Quality Issues in the Broads,* Broads Authority, Norwich, 1990, 7.
19. **Walsh, R. G.,** *Recreation and Economic Decisions,* Venture Publishing, Philadelphia, 1986.
20. **Cameron, J. and Abouchar, J.,** The precautionary principle: a fundamental principle of law and policy for the protection of the global environment. *Boston College International and Comparative Law Review,* 14, 1–27, 1991.
21. **O'Riordan, T. and Cameron, J., Eds.,** *Interpreting the Precautionary Principle,* Earthscan Publications, London, 1984.

# INDEX

## A

Active silicon, 39
Aeration, 125
Agriculture
    Dutch Wadden Sea and, 135
    Peel-Harvey Estuary and, 5, 8
    Shenzhen Bay and, 33
    Tolo Harbour and, 41, 45, 46, 52, 54
Algae, 182, 212, 213, 214, 229, see also specific types
    blue-green, 5, 13, 14, 177, 190
        fisheries and, 191, 197–198
    eutrophication and, 189
    fisheries and, 191, 193–198
    green, 13
    in lagoon of Venice, 67, 71, 72
    macro-, see Macroalgae
    micro-, 73–75, 175
    in Peel-Harvey Estuary, 191
    red, 182
    in Zandvlei Estuary, 122
Alkaline phosphatase, 212
Alumino-silicates, 211
Ammonia, 11, 86, 89
Ammonium, 23, 73, 134, 141, 181, 211
Australia, 151, 163, 166
    fisheries in, 193
    Peel-Harvey Estuary in, see Peel-Harvey Estuary
    Tuggerah Lakes System in, see Tuggerah Lakes
        System

## B

"Bar-built" type estuaries, 2, 5, see also specific estuaries
Baroclinic flow, 151, 152, 153–155, 163
    barotropic flow relation to, 159–162
Barometric tidal components, 6
Barotropic flow, 151, 152, 153, 155, 163–164, 171
    baroclinic flow relation to, 159–162
Barrier estuaries, 19, see also specific types
Barwon River, 206
BATHTUB model, 122
Bay of Fundy, 207
Benthic macrophytes, 209
Bernoulli's equation, 156
Bioaccumulation, 215
Bioavailability of nutrients, 212–217
Biological evolution, 64–69
Biological oxygen demand (BOD)
    in Shenzhen Bay, 33
    in Tolo Harbour, 48, 54
    in Tuggerah Lakes System, 26

Biomagnification, 215
Biomanipulation, 218, 229
Biomass
    of Dutch Wadden Sea, 144–145
    of lagoon of Venice, 64–69, 71, 72, 73, 75
    macroalgal, 72
    macrophyte, 75
    phytoplankton, 38–39
    zooplankton, 189
Biota, 98–100
Bioturbation, 209
Blue-green algae, 177, 190
    fisheries and, 191, 197–198
    in Peel-Harvey Estuary, 5, 13, 14
BOD, see Biological oxygen demand
Budgewoi Lake, 20, 22, 24, 25, 26

## C

Calcium, 49, 50, 142
Cape Province, South Africa's Zandvlei Estuary, see
        Zandvlei Estuary
Carbon
    annual fluxes of, 98
    in Ems Estuary, 95, 98, 102
    in lagoon of Venice, 71, 74
    nitrogen ratio to, 181
    phosphorus ratio to, 181
    in Tolo Harbour, 45, 49, 50
    in Zandvlei Estuary, 120
Catchments
    of Dutch Wadden Sea, 133–135
    of Ems Estuary, 85–88
    of lagoon of Venice, 61–63
    of Peel-Harvey Estuary, 7–9
    of Shenzhen Bay, 33
    of Tolo Harbour, 41, 45–46, 52, 53
    of Tuggerah Lakes System, 22
    of Zandvlei Estuary, 111–116, 123–124
Channels, 151, 153, see also specific types
    shape of, 158
    sills in, 168–170
    two-layer flow through, 155–159
Chemical oxygen demand (COD), 36
Chesapeake Bay, 176, 215
Chlorophyll a, 198
    in Dutch Wadden Sea, 144
    in Ems Estuary, 102
    in lagoon of Venice, 71
    in Tolo Harbour, 45, 47, 48, 52, 53, 54
    in Tuggerah Lakes System, 22, 26
    in Zandvlei Estuary, 116

Chlorophytes, 216
Climate
    for Dutch Wadden Sea, 131
    for Ems Estuary, 82–83
    for Tolo Harbour, 43
    for Zandvlei Estuary, 111
Coastal upwelling, 214
Cockburn Sound, 175, 176
COD, see Chemical oxygen demand
Competition, 180–181
Convection, 152
Cultural eutrophication, 189
Cyanobacteria, 5, 14, 27
Cyanophytes, 215

## D

Decomposition, 213
Denitrification, 141, 178, 212, 213
Diagenesis, 211
Diatom blooms, 216, 229
    in Ems Estuary, 93
    fisheries and, 189, 190, 191
    in Peel-Harvey Estuary, 12, 13, 14, 191
    in Shenzhen Bay, 36, 37
    in Tolo Harbour, 52
Diep River, 111
Diking, 81
DIN, see Dissolved inorganic nitrogen
Dinoflagellates, 52
DIP, see Dissolved inorganic phosphorus
Dissolved inorganic nitrogen (DIN), 34, 35–36, 39, 120
Dissolved inorganic phosphorus (DIP), 35–36, 39, 140
Dissolved oxygen, 23, 24, 33, 48, 53, 54
DO, see Dissolved oxygen
Dover Strait, 135
Drainage density, 9
Dredging, 124, 218
Drift currents, 83
Dutch Wadden Sea
    biomass of, 144–145
    catchments of, 133–135
    climate for, 131
    eutrophication of, 129–147
        sediment and, 143–144
        tides and, 132–133
        water exchange and, 135–139
    fish in, 145–146
    management of, 147
    morphology of, 132
    nutrients in, 134–135, 139–143
        enrichment of, 144–147
        in sediment, 141–142
    physical properties of, 131–133
    restoration of, 147

    sediment in, 143–144
    tides of, 132–133
    topography of, 132
    water exchange and movement in, 135–139

## E

Ecological models, 153
Effluents, 41, 45, 49, 54, 177
EIA, see Environmental impact assessment
Elbe Estuary, 217
Electric power stations, 19–20, 22, 25
Ems Estuary, 81–104
    catchments of, 85–88
    climate for, 82–83
    microbenthos in, 96–97
    morphology of, 83–84
    nutrients in, 86–88, 89–93, 104
    physical properties of, 82–85
    sediment in, 83, 93–101
        fluxes of, 98–100
        resuspension of, 96–97, 100–101
    suspended matter in, 95–96
    tides of, 84–86
    topography of, 83–84
    water movement in, 88–89
Environmental impact assessment (EIA), 103
Epiphytes, 174, 175, 178
Estuaries, see also specific estuaries by name
    "bar-built" type, 2, 5
    barrier, 19
    Elbe, 217
    Ems, see Ems Estuary
    Garigliano, 208
    Gironde, 209
    Great Bay, 209
    Harvey, 151, 190, 191, 197
    Mackenzie, 207
    Mersey, 207
    Murray-Serpentine, 215
    Neuse, 214
    Odawa, 213
    particle size distribution in, 207–208
    Patuxent, 215
    Pearl River, 34
    Peel-Harvey, see Peel-Harvey Estuary
    Potomac, 214, 215
    Rhone, 216
    River Ems, see Ems Estuary
    Scheldt, 217
    sediment transport in, 103–104
    Severn, 217
    Stommel-Farmer steady state model of, 170–171
    Swan, 190, 191, 192, 193, 194, 195, 196
    Tama, 213

Tiber, 213
Volturno, 208
water exchange between, see under Water exchange
Zandvlei, see Zandvlei Estuary
Euphrates River, 207
Eutrophication, 151
  consequences of, 14–15
  control of, 225–231
  cultural, 189
  of Dutch Wadden Sea, 129–147
    sediment and, 143–144
    tides and, 132–133
    water exchange and, 135–139
  economics of control of, 225–231
  experimental measures to control, 228–229
  fisheries and, 193–198
  food web and, 190
  of lagoon of Venice, 59–77
    biological evolution and, 64–69
    biomass distribution and, 64–69
    catchments and, 61–63
    history of, 59–61
    nutrients and, 69–75
    physical properties and, 61
    sediment and, 63–64, 65, 70–72
  of Norfolk Broads, 227–231
  of Peel-Harvey Estuary, 14–15
  sustainable economics of control of, 225–231
  of Tolo Harbour, 41, 47, 49, 52, 54
  of Tuggerah Lakes System, 27
  of Zandvlei Estuary, 122

**F**

Fertilizers, 8, 15, 135, 173, 175, 228
Fick's equation, 216
Fisheries, 189–201
  algae and, 191, 193–198
  blue-green algae and, 197–198
  eutrophication and, 193–198
  Leschenault Inlet and, 191, 192–193, 194, 195, 196
  macroalgae and, 193–197
  Peel-Harvey Estuary and, 5, 189, 190–192, 193, 194, 195, 196, 197
  phytoplankton and, 189, 190, 191, 192
  in southwestern Australia, 193
  Swan Estuary and, 190, 191, 192, 193, 194, 195, 196
  Tolo Harbour and, 41, 42, 47
  Tuggerah Lakes System and, 20
Fishing, 81
Flagellates, 52, 69
Free-floating macroalgae, 174
Friction, 154
Froude numbers, 155, 156, 158

**G**

Garigliano Estuary, 208
Geography of Zandvlei Estuary, 111–112
Gippsland Lakes, 166
Gironde Estuary, 209
Grazing, 175, 179, 180
Great Bay Estuary, 209
Great Lakes, 174, see also specific lakes
Green algae, 13
Gulf of Gaeta, 208

**H**

Harvey Estuary, 151, 190, 191, 197
Harvey River, 7, 164, 169
History
  of lagoon of Venice, 59–61
  of Tuggerah Lakes System, 20
Hong Kong's Tolo Harbour, see Tolo Harbour
Huangfu River, 206
Hydraulic control, 154
Hydrogen sulfide, 175

**I**

IJssel Lake, 102, 133, 135, 138, 140
IJssel River, 133
Inlet channels, 2, 6
Iron oxides, 142

**K**

Kester's ratio, 214
Keysers River, 111

**L**

Lagoon, defined, 2
Lagoon of Venice
  biological evolution of, 64–69
  biomass of, 64–69, 71, 72, 73, 75
  catchments of, 61–63
  description of, 59
  drainage area of rivers feeding, 61–62
  eutrophication of, 59–77
    biological evolution and, 64–69
    biomass distribution and, 64–69
    catchments and, 61–63
    history of, 59–61
    nutrients and, 69–75
    physical properties and, 61
    sediment and, 63–64, 65, 70–72
  history of, 59–61
  nutrients in, 62–63, 69–75

physical properties of, 61
  sediment in, 63–64, 65, 70–72
  water exchange and movement in, 63
Lake Budgewoi, 20, 22, 24, 25, 26
Lake IJssel, 102, 133, 135, 138, 140
Lake Macquarie, 19
Lake Merimbula, 19
Lake Michigan, 174
Lake Mummorah, 20, 22, 24, 25, 26
Lake Pambula, 19
Lake Songkhla, 153, 167
Lake Tuggerah, 20, 22, 25, 26
Lake Wallaga, 19
Lake Washington, 173
Lake Wingra, 173
Laurentian Great Lakes, 207
Lawrence Lake, 175
Legumes, 8
Leschenault Inlet, 191, 192–193, 194, 195, 196
Limpopo River, 206
Lock-exchange, 158, 159, 160, 161, 171
Low frequency tides, 166–167

**M**

Mackenzie Estuary, 207
Macquarie Lake, 19
Macroalgae, 177, 182, 190, 215
  in Ems Estuary, 102
  eutrophication and, 189
  fisheries and, 193–197
  free-floating, 174
  in lagoon of Venice, 67, 71, 72
  in Peel-Harvey Estuary, 11, 13, 14, 191
  in Tuggerah Lakes System, 22, 23, 24, 26, 27
Macrophytes, 175, 180–181
  benthic, 209
  biomass of, 75
  consequences of, 121–122
  fisheries and, 189, 190, 191, 192
  in Leschenault Inlet, 192
  loss of, 174
  in Peel-Harvey Estuary, 191
  in Swan Estuary, 192
  in Zandvlei Estuary, 120, 121–122
Malamocco-Marghera canal, 59, 61
Mandurah Channel, 163, 166–167, 169, 170
Manning's equation, 207
Marginal platforms, 6
Mathematical models, 47
Merimbula Lake, 19
Mersey Estuary, 207
Mersey River, 207
Mesocosms, 178
Michigan Lake, 174
Microalgae, 73–75, 175

Microbenthos, 96–97
Microcosms, 179, 180
Microfossils, 174
Microphytobenthos, 97, 100–101, 143–144
Mineralization, 90, 139, 178
Moriches Bay, 175
Morphology, 83–84, 132
Morphometry, 111
Mudpumping, 229
Mumford Cove, 178
Mummorah Lake, 20, 22, 24, 25, 26
Murray River, 7, 8, 9
Murray-Serpentine Estuary, 215

**N**

Nanoplankton, 189
National RIvers Authority, 226
Netherlands' Ems Estuary, see Ems Estuary
Neuse Estuary, 214
New South Wales, Australia's Tuggerah Lakes System,
  see Tuggerah Lakes System
Nile River, 206
Niobrara River, 207
Nitrates, 181
  in Dutch Wadden Sea, 134, 140, 142
  in Ems Estuary, 86, 89
  in lagoon of Venice, 73
  in Peel-Harvey Estuary, 11
  reduction in, 142
  in Tolo Harbour, 45, 52–53
  in Tuggerah Lakes System, 23
Nitrification, 11
Nitrites, 73
Nitrogen, 180
  carbon ratio to, 181
  dissolved inorganic, 34, 35–36, 39, 120
  in Dutch Wadden Sea, 134, 141
  in Ems Estuary, 87
  in lagoon of Venice, 71–72, 73, 74, 75
  in Peel-Harvey Estuary, 8
  phosphorus ratio to, 71–72, 75, 120, 181
  in sediment, 71–72, 206, 211
  in Shenzhen Bay, 34, 35–36
  in Tolo Harbour, 54
  total inorganic, 73, 116
  in Tuggerah Lakes System, 23, 24, 25, 26
  in Zandvlei Estuary, 73, 116, 120
Nitrogen cycle, 212
Nitrogen fixation, 8
Nonpoint source pollution, 218
Norfolk Broads, 173, 227–231
Norfolk and Suffolk Broads Authority, 226–227, 229
North Sea, 135, 136
Nutrient enrichment, 1–2, see also Nutrients
  consequences of, 52–53, 102, 146–147

in Dutch Wadden Sea, 144–147
in Ems Estuary, 102
in Peel-Harvey Estuary, 13–14, 177
plant changes after, 173–183
  in freshwater systems, 173–175, 179–182
  future research on, 182
  in marine systems, 175–182
in Shenzhen Bay, 37–39
species diversity and, 182
symptoms of, 13–14
  in Dutch Wadden Sea, 144–146
  in Ems Estuary, 102
  in Peel-Harvey Estuary, 13–14
  in Shenzhen Bay, 37–39
  in Tolo Harbour, 51–52
  in Tuggerah Lakes System, 26–27
  in Zandvlei Estuary, 120–122
in Tolo Harbour, 51–53
in Zandvlei Estuary, 120–122
Nutrients, 174, see also Nutrient enrichment; specific
    nutrients
accumulation of, 209–212
bioaccumulation of, 215
bioavailability of, 212–217
biomagnification of, 215
in Dutch Wadden Sea, 134–135, 139–143
  enrichment of, 144–147
  in sediment, 141–142
dynamics of, 221
in Ems Estuary, 86–88, 89–93, 104
enrichment with, see Nutrient enrichment
export of, 212
fluxes of, 135, 181, 219–220, 221
inactivation of, 218
inputs of, 143
in lagoon of Venice, 62–63, 69–75
loading of, 2, 8, 52, 174
  in Dutch Wadden Sea, 135
  in lagoon of Venice, 62–63
  reduction of, 178
  in Zandvlei Estuary, 113–116
metals and, 215
microalgae recycling of, 73–75
in North Sea, 135, 136
in Peel-Harvey Estuary, 10, 11, 13–14, 177
pools of, 209–215
recycling of, 73–75
release of, 142, 212–217
response to, 175–178
seasonal cycles of, 90, 141
in sediment, 70–71, 93, 120, 141–142, 206
  accumulation of, 209–212
  bioavailability of, 215–217
  dynamics of, 221
  forms of, 211–212
  pools of, 209–215

release of, 215–217
uptake of, 215–217
in Shenzhen Bay, 34–36, 37–39
sinks of, 206
sources of, 206
spatial distributions of, 69–71
tissue concentration of, 182
in Tolo Harbour, 41, 45, 46, 47–48, 51–53
transport of, 104
trapping of, 11, 12, 209–211
in Tuggerah Lakes System, 23–27
uptake of, 215–217
withdrawal of, 174–175
in Zandvlei Estuary, 113–116, 120–122

**O**

Odawa Estuary, 213
Organic carbon
  in Ems Estuary, 95, 98
  in lagoon of Venice, 71, 74
  in Tolo Harbour, 45, 49, 50
  in Zandvlei Estuary, 120
Orthophosphates, 23, 45, 48
Oxygen, 181
  biological demand for, 26, 33, 48, 54
  chemical demand for, 36
  consumption of, 214
  depletion of, 52, 214
  dissolved, 23, 24, 33, 48, 53, 54
  in Ems Estuary, 102

**P**

Pambula Lake, 19
Paralytic shellfish poisoning (PSP), 52
Particle size distribution, 206–208
Patos Lagoon, 153
Patuxent Estuary, 215
Pearl River Delta, 43
Pearl River Estuary, 34
Peel-Harvey Estuary, 5–16, 151, 163, 164, 166, 215
  catchments of, 7–9
  consequences of eutrophication in, 14–15
  eutrophication in, 14–15
  fisheries and, 5, 189, 190–192, 193, 194, 195, 196,
    197
  nutrients in, 10, 11, 13–14, 177
  physical properties of, 5–7
  sediment in, 11–13
  water exchange and movement in, 9–11
Peel Inlet, 5, 11, 14, 169, 170, 190
  algae in, 197, 198, 215
  barotropic ocean flushing in, 163
  blue-green algae in, 197, 198
  salinity of, 166

Periphyton, 189
pH, 23, 28, 32, 45, 47
Phosphates, 175
    accumulation of, 209
    in Dutch Wadden Sea, 134, 141
    in Ems Estuary, 86, 89, 94, 102
    ortho-, 23, 45, 48
    in Peel-Harvey Estuary, 11, 12, 13, 15
    reactive, 86
    release of, 13
    in sediment, 209
    stripping of, 228, 229
    in Tolo Harbour, 45, 48, 52, 53
    in Tuggerah Lakes System, 23
Phosphorus, 174, 177, 179, 180
    accumulation of, 209, 211
    carbon ratio to, 181
    dissolved inorganic, 35–36, 39, 140
    in Dutch Wadden Sea, 134, 140, 141, 142
    in Ems Estuary, 87
    in lagoon of Venice, 62, 71–72, 73, 74, 75
    nitrogen ratio to, 71–72, 75, 120, 181
    in Peel-Harvey Estuary, 8, 9, 12
    release of from sediment, 12
    in sediment, 71–72, 206, 209, 211
    in Shenzhen Bay, 35–36
    in Tolo Harbour, 45, 48, 49, 51, 54
    in Tuggerah Lakes System, 24, 25, 26
    in Zandvlei Estuary, 113–116, 120
Physical properties, see also specific properties
    of Dutch Wadden Sea, 131–133
    of Ems Estuary, 82–85
    of lagoon of Venice, 61
    of Peel-Harvey Estuary, 5–7
    of Shenzhen Bay, 31–33
    of Tolo Harbour, 43–44
    of Tuggerah Lakes System, 20–22
    of Zandvlei Estuary, 111
Phytobenthos, 141
Phytoplankton, 173, 174, 175, 182
    in Dutch Wadden Sea, 141, 144
    in Ems Estuary, 102
    fisheries and, 189, 190, 191, 192
    in lagoon of Venice, 67, 69, 71, 74
    in Peel-Harvey Estuary, 191
    in Shenzhen Bay, 33, 37, 38–39
    in Swan Estuary, 192
    in Tolo Harbour, 45, 47, 52
    in Tuggerah Lakes System, 22, 23, 26, 27
    in Zandvlei Estuary, 122
Plankton, 177
Plant group changes, 173–183
    in freshwater systems, 173–175, 179–182
    future research on, 182
    in marine systems, 175–182
Plover Cove Reservoir, 42, 47, 53

Pollution, see also specific types
    of Ems Estuary, 103–104
    nonpoint source, 218
    thermal, 19
    of Tolo Harbour, 41, 45, 46, 47, 49, 52, 54
Pontine Island Basin, 211
Potomac Estuary, 214, 215
Power stations, 19–20, 22, 25
Precautionary principle, 103, 230
Primary production, 144–145, 174, 176
PSP, see Paralytic shellfish poisoning

**R**

Reactive phosphates, 86
Reactive silicate, 86, 89, 94
Reactive silicon, 36, 37, 38, 48
Recreational sites, 5, 41, 52, 109
Red algae, 182
Redfield ratios, 181, 220
Red tides, 52, 53
Regeneration, 214–215
Remobilization, 213–214, 215
Resource limitation, 179–180
Resuspension of sediment, 208–209, 212–213
    in Dutch Wadden Sea, 143–144
    in Ems Estuary, 96–97, 100–101
    in Peel-Harvey Estuary, 11, 12, 209
Retention coefficient, 164–166
Reynolds numbers, 154
Rhone Estuary, 216
Riber Ems, see also Ems Estuary
River Ems Estuary, see Ems Estuary
River IJssel, 133

**S**

Salinity, 151
    of Dutch Wadden Sea, 140
    of Ems Estuary, 89
    of Peel-Harvey Estuary, 9, 10
    of Shenzhen Bay, 32
    of Tolo Harbour, 47
    of Tuggerah Lakes System, 23, 28
    of Zandvlei Estuary, 119, 120, 124
Sand River, 111
Scheldt Estuary, 217
Seagrasses, 2, 24, 27, 28, 192
Secondary production, 145
Sediment, 205–221
    characterization of, 206–209
    composition of, 93–95, 97, 143
    in Dutch Wadden Sea, 143–144
    in Ems Estuary, 83, 93–101
        fluxes of, 98–100
        resuspension of, 96–97, 100–101

fluxes of, 98–100
in lagoon of Venice, 63–64, 65, 70–72
metals in, 216–217
nitrogen in, 71–72
nutrients in, 70–71, 93, 120, 141–142, 206
    accumulation of, 209–212
    bioavailability of, 215–217
    dynamics of, 221
    forms of, 211–212
    pools of, 209–215
    release of, 215–217
    uptake of, 215–217
particle size distribution in, 206–208
in Peel-Harvey Estuary, 11–13
phosphorus in, 71–72
pools of nutrients in, 209–215
properties of, 143–144
resuspension of, 208–209, 212–213
    in Dutch Wadden Sea, 143–144
    in Ems Estuary, 96–97, 100–101
    in Peel-Harvey Estuary, 11, 12, 209
in Shenzhen Bay, 37
sorting of, 83
in Tolo Harbour, 49–51
transport of, 103–104, 219
in Tuggerah Lakes System, 25–26, 28
in Zandvlei Estuary, 120
Sedimentation, 206–207, 211
Sediment traps, 11, 12, 218
Serpentine River, 7
Severn Estuary, 217
Sewage, 25, 47, 52, 54, 178
Sewage effluents, 177
Sewage outfalls, 175
Sewage treatment plants, 47, 52, 54, 62
Shenzhen Bay, 31–39
    catchments of, 33
    nutrients in, 34–36, 37–39
    physical properties of, 31–33
    sediment in, 37
    species components in, 37–38
    water exchange and movement in, 34, 39
Shenzhen River, 33, 34
Shoe Lake, 174
Silicate, 23, 86, 89, 94, 141, 211
Silicon
    active, 39
    reactive, 36, 37, 38, 48
    in Shenzhen Bay, 36, 37, 38, 39
    in Tolo Harbour, 48
Sills, 168–170
Songkhla Lake, 153, 167
South Africa's Zandvlei Estuary, see Zandvlei Estuary
South China Sea's Shenzhen Bay, 31–39
Species components, 37–38, 147
Species diversity, 182

Stoichiometry, 214
Stokes' Law, 208
Stommel-Farmer steady state estuarine model, 170–171
Suspended matter
    accumulation of, 95–96, 143
    in Dutch Wadden Sea, 132, 143
    in Ems Estuary, 95–96
    sedimentation and, 206
    transport of, 95–96, 132, 143
Swan Estuary, 190, 191, 192, 193, 194, 195, 196

**T**

Tama Estuary, 213
TDS, see Total dissolved solids
Temperature
    of Ems Estuary, 82, 96
    of Shenzhen Bay, 32
    of Tolo Harbour, 43, 48
    of Tuggerah Lakes System, 23, 28
    of Zandvlei Estuary, 111
Thermal pollution, 19
Tiber Estuary, 213
Tides, 6, 151, 153, 163, 207, 209
    in Dutch Wadden Sea, 132–133
    in Ems Estuary, 84–86, 88
    exchange of, 153
    low frequency, 166–167
    in Tolo Harbour, 41, 43, 44, 47, 52
    in Tuggerah Lakes System, 28
Time scales, 88–89, 139
TIN, see Total inorganic nitrogen
Tokuyama Bay, 217
Tolo Harbour, 41–54
    catchments of, 41, 52, 53
    map of, 43
    nutrients in, 41, 45, 46, 47–48, 51–53
    physical properties of, 43–44
    sediment in, 49–51
    water exchange and movement in, 46–47
Topography, 83–84, 114, 132, 154
Total dissolved solids (TDS), 116
Total inorganic nitrogen (TIN), 73, 116
Tuggerah Lake, 20, 22, 25, 26
Tuggerah Lakes System, 19–28
    catchments of, 22
    history of, 20
    nutrients in, 23–26, 23–27
    physical properties of, 20–22
    sediment in, 25–26, 28
    temperature in, 23, 28
    urban inflow in, 24–25
    water exchange and movement in, 22–23, 26, 28
    water quality in, 23
Turbidity, 208
    of Peel-Harvey Estuary, 11

of Tolo Harbour, 43, 44
of Tuggerah Lakes System, 23, 24, 26
Two-layer flow, 155–159, 162

## U

Urban runoff, 24–25, 26, 27, 47, 117, 189

## V

Van River, 207
Vittorio Emanuele canal, 59, 61
Volturno Estuary, 208

## W

Wadden Sea, 83, see Dutch Wadden Sea
Wallaga Lake, 19
Wallis Lakes, 19
Washington Lake, 173
Wastewater plants, 102
Water exchange
    baroclinic, see Baroclinic flow
    barotropic, see Barotropic flow
    in Dutch Wadden Sea, 135–139
    in lagoon of Venice, 63
    in Peel-Harvey Estuary, 9–11
    between shallow estuaries, 151–171
        baroclinic, 151, 152, 153–155, 159–162, 163
        barotropic, 151, 152, 153, 155, 159–162, 163–164
        channels and, 153
        low frequency tides and, 166–167
        retention coefficient and, 164–166
        two-layer flow in, 155–159, 162
    in Shenzhen Bay, 34, 39
    in Tolo Harbour, 46–47
    in Tuggerah Lakes System, 22–23, 26, 28
    in Zandvlei Estuary, 116–119
Water movement
    in Dutch Wadden Sea, 135–139
    in Ems Estuary, 84, 88–89

in lagoon of Venice, 63
in Peel-Harvey Estuary, 9–11
in Shenzhen Bay, 34
in Tolo Harbour, 46–47
in Tuggerah Lakes System, 22–23
in Zandvlei Estuary, 116–119
Water quality, 218
    modeling of, 47
    monitoring of, 126
    in Tuggerah Lakes System, 23, 24
    in Zandvlei Estuary, 121, 123, 126
Water recreation, 5, 41, 52, 109
Wave action, 207
Western Australia's Peel-Harvey Estuary, see
        Peel-Harvey Estuary
Westlake Wetland, 111
West Looe Estuary, 217
Wind forcing, 152
Wind mixing, 152
Wingra Lake, 173
"Wise use principle", 103
Wyong River, 22

## Y

Yarra Estuary, 216
Yellow River, 206
Yoshii Estuary, 208

## Z

Zandvlei Estuary, 109–126, 212
    catchments of, 111–116, 123–124
    eutrophication of, 122
    geography of, 111–112
    management philosophy for, 122–126
    nutrients in, 113–116, 120–122
    physical properties of, 111
    sediment in, 120
    water exchange and movement in, 116–119
Zooplankton, 39, 189